SOILS AND ENVIRONMENT

Lying at the interface of the geosphere, hydrosphere, biosphere and atmosphere, soils are of major importance to an understanding of both natural environmental processes and those affected by human activity. They influence water movement, ground stability, near-surface climate, and plant and animal distributions, as well as determining the agricultural potential of an area, affecting water quality and featuring strongly in many aspects of landuse planning and management. Soil will continue to play a vital role in safeguarding the future of our environment, yet is one of the most undervalued and abused of the Earth's resources.

Soils and Environment examines the ways in which soils both influence, and are influenced by, the environment, adopting a range of approaches – micro and macro-scale, pure and applied, spatial and temporal, natural and human-related. Assuming only a basic scientific knowledge, the book analyses the constituents and properties of soils, the processes and pathways of soil formation, and the range of soil types found globally. It discusses the environmental factors which influence soil development and the resulting characteristics of soils, and considers the influence of environmental change on soils and the role of soils in environmental reconstruction. It then examines soils as components of natural environmental systems before considering soil–human interactions in landuse systems, environmental problems and landuse management, and soil survey and land evaluation.

Supported by a glossary of terms and comprehensive references, this fully illustrated book breaks down the barriers of scientific jargon for readers new to the subject, but also provides important insights for more advanced study. *Soils and Environment* offers readers across a wide spectrum of environmentally related subjects a broad, balanced coverage of an area of major human and environmental significance.

S. Ellis is a lecturer in Geography and Earth Resources at the University of Hull. **A. Mellor** is a senior lecturer in Geography and Environmental Management at the University of Northumbria at Newcastle.

A VOLUME IN THE
ROUTLEDGE PHYSICAL ENVIRONMENT SERIES
Edited by Keith Richards
University of Cambridge

The Routledge Physical Environment series presents authoritative reviews of significant issues in physical geography and the environmental sciences. The series aims to become a complete text library, covering a range of themes in physical geography and environmental science, including specific processes and environments, environmental change on a variety of scales, policy and management issues, as well as developments in methodology, techniques and philosophy.

Other titles in the series:

ICE AGE EARTH
Late Quaternary Geology and Climate
Alastair G. Dawson

WATER RESOURCES IN THE ARID REALM
E. Anderson and C. Agnew

ENVIRONMENTAL HAZARDS
K. Smith

MOUNTAIN WEATHER AND CLIMATE
2nd edition
R.G. Barry

THE GEOMORPHOLOGY OF DESERT DUNES
N. Lancaster

Forthcoming

GLACIATED LANDSCAPES
M. Sharp

HUMID TROPICAL ENVIRONMENTS AND LANDSCAPES
R. Walsh

PROCESS, ENVIRONMENT AND LANDFORMS
Approaches to Geomorphology
K. Richards

SOILS AND ENVIRONMENT

S. Ellis and A. Mellor

London and New York

First published 1995
by Routledge
11 New Fetter Lane, London EC4P 4EE

Transferred to Digital Printing 2003

Simultaneously published in the USA and Canada
by Routledge
29 West 35th Street, New York, NY 10001

Typeset in Garamond by
Solidus (Bristol) Limited

British Library Cataloguing in Publication Data
A catalogue record for this book is available from the British Library

Library of Congress Cataloguing in Publication Data
Ellis, S.
Soils and Environment / S. Ellis and A. Mellor.
    p.   cm. – (Routledge Physical Environment series)
Includes bibliographical references and index.
1. Soils–Environmental aspects.   I. Mellor, A.
II. Title. III. Series.
S596.E38   1995
631.4–dc20   94–23894

ISBN 0–415–06887–8 (hbk)
ISBN 0–415–06888–6 (pbk)

CONTENTS

PLATES

FIGURES

TABLES

PREFACE

Soil is undoubtedly one of the most important components of the environment, yet it is perhaps one of the most undervalued, misused and abused of the Earth's resources. Lying at the interface of the geosphere, hydrosphere, biosphere and atmosphere, soils represent the end product of a complex set of interacting processes, operating over a vast range of time-scales. To a large extent soils determine the agricultural potential of an area, they influence many geomorphological and hydrological processes, and they also feature strongly in many aspects of rural and urban planning, including mineral extraction, construction, waste disposal and conservation. An understanding of soils within an environmental context is therefore important to many disciplines, and this is increasingly the case as environmental problems, such as pollution, acidification, erosion and climatic change, become a matter of increasing public concern.

In many academic bookshops you may find at least half a dozen books dealing with various aspects of soils within an environmental context, so why another to add to the problem of choice? The answer lies in the fact that existing texts tend to concentrate on particular areas of the subject, some at an elementary level and others in a more advanced way, rather than providing the breadth of coverage which many students of environmentally related subjects are likely to encounter during the course of their training. It is therefore our intention to provide a wide range of material which is presented in such a way that it can be used by students in a variety of environmentally based subjects and at a variety of levels.

In this book we examine the constituents and properties of soils, the processes and pathways of soil formation, and the range of soil types and their classification. The environmental factors that influence their development and the resulting characteristics of soils at the global scale are then addressed. We go on to discuss the influence of past environmental conditions on soils, the role of soils in reconstructing and dating past environments, and soils as components of natural environmental systems. Finally, we concentrate on soil–human interactions through the consideration of soils in landuse systems, soils and environmental problems, and soil survey and land evaluation. Throughout the book we adopt a range of approaches – micro- and macro-scale, pure and applied, spatial and temporal, and natural and human-related. The text is designed for students of geography, soil science and environmental science at undergraduate and masters level, but it will also be of value to those studying other disciplines, such as agriculture, ecology, geology, hydrology, archaeology, land management, landuse planning and conservation.

Students of the environment do not always possess a strong science background and can therefore be put off soils at an early stage, often regarding them as the domain of the pure scientist. Consequently we have aimed to present the material in a way that assumes only a rudimentary knowledge of science. The jargon associated with soils may also create unnecessary barriers, so we have been careful to introduce terms in an explicit and hopefully painless manner in order to make reading a rewarding, rather than an unnecessarily challenging, experience; this is reinforced through the provision of a glossary.

We are the first to admit that with the breadth of coverage provided, a certain degree of depth must inevitably be sacrificed in the interests of length and pricing. We hope, however, that this will not be seen as a major problem in the context within which the book has been devised, and that the benefits of presenting a wide range of material within a single volume will outweigh the problems of the omissions that have undoubtedly occurred. In short, we have aimed to provide a broad and balanced text which will be of value to students and teachers alike across a spectrum of environmentally related courses.

In preparation of this work we would like to thank Keith Richards, Tristan Palmer and Sarah Lloyd for their editorial advice and assistance, John Garner for photographic work, Gary Haley for production of the diagrams, and Ian Fenwick and Phil O'Keefe for comments on the manuscript. We are also indebted to our families for their unfailing support, and to the many students who over the years have helped us clarify our ideas about teaching and presentation.

Steve Ellis and Tony Mellor
March 1995

ACKNOWLEDGEMENTS

The authors and publisher wish to thank the following for permission to reproduce copyright material. Every effort has been made to contact copyright holders and we apologise for any inadvertent omissions:

Academic Press – Fig. 2.27; Agricultural Institute of Canada – Fig. 4.7, Table 3.1; American Society of Agronomy – Fig. 8.25; Atmospheric Research and Information Centre, Manchester Metropolitan University – Fig. 8.16; Blackwell Scientific Publishers Ltd – Figs 2.11, 2.19, 2.23, 2.26, 4.1, 4.4, 4.11, 5.9, 5.10, 5.14, 7.5, 8.13, 8.14, 8.19, 8.21, 8.26, Tables 4.2, 8.12; Butterworth Heinemann – Figs 2.14, 7.7; CAB International – Figs 7.14, 7.17, 7.18, 8.3, 8.8, 8.20, 8.27, 9.5, Tables 8.4, 8.6; Cambridge University Press – Figs 6.16, 7.4, 9.1; Central Soil and Water Conservation Research and Training Institute – Fig. 7.11; Chapman & Hall – Figs 2.7, 4.10, 5.12, 5.15, 6.4; Duckworth – Table 4.1; Elsevier – Figs 4.15, 5.6, Tables 2.2, 8.5; Gebrueder Borntraeger – Figs 6.9, 6.10, 6.13; Geologiska Föreningens i Stockholm Förhandlingar – Fig. 5.11; L. Heathwaite – Fig. 5.5; HMSO – Fig. 7.6 (Crown copyright is reproduced with the permission of the Controller of HMSO); Hodder & Stoughton Ltd – Figs 3.10, 8.12; M. Hornung – Table 8.7; Institute of Grassland and Environmental Research – Fig. 7.8; International Society of Soil Science – Fig. 3.15; Journal of Terramechanics – Fig. 8.11; J.P. Legros and Prof. Ozenda – Fig. 4.17; Longman – Figs 2.20, 2.21, 4.2, 7.1, 7.2, 8.4, 8.9, 9.2, Table 7.1; McGraw-Hill – Figs 6.2, 6.5, 6.20; *New Scientist* – Fig. 8.18; M. Newson – Fig. 6.14; New Zealand Society of Soil Science – Table 3.2; Oxford University Press – Figs 3.16, 3.17, 4.3, 4.9, 6.15, 8.17, 9.4; Philip Allen – Figs 7.13, 7.16, Table 8.10; Rothamsted Experimental Station – Fig. 7.9; Royal Geographical Society – Fig. 9.6; Science and Technology Letters – Table 7.2; Simon & Schuster, Inc. – Figs 3.1, 6.1, Tables 4.3, 6.3 (reprinted with permission, from the Macmillan College texts. Copyright by Macmillan College Publishing Company, Inc.); Society of Chemistry and Industry – Fig. 7.15; Soil Science Society of America – Fig. 4.5; Soil Survey and Land Research Centre – Figs 2.12, 7.3, Table 2.4; Soil and Water Conservation Society – Fig. 8.2; Springer Verlag – Fig. 8.1, Tables 8.2, 8.8, 8.13; Van Nostrand Reinhold – Fig. 3.5; B. Van Vliet-Lanoë – Fig. 5.3; John Wiley & Sons Ltd – Figs 2.17, 2.25, 4.16, 4.18, 6.19, 7.10, 7.12, 8.5, Tables 6.2, 8.3; Williams & Wilkins Co. – Figs 3.13, 4.3.

UNITS OF MEASUREMENT

Physical quantity	Unit(s)	Symbol	Value
Length	metre	m	
	Ångstrom	Å	10^{-10} m
Mass	gram	g	
	tonne	t	1000 kg
Area	hectare	ha	10,000 m^2
	square metres	m^2	
Volume	litre	l	
Density	grams per cubic cm	g cm^{-3}	
Temperature	degrees celsius	°C	
Pressure/tension/suction	atmospheres	atm	
	bar	bar	1.02 atm or 10^5 Pa
	Pascal	Pa	1N m^{-2}
	Newton	N	1kg m s^{-2}
Cation exchange capacity	milliequivalents per 100 g	me/100 g	
	centimoles positive charge per kg	cmol(+) kg^{-1}	
Concentration	molar solution	M	
	moles per litre	mol l^{-1}	
	mg per kg	mg kg^{-1}	
Electrical conductivity	millimho per cm	mmho cm^{-1}	
	deciSiemens per m	dS m^{-1}	
Radionuclide activity	Becquerels per square cm	Bq cm^{-2}	

Unit prefixes

kilo-	k	10^3	milli-	m	10^{-3}
deca-	da	10^1	micro-	μ	10^{-6}
deci-	d	10^{-1}	nano-	n	10^{-9}
centi-	c	10^{-2}	pico-	p	10^{-12}

ABBREVIATIONS

AEC   Anion exchange capacity
BP   Before Present. If used in the context of radiocarbon dating, the *present* is taken to be the year AD 1950
CEC   Cation exchange capacity
CERCLA   Comprehensive Environmental Response Compensation and Liability Act (USA)
CFDP   Community Forestry Development Programme
COLE   Coefficient of linear extensibility
DDM   Digestibility of dry matter
DDT   Dichlorodiphenyltrichloroethane
DoE   Department of the Environment (UK)
EC_e Electrical conductivity of the saturation extract
EDTA   Ethylenediaminetetraacetic acid
ESP   Exchangeable sodium percentage
FAO   Food and Agriculture Organization
FYM   Farmyard manure
GIS   Geographical information systems
ICRCL   Interdepartmental Committee on the Redevelopment of Contaminated Land (UK)
ICRISAT   International Crop Research Institute of the Semi-Arid Tropics
MAFF   Ministry of Agriculture, Fisheries and Food
NAA   Nitrate Advisory Area (UK)
NSA   Nitrate Sensitive Area (UK)
PAM   Polyacrylamide
PCB   Polychlorinated biphenyl
PEG   Polyethylene glycol
PVA   Polyvinyl alcohol
SAR   Sodium adsorption ratio
SMD   Soil moisture deficit
SSA   Specific surface area
UN   United Nations
Unesco   United Nations Educational, Scientific and Cultural Organization
USDA   United States Department of Agriculture

1

INTRODUCTION

1.1 INITIAL CONCEPTS

Being at the interface of the Earth and its atmosphere, soil is a complex medium which both responds to, and influences, environmental processes and conditions. It is vital to human existence since, via plant growth, it forms the basis of much of our food supply, and it can also affect our activities by its influence on factors such as ground stability, drainage and water supply. Not only is an understanding of soil therefore important in the study of human–environment interactions, but it also allows a greater insight into a wide range of processes operating within the natural environment, both above and below the surface. Before considering these aspects in more detail, it is useful to start by asking a few basic questions – what do we mean by the terms *soil* and *environment*, how do the two interact and how can we go about studying this interaction?

There is no standard definition of soil, but in its broadest sense it is simply weathered mineral material at the Earth's surface, which may or may not contain organic matter, and often also contains air and water. It may range in thickness from a few millimetres to many metres, and is present over most of the Earth's land surface. Soils often comprise a series of layers aligned roughly parallel with the surface, which can be distinguished from one another on the basis of certain physical or chemical characteristics. These layers are termed *horizons*, and the combined, vertical sequence of horizons is known as a *profile*. The number of horizons can vary greatly between profiles, but in many profiles three basic horizons can be recognised, usually referred to by the letters A, B and C (Figure 1.1). The uppermost, or A, horizon contains organic matter, sometimes exclusively, but often mixed with mineral material. The underlying B horizon is usually a more mineral-rich zone, into which material is often moved, vertically or laterally, from elsewhere in the soil. The combination of A and B horizons is referred to as the *solum*. The B horizon overlies the C horizon, which represents the little altered form of the material from which the soil derives, known as the *parent material*. In some soils this material may occur as a geologically recent, superficial deposit, having been laid down by, for example, a river, a glacier,

1

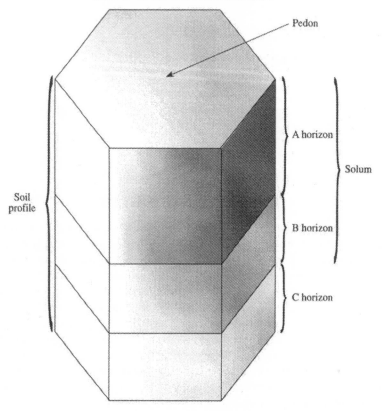

Figure 1.1 A soil profile and pedon, showing soil horizons. The profile is a two-dimensional unit, while the pedon shows soil characteristics in three dimensions

the wind or the sea. In others it comprises the underlying bedrock, which may be referred to as the D or R horizon.

It is important to recognise that although soils are often represented diagrammatically in two dimensions, they are a three-dimensional medium whose properties can vary rapidly both vertically and laterally. The smallest unit of soil which can encompass such variation is known as a *pedon*, which, in contrast to a soil profile, takes a three-dimensional form (Figure 1.1). The size of a pedon will therefore depend on the scale of soil variability, and can range from 1 to 10 m². Similar, adjacent pedons can be grouped into larger units known as *polypedons*.

The processes by which soils form can be divided into four major groups – the addition of material, both organic and inorganic, to the soil, the transformation of this material via the processes of organic matter decomposition, weathering and clay mineral formation, its transfer within the soil

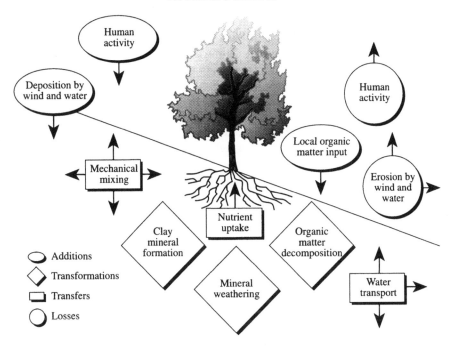

Human
activity

Deposition by
wind and water

Human
activity

Local organic
matter input

Mechanical
mixing

Erosion by
wind and
water

Nutrient
uptake

Clay
mineral
formation

Organic
matter
decomposition

Additions

Transformations

Mineral
weathering

Water
transport

Transfers

Losses

Figure 1.2 The major soil-forming processes (based on Simonson 1978)

by water or by mechanical means, and its loss from the soil via either the surface or subsurface (Figure 1.2). The different ways in which these processes operate between soils produces different horizon sequences and other profile characteristics, and therefore different soil types.

The environment can be defined simply as the surroundings in which we live, and consists of four components – the *geosphere, hydrosphere, atmosphere* and *biosphere*, which interact in a complex manner. The internal structure of the Earth is usually divided into three main units – the core, mantle and crust, and the geosphere occupies the solid, upper part of the crust in which soils are formed. Below the crust and upper mantle, the Earth possesses a more plastic character due to higher temperatures and pressures. The crust consists predominantly of eight elements, with oxygen and silicon occurring in the greatest quantity (around 74 percent by weight), and the others being, in decreasing content, aluminium (around 8 percent), iron, calcium, sodium, potassium and magnesium (around 16 percent in total). There are many other elements also present, but these account in total for less than 2 percent by weight and are known as *trace elements*. The elements in the crust make up the minerals and rocks, which interact with the other environmental components via the processes of weathering and erosion, to produce the wide variety of relief forms at the Earth's surface.

The hydrosphere refers to the many forms in which water can occur both at and below the Earth's surface. Surface water occurs as rivers, lakes or oceans, while ground water can range from water held in individual soil pores to large underground streams or reservoirs within the bedrock. The hydrosphere comprises predominantly hydrogen and oxygen, but can contain other elements in varying quantities and combinations. Sea water comprises predominantly sodium and chloride ions, which together account for approximately 85 percent by weight, and together give a mean salinity value of 35 g kg^{-1}. In contrast, the chemical composition of terrestrial water varies widely, depending on factors such as rock type, dilution by precipitation and concentration by evaporation.

The atmosphere is the air surrounding the Earth, and in its lower part comprises two predominant gases – nitrogen (around 78 percent by weight of dry air) and oxygen (21 percent). The air can also contain water vapour in variable amounts, along with suspended particles and small quantities of other gases such as argon and carbon dioxide. The lowest layer of the atmosphere is known as the *troposphere*, up through which temperature decreases at a rate of around 6.5°C km^{-1} to a height of approximately 16 km at the equator and 8 km at the poles; above this lies the *stratosphere*, extending to an altitude of around 50 km and up through which temperatures increase due to the absorption of ultraviolet radiation from the sun by ozone. These two layers have an important influence on climatic processes, which vary both laterally and vertically in response to complex atmospheric interactions.

The biosphere represents the fauna and flora which live above, at and below the Earth's surface, along with organic material which is no longer alive. It comprises primarily hydrogen, oxygen and carbon which, together with other elements, produce a massive variety of organic compounds and life forms. These forms both influence, and are influenced by, atmospheric, geospheric and hydrospheric conditions. This zone of interaction is known as the *ecosphere* and the interactive processes can be modelled as an ecological system or *ecosystem*. Within this zone are various levels of organisation and interaction. At the lowest level is the cell which makes up individual organisms, and the members of any one species combine to form a population. Populations of different species live as a community, many of whose members are dependent on each other, either directly or indirectly, via the food chain. At the top of the chain are the carnivores. These prey on other carnivores, or on the herbivores, which in turn eat vegetation. There are also groups of organisms which decompose dead organic matter produced either by plants or animals.

Because soil contains rock material, water, air and biota, it is the interface at which all the environmental components interact and is therefore arguably the most complex medium within environmental systems, both influencing

4

and responding to their operation. For example, the geosphere determines the parent material from which a soil develops, the hydrosphere determines the presence of water which is vital for the operation of many of the processes of soil formation, the atmosphere determines the climatic conditions which influence their rate of operation, and the biosphere determines which fauna and flora are available for participation in these processes. Conversely, soil will influence the geosphere by controlling weathering and the transport of weathered material, and therefore the nature of surface relief. It also controls the movement, storage and composition of water in the hydrosphere, the composition of the atmosphere below, and immediately above, the ground surface, and the climate within these parts of the atmosphere. The ability of biota to inhabit a particular area of the Earth's surface is also dependent on the soil, via its influence on factors such as nutrient type and availability, water supply, drainage and ground stability.

1.2 APPROACHES TO SOIL-ENVIRONMENTAL STUDY

The importance of soils in the environment has been recognised since the earliest times, particularly in relation to agriculture. For example, the value of terracing soils on steep slopes and of irrigating soils in seasonally dry climates was known to many ancient civilisations. The agricultural value of soils was recorded some 2,000 years ago by the Romans. These ideas were later synthesised in medieval Italy and were subsequently developed here and elsewhere in Europe, along with experiments to identify the factors responsible for plant growth. This research led to the establishment of modern agricultural science in the nineteenth century, and to the foundation of the allied discipline of soil science, involving the study of the physics, chemistry and biology of soils.

The influence of soils on plant growth, established by the work of agriculturalists, was also utilised in the development of biogeography and ecology, which considered soil and vegetation within a broader environmental framework. Through the increase in travel and overseas exploration by scientists during the eighteenth and nineteenth centuries, it became recognised that both global and regional variations in vegetation distribution could be explained in terms of environmental factors, climate and soil being considered as particularly important controls. During the nineteenth century, attempts were also being made to study the processes by which soils formed, the ways in which these processes were influenced by environmental factors, the resulting soil types produced and their geographical distribution; this marked the beginning of the discipline known as *pedology*.

Around the same time, soils were beginning to be studied within a geological context. For example, where soil profiles had been preserved by being buried beneath sediment, they could be used to identify periods of landscape stability within geological sequences during which soil development occurred. In the twentieth century, the importance of soils became increasingly recognised in the study of landform development or *geomorphology*, particularly in the context of slope processes. Here they were used to identify periods of landscape stability and erosion, and also to examine in detail the processes of weathering and sediment transport; allied to geomorphology was the study of soil by engineers from a viewpoint of its mechanical behaviour. The twentieth century also saw the study of soil by hydrologists and climatologists. Hydrologists recognised that soils controlled the retention and movement of water in the uppermost part of the geosphere, and hydrological models were developed based on studies of soil physics. Soil physics also featured in the development of microclimatology, with respect to the influence of soils on temperature, moisture and air movement.

The use of soils in providing information about past environmental conditions has been developed in more recent decades through a number of disciplines. Soils can preserve fossils such as plant remains or artefacts which can be used by palaeo-ecologists and archaeologists to determine the nature of vegetation and climate, or human activity. Fossil soil developmental characteristics can also be used by pedologists and geologists to determine the nature of the environment in which a soil formed in the past. Soils may also contain material which allows them to date past environmental conditions or processes, which can be useful to many disciplines.

In more recent times, soils have become involved increasingly in environmental management. For example, the agricultural potential of a soil is an important factor in urban planning, while its hydrological properties may be important in waste disposal. The increasing awareness of the dangers of environmental degradation has also emphasised the fundamental role of soils through recognition of problems such as pollution, soil erosion, acidification and desertification, and the role of soils in sustainable land use is now becoming widely recognised. This takes us full circle by illustrating the importance of soils in sustaining life, equally relevant now as to the earliest civilisations.

1.3 AIMS AND SCOPE OF THIS BOOK

This book aims to provide an understanding of the ways in which soils are influenced by the environment and, in turn, the ways in which the environment is affected by soil conditions, using the various disciplinary approaches outlined in the previous section (Figure 1.3). There are, however, problems in designing a book from a multidisciplinary viewpoint, in that each discipline has its own approaches and terminology which, although second nature to its exponents, may be unfamiliar to others. We have therefore assumed only a basic scientific knowledge and have attempted to

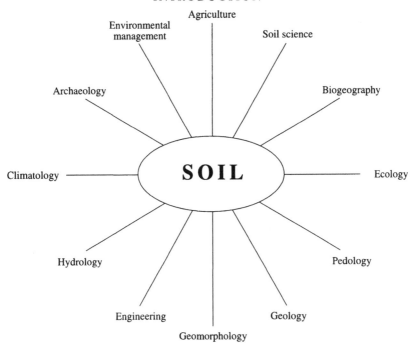

Figure 1.3 Approaches to soil-environmental study

keep jargon to a minimum; a glossary of commonly encountered terms is provided at the end of the book.

Our coverage starts by examining the basic constituents and properties of soil, which are fundamental to the study of soil-environmental relationships from any disciplinary angle. We then proceed to consider how soils form in terms of the processes responsible for horizon and profile development and the ways in which the resulting soil types can be classified. This is followed by an examination of the factors which influence soil development and the principal characteristics of soils formed under different environmental conditions. Attention is then focused on soils in the context of environmental history from three principal viewpoints – the ways in which soils have formed under environmental conditions different from those of today, the uses of soils in providing information about environmental change, and the ways in which soils can be used to date the past. Next we consider the role of soil in natural environmental systems, looking at how it influences the hydrosphere, atmosphere, geosphere and biosphere. The important topic of soils and land use is then examined in terms of the ways in which soils determine and respond to various landuse practices. This is followed by a consideration of the role of soils in some of our major environmental problems, together with approaches to their management and remediation. Having examined soil-environmental interrelationships in both natural and human systems, methods of recording the distribution of soils, and of evaluating them for particular purposes, are discussed, and the applications of these techniques considered. Finally, in our concluding chapter we present an overview of the present state of knowledge of soils within an environmental context, and consider possible developments for the future.

Clearly there are many other ways in which the material could have been presented, but we have adopted this scheme with two important but contrasting requirements in mind – it can be followed sequentially by a non-specialist in one of a variety of disciplines in order to obtain a wide range of information and knowledge in a logical and cumulative manner; it can also be used in such a way that more specialist readers can refer to one or more chapters in any order to obtain a synthesis of a particular part of the subject area, not only their own, and in this way it is hoped that interdisciplinary recognition and collaboration may be encouraged in a subject which is not only fascinating from an academic viewpoint but is also of major human significance.

2

SOIL CONSTITUENTS AND PROPERTIES

2.1 INTRODUCTION

In terms of its constituents, the soil can be viewed as a three-phase system comprising solid, liquid and gaseous constituents (Bonneau and Souchier 1982). The solid phase consists of both mineral and organic material; the mineral fraction is derived largely from the parent material and the organic fraction largely from vegetation growing in and above the soil. Individual particles and fragments often join together to form larger units known as *aggregates* or *peds*, and these play a crucial role in the development of many physical and chemical characteristics of soils. Between the solid material there are usually spaces known as *pores* or *voids*, and these are occupied by the liquid and gaseous phases. The liquid component, or *soil water*, derived from precipitation and ground-water sources, is able to transport material through the soil in both suspended and dissolved forms, and is often referred to as the *soil solution*. The gaseous component, known as the *soil atmosphere* or *soil air*, consists of a mixture of gases derived from the above-ground atmosphere and from the respiration of soil organisms. The relative proportion of constituents in a typical topsoil is shown in Figure 2.1, although the proportions can vary widely depending on the time of sample collection, soil type and environmental conditions.

The constituents are intimately associated and interact in a variety of ways, which gives a soil a wide range of properties, from relatively simple morphological characteristics recognisable in the field to complex chemical properties recognised only by laboratory analysis, and many of the properties are strongly interrelated (Figure 2.2). There are a vast number of soil properties, which have been the subject of textbooks in their own right (e.g. Baize 1993), and this chapter will therefore focus only on those properties most commonly encountered in the context of soil-environmental study. Although technical and analytical procedures will be examined briefly, more specialised texts should be consulted for further details (e.g. Rowell 1994). The chapter starts by examining the major soil constituents, as these are fundamental to an understanding of soil properties, which will form the subject of the remainder of the chapter.

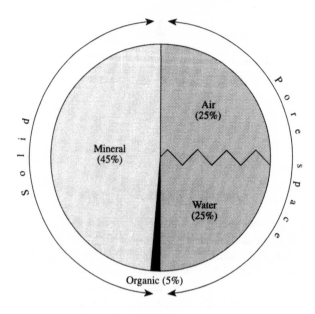

Figure 2.1 Composition by volume of a typical topsoil

2.2 SOIL CONSTITUENTS

2.2.1 Mineral material

The mineral fraction of soils is derived largely from weathering of the underlying parent material (section 3.2.2), which may consist of consolidated bedrock or unconsolidated superficial deposits. Consolidated bedrock can be classified into one of three categories – igneous, sedimentary or metamorphic (Table 2.1). Igneous rocks are derived from the consolidation of molten magma, either within the Earth's crust or at its surface, whereas sedimentary rocks are the product of cycles of weathering, erosion and deposition operating at the surface, and metamorphic rocks originate from the alteration of any other rock type by high temperature and/or pressure but without melting. Unconsolidated superficial deposits are just as variable as solid bedrock materials and are often classified according to the nature of their depositional environment, for example riverine alluvium, glacial till and lacustrine clays. These deposits are often indurated or cemented to some extent by calcareous, siliceous or iron-rich components. In addition to the parent material origin of mineral material, it can also be added to soils by movement from upslope, by aeolian transport or by atmospheric fall-out of materials such as volcanic ash.

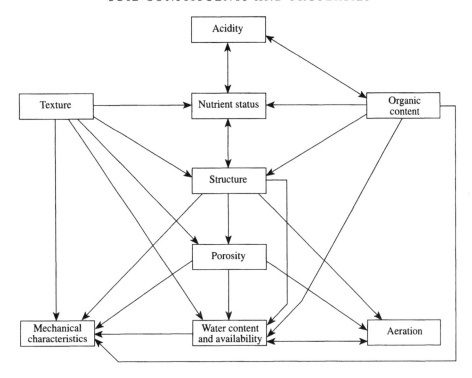

Figure 2.2 Interrelationships between selected soil properties

Table 2.1 Geological classification of the more common bedrock parent materials

Igneous	*Sedimentary*	*Metamorphic*
Granite	Grit	Quartzite
Felsite	Sandstone	Marble
Rhyolite	Siltstone	Hornfels
Diorite	Mudstone	Slate
Andesite	Shale	Phyllite
Gabbro	Limestone	Schist
Dolerite	Dolomite	Gneiss
Basalt	Chalk	
	Ironstone	
	Evaporite	

Soil minerals occur in a wide variety of forms, but these can be arranged into groups on the basis of their chemical composition and the structural arrangement of their constituent elements. By far the most abundant group in soils and their parent materials is the *silicate* group. The fundamental

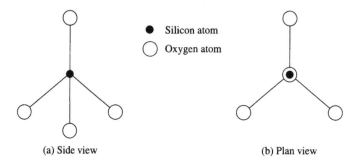

Figure 2.3 Silicon-oxygen tetrahedron, the basic building block of silicate minerals

building block of all silicate minerals is the silicon-oxygen (Si-O) tetra-hedron, which consists of a central silicon ion (Si^{4+}) surrounded by four closely spaced oxygen ions (O^{2-}) (Figure 2.3). Because the silicon ion has four units of positive charge and each oxygen ion has two units of negative charge, each discrete tetrahedron possesses four units of negative charge. This allows the tetrahedra to link together in a variety of characteristic structural arrangements which forms the basis of their classification. The potential range of silicate minerals is further widened by substitution of one ion for another within the mineral structure, a process known as *isomorphous substitution*. One of the most common substitutions involves replacement of some Si^{4+}, in Si-O tetrahedra, by aluminium ions (Al^{3+}). Other examples include the substitution of Al^{3+} by magnesium ions (Mg^{2+}) or iron (Fe^{3+} or Fe^{2+}). When ions with the same charge or valency are exchanged, the mineral structure remains electrically neutral. If, however, ions with different valencies are exchanged, there will be a charge imbalance; frequently this results in an excess negative charge, as in the examples above. In this situation, electrical neutrality is achieved either by incorporation of addi-tional cations such as calcium, magnesium, potassium or sodium (Ca^{2+}, Mg^{2+}, K^+, Na^+) into the crystal lattice, or by structural rearrangements that allow internal compensation of charge (White 1987). The most common silicate minerals and their associated structures are discussed briefly below.

Framework silicates or *tectosilicates* consist of a three-dimensional lattice of Si-O tetrahedra linked through their corners. The two main mineral groups within this category are quartz and feldspars, which are common minerals in many soils. Quartz consists simply of Si-O tetrahedra linked through the O^{2-} ions, and consequently there are twice as many O^{2-} ions as Si^{4+} ions in the mineral structure, which has the general formula $(SiO_2)_n$. Unlike quartz, feldspars contain significant amounts of Al^{3+} ions, originating from isomorphous substitution of Si^{4+} ions, and base cations, especially

Ca^{2+}, Na^+ and K^+, which satisfy the residual negative charges. There are two main groups of feldspar minerals – potassium feldspars such as orthoclase and microcline, and plagioclase feldspars which form a continuous series of minerals between the sodium end member, albite, and the calcium end member, anorthite.

Chain silicates or *inosilicates* comprise Si-O tetrahedra linked together to form a continuous chain structure of which there are two types. First there is the single chain where the Si-O tetrahedra are linked by sharing two out of the three basal O^{2-} ions, and second there is the double chain where tetrahedra are linked by sharing all three basal O^{2-} ions to form a hexagonal arrangement (Figure 2.4). Common minerals in these two categories are the

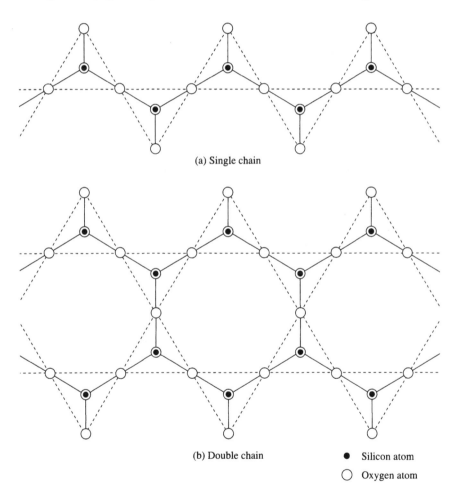

(a) Single chain

(b) Double chain

● Silicon atom
○ Oxygen atom

Figure 2.4 Chain silicate mineral structures (plan view)

13

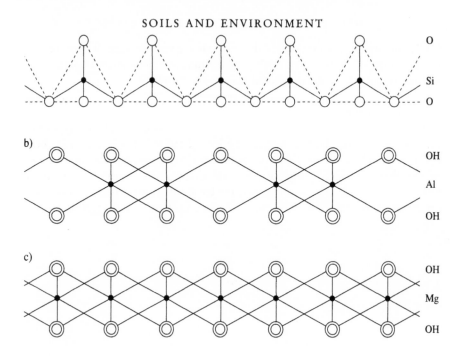

Figure 2.5 Sheet silicate mineral components: (a) siloxane sheet, (b) gibbsite sheet, (c) brucite sheet

pyroxene family (e.g. augite) and the amphibole family (e.g. hornblende) respectively. In chain silicates the chains themselves are linked together by a variety of cations including Ca^{2+}, Mg^{2+}, Fe^{2+}, Na^+ and Al^{3+}. Minerals in this group are very reactive and more easily weathered than the framework silicates. *Orthosilicates* and *ring-silicates* display a wide range of structures but, in comparison with other groups, their occurrence in soils is relatively minor. Some minerals in this group are very reactive and therefore particularly susceptible to weathering (e.g. olivine), whereas others are very resistant (e.g. zircon). This category can be divided into two groups – *neso-silicates* and *soro-silicates*. In the first group the Si-O tetrahedra occur as separate units, with no shared O^{2-} ions, and are linked by metallic cations. Examples include olivine and garnet. In soro-silicates the tetrahedra form separate groups in which they share one or more of their O^{2-} ions; when the group is formed by sharing O^{2-} ions by more than two Si-O tetrahedra, a ring structure results. Examples of soro-silicates are beryl and cordierite.

Sheet silicates or *phyllosilicates* are perhaps the most important group of minerals in soils, playing an important role in the development of many physical and chemical soil characteristics. Unlike the other mineral groups, many sheet silicate minerals occur in the smallest grain-size categories;

consequently they are often known as the *clay minerals*. Minerals in this group are made up of various combinations of three fundamental sheet structures: (a) Si-O (siloxane) sheet (Figure 2.5a) in which the Si-O tetrahedra are linked in a hexagonal arrangement, (b) Al-OH (gibbsite) sheet in which the Al^{3+} ions are surrounded by six closely packed hydroxyl (OH^-) ions (Figure 2.5b) to form an octahedral structure, and (c) Mg-OH (brucite) sheet whose structure is similar to that of the gibbsite sheet but contains Mg^{2+} ions rather than Al^{3+} (Figure 2.5c).

Classification of sheet silicate minerals is based on three criteria – the ratio of the above sheets within a unit layer of the mineral structure, the interlayer or basal spacing between unit layers, and the interlayer components or species (Brindley and Brown 1980, Bonneau and Souchier 1982) (Figure 2.6). The sheet silicate minerals most commonly found in soils include the micas, illite, chlorite, kaolinite, vermiculite and the smectites. The most frequently occurring micaceous minerals are muscovite and biotite. Both of these are 2:1 layer silicates (two siloxane sheets to one gibbsite sheet in the case of muscovite, and two siloxane sheets to one brucite sheet for biotite) with an interlayer spacing of approximately 10 Å (1.0 nm) and K^+ as the dominant interlayer component (Figure 2.6a). Illite is very similar in composition and structure to micaceous minerals and is sometimes known as hydrous mica.

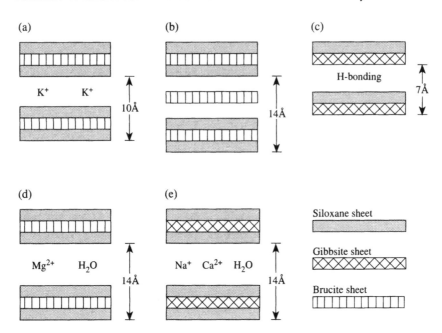

Figure 2.6 Diagrammatic representation of clay mineral structures: (a) biotite, (b) chlorite, (c) kaolinite, (d) vermiculite, (e) montmorillonite

Chlorite has a 2:1:1 layer structure (two siloxane sheets to one brucite sheet, forming a biotite mica layer, to one further brucite sheet which occupies the interlayer position) with an interlayer spacing of about 14 Å (1.4 nm) (Figure 2.6b). Kaolinite has a 1:1 layer structure (one siloxane sheet to one gibbsite sheet) and an interlayer spacing of around 7 Å (0.7 nm) (Figure 2.6c). Layer units of kaolinite are attracted by hydrogen bonding and the crystals have a characteristic hexagonal appearance. Vermiculite is a 2:1 layer silicate with a basal spacing of about 14 Å (1.4 nm) and interlayer positions occupied by Mg^{2+} ions and water molecules (Figure 2.6d). The most common smectitic mineral in soils is montmorillonite. This is a 2:1 layer silicate with interlayer positions dominated by Na^+ and Ca^{2+} ions, and water molecules (Figure 2.6e). Although its basal spacing is approximately 14 Å (1.4 nm), this varies somewhat due to its shrink–swell characteristics which occur on wetting and drying.

Other sheet silicate minerals sometimes found in soils include halloysite, imogolite and allophane, and mixed layer or interstratified minerals. Halloysite is similar in many ways to kaolinite, but its crystal layers are curved to form a tubular structure. Imogolite and allophane are gel-like hydrated alumino-silicate minerals with a poorly crystalline tube- or thread-like structure. Mixed layer minerals result from the interlayering of more than one sheet silicate mineral. Interlayering may be regular, where there is a regular repetition of the different mineral layers, or random where there is no clear pattern of repetition. These minerals may exist as intergrades which represent stages of transition between one mineral and another during weathering (e.g. Wilson and Nadeau 1985). Examples of mixed layer minerals are mica- or illite-vermiculite (hydrobiotite), illite-montmorillonite and kaolinite-montmorillonite.

In addition to the silicate minerals are the *non-silicate minerals*. These are often of only minor occurrence in soils and are usually referred to as *accessory minerals*, although they play a significant role in the development of some soil characteristics. Among the most common non-silicate minerals in soils are the free oxides and hydroxides of iron, aluminium and manganese; these may exist as crystalline or amorphous forms (Baize 1993). Iron minerals include goethite, ferrihydrite, hematite, lepidocrocite, maghemite and magnetite. Aluminium minerals include gibbsite and boehmite, while examples of manganese minerals are birnessite and pyrolusite. Other non-silicate minerals include calcite ($CaCO_3$), which occurs in soils developed from calcareous parent materials, and anatase (TiO_2) and amorphous silica, which are often found in soils developed on recent volcanic deposits.

16

2.2.2 Organic components

Soil organic matter is derived from a number of sources (Figure 2.7), of which the most important is usually plant litter. This consists of a variety of plant debris including leaves, stems, flowers, twigs, bark, and the larger branches and trunks of trees. Other organic components include plant roots, root exudates, soil organisms together with their faecal remains and metabolites, and organic substances washed into the soil from vegetation.

Soil organisms are responsible for the physical comminution and biochemical decomposition, and incorporation of organic matter. They exhibit tremendous diversity in terms of their numbers, size and morphology, and also vary dramatically in their function, mode of nutrition and environmental tolerance. Organisms can be referred to as *producers, consumers* or *decomposers*. Producers, such as plants, fix carbon from atmospheric carbon dioxide (CO_2) during photosynthesis, consumers feed on plants and other organisms, and decomposers utilise carbon from organic material, returning it to the atmosphere as CO_2 and other gaseous by-products and mineralising nutrients to their original ionic form. Soil organisms can also be classified according to their size – *micro-organisms* ($< 200\,\mu$m), which include both microflora and microfauna, *mesofauna* (200–$1,000\,\mu$m) and *macrofauna* ($> 1,000\,\mu$m). The classification of soil organisms according to their mode of nutrition is based on their sources of carbon and energy. Essentially, they obtain carbon from either organic material or inorganic sources (largely CO_2); these two groups are known as *heterotrophs* and *autotrophs* respectively. They can also obtain energy either from light (photoheterotrophs or photoautotrophs) or from chemical oxidation (chemoheterotrophs or chemoautotrophs). In addition to their carbon and energy sources, soil organisms, especially bacteria, vary in their oxygen requirements. *Aerobes* have a direct

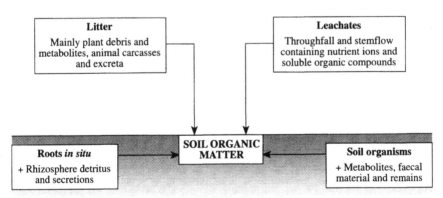

Figure 2.7 The main sources of soil organic matter (from White *et al.* 1992)

requirement for oxygen, *facultative anaerobes* normally require oxygen but may adapt to oxygen deficit by utilising nitrate and other inorganic compounds as electron receptors, and *obligate anaerobes* require an absence of oxygen as it is toxic to them. Before looking at the composition of organic material, the various groups of micro-organisms, mesofauna and macrofauna will be briefly considered.

Four main groups of micro-organisms can be recognised – bacteria and actinomycetes, fungi, algae and protozoa. Bacteria are very small organisms (1–5 μm) and are often round or rod-like in shape. They tend to exist in thin films of water surrounding soil particles, often reproducing with tremendous rapidity, with numbers in the range of 10^6–10^9 g^{-1} having been widely reported. Many species secrete polysaccharide gums, which facilitate aggregation (section 2.3.2) (Molope *et al.* 1987). Bacteria demonstrate great biochemical versatility in their ability to decompose a wide range of materials under a variety of conditions. For example, *Pseudomonas* sp. can metabolise a range of chemicals including pesticides, *Nitrobacter* sp. derives its energy from the oxidation of nitrite to nitrate, *Thiobacillus ferrooxidans* acquires energy from oxidation of reduced sulphur compounds and Fe^{2+}, and *Rhizobium* sp. forms nitrogen-fixing nodules on the roots of leguminous plants (Wood 1989). Actinomycetes consist of fine branching filaments (about 1 μm in diameter) during their vegetative stage, when they are similar in appearance to fungi. During reproduction, however, the filaments undergo fragmentation, sometimes forming dense colonies. For this reason actinomycetes are often classified as highly evolved and complex bacteria.

Fungi produce filamentous structures (hyphae) which are about 0.5–1.0 μm in diameter and which grow into a dense network or mycelium. They are generally less numerous than bacteria in soils (1–4 × 10^5 g^{-1}) (Wilhelmi and Rothe 1990), although they are common in acidic soils (section 2.4.3) where they can be responsible for 60–80 percent of organic matter decomposition. All fungi are heterotrophic, living mostly in the surface layers where they play an important role in the development of soil structure (Molope *et al.* 1987, Miller and Jastrow 1990). Mycorrhizal fungi live symbiotically in plant tissue, removing carbon in return for nutrients such as phosphorus. Algae are photosynthetic organisms and are therefore confined largely to the soil surface. They include Cyanophaceae (blue-green algae) and Chlorophaceae (green algae), the former playing a major role in development of the nitrogen status of soils. Algal numbers are usually intermediate to those of bacteria and fungi at around 10^5–3 × 10^6 g^{-1}. Protozoa are microfauna, unlike the previous groups which are microflora. They are uni- or non-cellular organisms of 5–40 μm in length and live in thin water films surrounding soil particles. They can be divided into amoeboid forms with silica or chitin sheaths, and rotifers with better differentiated cell structure. Protozoa are very numerous (often > 10^4 g^{-1}) and are particularly important in controlling the numbers of bacteria and fungi, on which they feed.

18

As in the case of micro-organisms, four main groups of mesofauna can be recognised – nematodes, arthropods, annelids and molluscs (Figure 2.8). Nematodes are unsegmented worms (eelworms or roundworms), and next to protozoa are the smallest of the soil fauna, being 0.5–1.0 mm in length. They are plentiful in soil and litter, feeding on plant remains, roots, bacteria and sometimes on protozoa. Arthropods can be divided into a number of categories, the most significant numerically being acari (mites) and collembola (springtails). Both of these feed on plant litter, bacteria and fungi, with acari being particularly common in acidic litter where they may constitute 80 percent of soil fauna. Other arthropod groups include myriapods, which are dominated by chilopods (centipedes) which are carnivorous, and diplopods (millipedes) which are herbivorous. Arthropods also include isopods (woodlice), beetles, insect larvae, ants and termites. Termites are prevalent in tropical soils and are particularly active soil mixers, producing structures at the surface (termitaria) up to several metres in height.

Figure 2.8 Some of the more common types of soil mesofauna (not necessarily to scale): (a) nematodes, (b) enchytraeid worms, (c) earthworms, (d) molluscs, (e) myriapods, (f) isopods, (g) ants, (h) beetles and larvae, (i) dipterous larvae, (j) spiders, (k) springtails and (l) mites (based on Pitty 1979)

Arthropod populations can be particularly high, for example 220,000 m^{-2} in old grassland soils (Wood 1989).

Annelids consist largely of enchytraeid worms (potworms) and lumbriscid worms (earthworms). Enchytraeids are small (0.1–5.0 cm in length) and have a thread-like appearance. They feed on algae, fungi, bacteria and soil organic matter, and can occur in populations as high as 200,000 m^{-2} (Wood 1989). Lumbriscid worms are probably more important than any other soil invertebrate in the decomposition and mixing of organic matter, and their numbers can exceed 800 m^{-2}, although they do not tolerate highly acidic conditions, and during prolonged drought or heat they burrow deeply into the soil and become inactive. Some species, such as *Lumbricus rubellus*, live only in the surface litter layer, but most migrate between organic and mineral horizons, ingesting both constituents and ensuring thorough mixing of the soil. Large populations under grassland can consume 90 t ha^{-1} a^{-1}, and the casts which are produced are important in the development of soil structure (section 2.3.2). The main mesofaunal mollusc groups are slugs and snails (gastropods) which feed on plants, fungi and faecal remains, but these are often limited in biomass to 20–45 g m^{-2} (Wood 1989). Macrofauna include larger molluscs, beetles and larger insect larvae, together with larger vertebrates such as moles, rabbits, foxes and badgers. These often burrow deeply into the soil, feeding on smaller organisms in their path. Moles in particular are voracious feeders and consume large numbers of earthworms, insect larvae and slugs.

In terms of its composition, organic material consists predominantly of carbon, hydrogen, oxygen and nitrogen, with carbon providing the framework for organic structures; most organic residues in soils contain around 45–55 percent carbon by weight. The elements which constitute organic material are arranged to form a variety of compounds, some of which have very complex structures. The basic building block of many organic compounds, however, is the carbon tetrahedron, where a central carbon atom with a valency of four links with other elements, notably hydrogen, oxygen and nitrogen (Figure 2.9a); this structural arrangement is very similar to the basic building block of silicate minerals (section 2.2.1). In some instances, numerous tetrahedra link together to form extremely large molecules known as *polymers*. Linkages can produce both linear and cyclic molecules (Figure 2.9b to e), and the number of potential combinations and structural arrangements is almost endless. Consequently, organic structures and their variety are considerably more complex than those of mineral material and are less well understood, although the majority of organic compounds can, however, be isolated. The main groups of compounds are carbohydrates, proteins and amino acids, and lignin, along with smaller amounts of fats, waxes, pigments and resins. The proportions of these compounds vary according to the type of organic matter and its stage of decomposition (Figure 2.10).

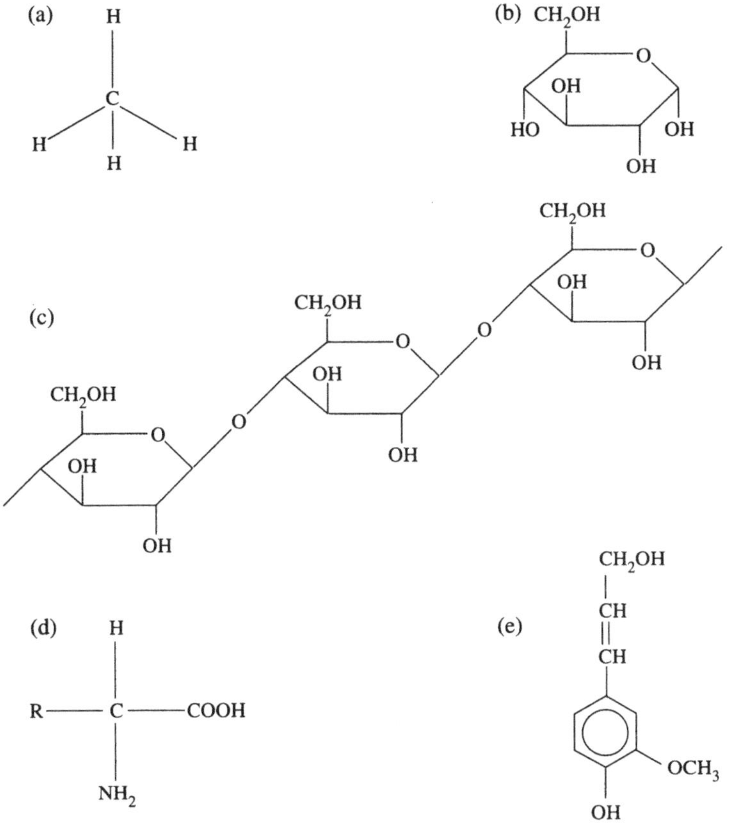

Figure 2.9 The composition and structure of selected organic compounds:
(a) carbon tetrahedron (methane), (b) glucose, (c) polysaccharide (cellulose),
(d) amino acid, (e) coniferyl alcohol, a building block of lignin.
Unlabelled corners of hexagons are occupied by carbon atoms.

Essentially, carbohydrates are hydrates of carbon with a general formula
$C_x(H_2O)_y$, although some contain other elements such as nitrogen and
sulphur. Carbohydrates are the main constituent of plant material at 60–90
percent of dry mass, and in soils 5–30 percent of the carbon exists in this
form. The simplest and most easily broken down carbohydrate compounds
in soils are the sugars and starches. These comprise relatively small molecules
such as glucose (a simple sugar), which is a monosaccharide (Figure 2.9b).
More complex, and thus more resistant to breakdown, are hemi-cellulose (a
pentose sugar polymer) and cellulose (a β(1–4) glucose polymer). These are
polysaccharides which consist of monosaccharide units joined by C-O-C
links (Figure 2.9c). Cellulose is the most abundant carbohydrate in plants,
where it may constitute more than 40 percent of the carbon. It forms the
resistant fibres of plants, the breakdown of which is catalysed by enzymes

21

Percentage composition

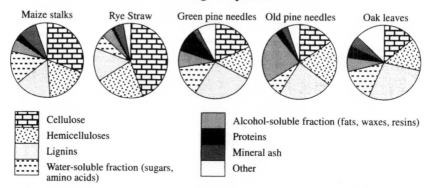

Maize stalks   Rye Straw   Green pine needles   Old pine needles   Oak leaves

Cellulose

Hemicelluloses

Lignins

Water-soluble fraction (sugars, amino acids)

Alcohol-soluble fraction (fats, waxes, resins)

Proteins

Mineral ash

Other

Figure 2.10 Organic composition (%) of a range of plant materials (from Ross 1989)

(cellulases) secreted by micro-organisms. Breakdown ultimately converts the polysaccharide compounds into monosaccharide compounds which can then be assimilated directly by the micro-organisms. Chitin is a polysaccharide of similar composition to cellulose, but it also possesses some nitrogen in the form of amino groups. It is a common constituent of insect cuticles and is also often present in fungi.

Proteins and amino acids are composed of about 50–55 percent carbon, 20–25 percent oxygen, 15–20 percent nitrogen and 6.5–7.5 percent hydrogen, although some contain small amounts of phosphorus and sulphur (Figure 2.9d). They are an important source of nitrogen, with around 20–50 percent of all organic nitrogen in soils existing as amino acids. Although they are rapidly metabolised in soils, they can persist for long time periods due to adsorption onto other constituents, such as clay, or by combination with more resistant organic components such as lignin or tannin. Lignin is a particularly resistant component of soil organic material, and constitutes 15–35 percent of the supporting tissues of plants. Essentially it is a complex phenolic polymer which displays a high degree of aromaticity in the development of cyclic, benzene ring structures (Figure 2.9e). The aromatisation and large number of C-C bonds are the main cause of its resistance to decomposition.

As a result of organic matter decomposition, fresh organic components are gradually converted, via a range of fermented products, to a material known as *humus*, which represents the end-product of this process (section 3.2.2). Humus is finely divided and amorphous with no cellular structure. Essentially it consists of insoluble, heterogeneous polymers and has an approximate composition, on an ash-free basis, of 44–53 percent carbon, 40–47 percent oxygen, 3.5–5.5 percent hydrogen and 1.5–3.5 percent nitrogen. The character of humus is, however, particularly complex and remains the subject of on-going research (Newman and Tate 1991, Reid *et al.* 1991).

22

2.2.3 Water

Soil water is derived from two principal sources – precipitation and ground water. Precipitation arrives at the soil surface in various forms; although rain, snow and hail are the main types, fog and mist may provide significant amounts of moisture, particularly in coastal and upland areas. The proportion of precipitation that reaches the ground surface depends largely on the nature and density of vegetation cover. On surfaces devoid of vegetation, precipitation reaches the soil directly, but on vegetated surfaces a significant proportion of precipitation can be intercepted; much of this ultimately reaches the soil as canopy throughfall and stemflow, while some is returned to the atmosphere by evaporation. On reaching the surface, water can either infiltrate the soil or, if the rate of arrival of water at the surface exceeds the rate of infiltration, it will run off over the surface. The precipitation that infiltrates the soil will either be lost via evapotranspiration or drainage, or will be retained by forces which hold it in place or allow it to move only very slowly (sections 2.3.4 and 6.2). Ground water can be derived by lateral movement from upslope, or by upward movement from the underlying rock strata. The various components of soil water movement will be examined in more detail in section 6.2.

The composition of soil water is a particularly dynamic characteristic, varying dramatically even over short time periods. This behaviour arises from the intimate association between the water, small mineral and organic particles (clay and humus) and plant roots, which can involve the exchange of ions between these components (section 2.4.2). Soil water contains a number of dissolved solid and gaseous constituents, many of which exist in mobile ionic form, and a variety of suspended solid components (Table 2.2). Base cations (Ca^{2+}, Mg^{2+}, K^+, Na^+, NH_4^+) may be derived from a number of sources. They are present in the atmosphere and are later dissolved in precipitation; in maritime areas, for example, precipitation often contains significant quantities of marine salts which are particularly rich in sodium and chloride ions (Na^+ and Cl^-) (Reynolds *et al.* 1987). As well as being deposited in the soil directly, ions accumulate on the surfaces of vegetation and are subsequently washed off into the soil. Cations can also be derived from mineral weathering and organic matter decomposition, entering the soil directly or via the soil exchange system; these processes play an important role in the buffering of soil acidity (section 2.4.3). In agricultural systems, lime and fertilisers provide yet another source of base cations, particularly Ca^{2+}, K^+ and NH_4^+.

The concentration of hydrogen ions (H^+) in soil water is a measure of its acidity, which is expressed in terms of *pH* (section 2.4.3). A major source of such acidity is carbon dioxide (CO_2) which is derived from the atmosphere, where it is dissolved in precipitation, and from the soil air where it is a product of soil organism respiration. Carbon dioxide dissolved in water

Table 2.2 Mean concentrations of major solutes in soil water from a podzolic soil profile in Wales

		Horizons			
	L	O	E	B	C
pH	3.8	3.7	4.0	4.0	4.0
$(\mu mol\ 1^{-1})$					
H^+	153	213	101	91	107
Al (inorganic)	20	91	277	317	268
Al (organic)	4	8	8	9	8
Cl^-	334	498	423	518	303
SO_4^{2-}	240	264	262	280	243
NO_3^-	68	9	31	34	55
Na^+	337	445	320	399	248
K^+	57	13	19	7	5
Ca^{2+}	41	20	31	25	19
Mg^{2+}	100	98	72	69	36
SiO_2	21	68	107	78	71
$(mg1^{-1})$					
DOC	21	15	7	10	8

Source: Soulsby and Reynolds 1992
Note: L = surface litter layer; O = peaty organic layer which is partially humified; E = leached horizon with bleached appearance; B = horizon characterised by humus and sesquioxide deposition; C = relatively unaltered parent material

produces carbonic acid (H_2CO_3) which undergoes dissociation to a greater or lesser extent, depending on acidity, to form hydrogen and bicarbonate ions (H^+ and HCO_3^-):

$$H_2O + CO_2 \Leftrightarrow H_2CO_3 \Leftrightarrow H^+ + HCO_3^-$$

Unpolluted rain water in equilibrium with atmospheric CO_2 has a pH of about 5.6, whereas soil water in equilibrium with CO_2 in soil air is usually more acidic, with pH values often below 5.0; this is because CO_2 levels in soil air are considerably greater than in the atmosphere (section 2.2.4). Another source of acidity in soil water derives from industrial and urban emissions (section 8.3.1). In addition to these pollutants, organic acids derived from decaying organic material are an important source of soil acidity. H^+ is also released by plants in exchange for nutrient base cations, and as part of the process of nitrification where NH_4^+ is converted to NO_3^- (section 8.3.1).

Iron and aluminium are also important constituents of soil water, particularly under acidic conditions. They are derived from mineral weathering and may exist in the soil solution either as ions (Fe^{2+} and Al^{3+}) or in the form of soluble organo-metallic complexes; significant quantities of dissolved organic carbon may also exist in this form. In some circumstances aluminium may be mobilised by mineral acids, such as sulphuric and nitric acid deposited in acid precipitation.

As in the case of cations, the anions in soil water are derived from a number of sources. Nitrate and phosphate ions (NO_3^- and PO_4^{3-}) are produced by mineralisation processes during organic matter decomposition (section 3.2.2), and are also derived from fertilisers. Chloride ions (Cl^-) and, to a lesser extent, sulphate ions (SO_4^{2-}) originate from atmospheric sources, including airborne marine salts and acid deposition. Bicarbonate ions (HCO_3^-) originate largely from the dissociation of H_2CO_3 as mentioned above, and are associated with mineral weathering, especially in soils developed from carbonate-rich parent materials.

In addition to the major dissolved constituents of soil water, there are other dissolved components although these are usually relatively minor and local in their occurrence. These include organic material and silica, together with a number of pollutants such as heavy metals (e.g. lead, zinc and cadmium) and radionuclides (e.g. caesium) (section 8.3.4).

Soil water contains not only dissolved solids but also a number of suspended constituents. These include small particles of mineral and organic material, which often result in discoloration and increased turbidity of soil water. Similarly, precipitates may accumulate in soil water, usually as a result of chemical changes as the water migrates through the soil. For example, orange-brown precipitates of insoluble ferric iron compounds can sometimes be observed where water, in which iron compounds usually exist in the soluble ferrous form, becomes more oxygenated. The chemical characteristics of water can also influence the behaviour of fine sediments, particularly clays, in suspension. In waters with low solute concentrations, or significant quantities of monovalent cations such as Na^+, the clays tend to remain in a dispersed state and can be held in suspension for long periods of time. In waters with high solute concentrations, however, particularly where concentrations of divalent cations such as Ca^{2+} are high, the clays may undergo rapid flocculation and settling.

2.2.4 Air

Air and water have a reciprocal arrangement in terms of their occupancy of soil pore space; in saturated soils, air content is low, whereas in dry soils the pore spaces are largely air-filled. Changes in water and air content are particularly dynamic because much of the water present in a saturated soil drains away rapidly, while heavy rainfall can quickly bring the soil back to saturation. The gaseous constituents of soil air are derived largely from the atmosphere, the respiration and metabolism of soil organisms, and from the evaporation of soil moisture. Soil air is continuous with the atmosphere provided that the soil surface is not sealed due to compaction or crusting, and such continuity ensures the free movement and exchange of gases. The gases move along gradients of partial pressure, so that oxygen will tend to migrate from the atmosphere where its partial pressure is high into the soil where it is low. Conversely, carbon dioxide and water vapour will tend to migrate from the soil into the atmosphere. Movement of gases may occur by diffusion, mass flow or in dissolved form, with diffusion being by far the most significant of these processes.

The atmosphere contains approximately 78 percent nitrogen, 21 percent oxygen and 0.03 percent carbon dioxide by volume. In comparison, soil air contains similar amounts of nitrogen, slightly less oxygen (15–20 percent) and more carbon dioxide (0.25–5.0 percent). The balance between levels of oxygen and carbon dioxide in soil air depends largely on the rate of respiration of soil organisms and on the diffusivity characteristics of the soil. Respiration rates often vary seasonally in response to the availability of organic substances for consumption, and to variations in temperature and moisture conditions. Diffusion of gases occurs most effectively in dry soils with large and interconnected pores. The presence of water tends to reduce pore continuity and consequently the soil becomes increasingly oxygen deficient or *anaerobic*. In addition to carbon dioxide, organisms release other gases into the soil, including methane (CH_4) and hydrogen (H_2), as a result of organic matter decomposition (section 3.2.2).

2.3 SOIL PHYSICAL PROPERTIES

2.3.1 Mineral particles

The principal properties of soil mineral particles in an environmental context are their size, shape, nature of surface, orientation and mineralogy. The mineral fraction of soils consists of particles that vary dramatically in size from large boulders (several m in diameter), through cobbles and pebbles (several cm in diameter) to sand, silt and clay (< 2 mm in diameter). Most studies of soil are concerned with the < 2 mm size range, which is often referred to as the *fine fraction* or *fine earth*. It is the proportion by weight of the size categories within the fine fraction which defines the particle size distribution or *texture* of a soil. Examples of commonly used textural classification systems are shown in Figure 2.11. The particle size distribution of a soil is usually recorded numerically on a percentage by weight basis, or graphically using either a triangular graph, on which only three size categories can be shown, or a cumulative frequency curve which allows a fuller range of size classes to be recorded (Figure 2.12). Most soils comprise a continuous spectrum of particle sizes, and the width of this spectrum is defined by the degree of *sorting*. Poorly sorted soils possess a wide range of particle sizes, whereas well sorted soils have a narrow range (Figure 2.12b). Texture can be estimated in the field simply by rubbing the soil between thumb and forefinger. Sand grains are easily distinguished by their coarseness, while silt has a distinctive soapy feel and clay is characteristically plastic and mouldable when moist. For a more detailed indication of texture,

26

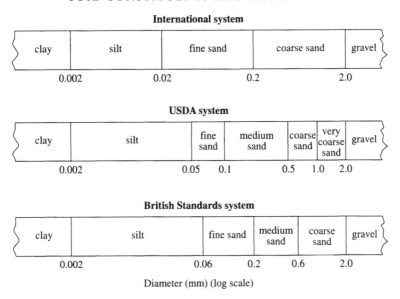

Figure 2.11 Examples of commonly used soil textural classification systems
(from White 1987)

however, laboratory analysis of the soil is required. This can be effected by the traditional method which involves sieving and sedimentation (Rowell 1994), or by more rapid methods using specialised instrumentation such as a Coulter Counter or laser diffractometer (Baize 1993).

The morphology of mineral particles relates to their shape and surface characteristics. For large particles this can be determined in the field, but for smaller material it is necessary to use a microscope. Particle shape can be expressed in two or three dimensions. Common terms adopted in two-dimensional expression include roundness, angularity and elongation, while in three-dimensional expression particles are recorded in terms of the extent of sphericity and flatness, using descriptions such as sphere, disc, rod, cube and prism. Such descriptions can be qualitative, or by reference to a comparison chart semi-quantitative estimates can be made. A number of indices have also been devised in order to quantify shape more accurately (Briggs 1977a) (Table 2.3). The nature of the surface of a particle is usually referred to as its *surface texture*, although this in no way relates to the texture of a soil as an expression of its particle size distribution. Surface texture is most commonly evaluated for the fine fraction, using a scanning electron microscope, which allows the grain surfaces to be viewed in three dimensions at very high magnification (Figure 2.13). Specific surface features, or combinations of features, are often characteristic of a particular

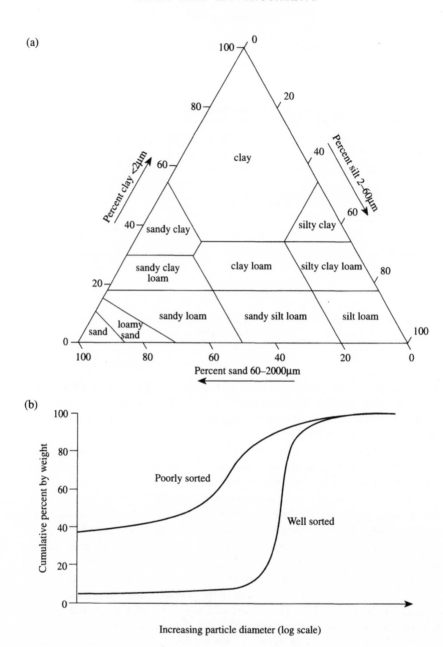

Figure 2.12 Depiction of soil textural data: (a) textural triangle (based on Hodgson 1976), (b) cumulative frequency curves

Table 2.3 Indices of particle shape

Dimensional expression of shape	Index	Formula
2-D	Cailleux's roundness (R)	$R = (2r/a) \times 1,000$
	Powers' scale	Visual comparison chart
3-D	Zingg's classification	Based on ratios of b/a and c/b
	Krumbein's sphericity (S)	$S = 3 \sqrt{(bc/a^2)}$
	Cailleux's flatness (F)	$F = ((a + b)/2c) \times 100$

Note: a = long axis; b = intermediate axis; c = short axis ; r = minimum radius of curvature at the end of the longest axis of the particle in its plane of maximum projection (measured by comparison with a set of standard concentric semi-circles)

genetic environment (Krinsley and Doornkamp 1973, Whalley and McGreevy 1983).

Orientation of mineral particles describes the disposition of the particle in three dimensions, and this can be a useful indicator of the direction of movement of soil material. The measurement of orientation is usually made with reference to particle longest axes, either in the horizontal plane as a compass bearing, or in the vertical plane as an angle of dip or plunge. In some cases, both sets of measurement are made and the results combined. Orientation can be depicted in a variety of ways including bar charts, rose diagrams and polar scattergrams (Briggs 1977a) (Figure 2.14).

Minerals differ markedly in their composition (section 2.2.1) and consequently many physical and chemical characteristics of soils, including texture, acidity and nutrient status, can be related to their mineralogy. Mineralogical studies are also important in the examination of weathering in soils (section 3.3.1). The mineralogical characteristics of large particles can be evaluated in the field by the inspection of hand specimens, but for smaller particles laboratory methods are required. Sand- and coarse silt-sized particles are usually examined using a petrological microscope, but clay and fine silt are too small to be easily resolved in this way and their mineralogy is commonly identified using the technique of X-ray diffractometry (Baize 1993). This allows the crystal lattice dimensions to be measured and is particularly useful in the identification of sheet silicate minerals (section 2.2.1). Samples are scanned by an X-ray beam and when a dominant lattice spacing is encountered, diffraction occurs and a peak is registered on a moving chart recorder (Figure 2.15). Each mineral has a characteristic set of peaks, although in some cases more than one mineral may have a peak in the same position. However, it is possible to distinguish between these minerals using simple chemical or heat pre-treatments which produce characteristic changes in the lattice spacing of one mineral but not in another (Brindley and Brown 1980).

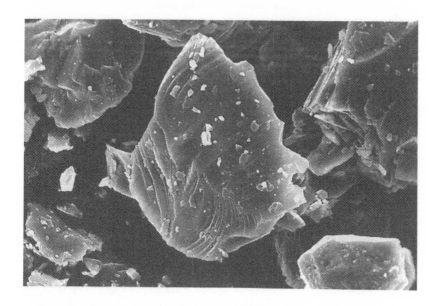

(a)

(b)

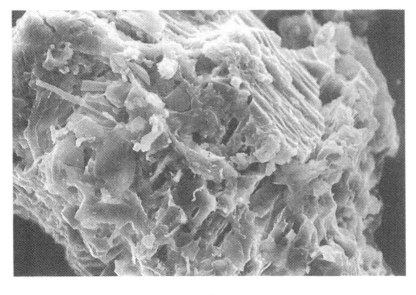

(c)

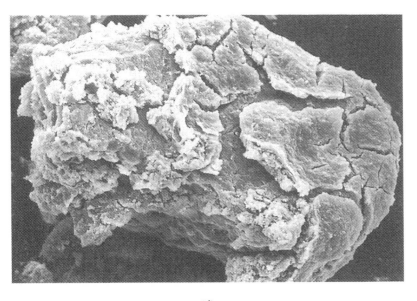

(d)

Figure 2.13 Scanning electron photomicrographs showing a variety of surface textures on fine sand grains: (a) glacially modified quartz grain, (b) incipient etching of feldspar grain along crystallographic weaknesses, (c) advanced etching of feldspar grain, (d) mineral grain with illuviation coating

31

(a)   (b)

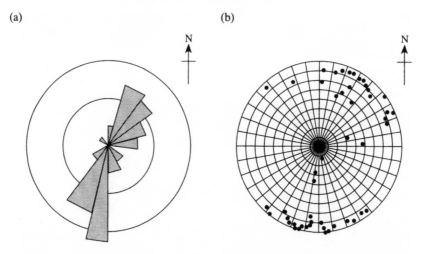

Figure 2.14 Depiction of particle orientation data: (a) rose diagram showing percentage of particles in each azimuthal class, (b) polar scattergram showing orientation and dip of each particle; particle dip is plotted on circular scale and decreases with increasing distance from centre of circle (based on Briggs 1977a)

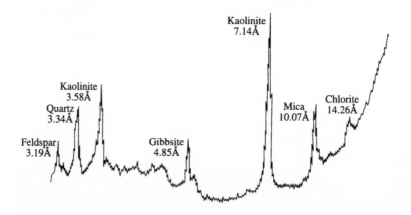

Figure 2.15 An X-ray diffraction trace showing the clay mineralogy of a soil B horizon from northeast Scotland

2.3.2 Aggregates

Aggregation in soils is promoted by a number of physical, chemical and biotic forces. Physical forces include expansion and shrinkage associated with wetting and drying, and compaction by raindrop impact, animal trampling and agricultural machinery. Chemical forces are largely electro-

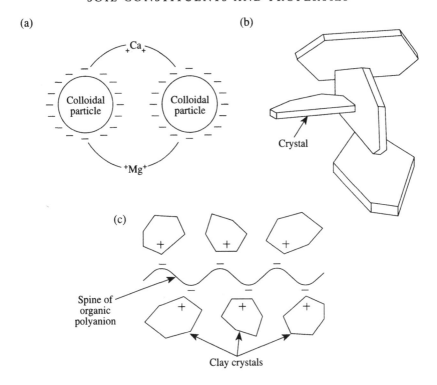

Figure 2.16 Interparticle attraction in aggregate formation: (a) cation bridging, (b) edge-face attraction of kaolinite crystals, (c) attraction between kaolinite crystals and an organic polyanion to form a 'string of beads' arrangement

static in character and often depend on the presence of adsorbed cations in association with the negative surface charge of colloidal particles such as clay and humus. The generation of negative surface charge on colloidal particles and the processes of cation adsorption are discussed in more detail in section 2.4.2. Multivalent cations in particular, such as Ca^{2+}, Mg^{2+} and Al^{3+}, have the ability to form an attachment with more than one colloidal particle, a process known as *cation bridging* (Figure 2.16a). Similarly, attraction may occur between positive charges on the broken edges of sheet silicate particles and the negative charges on the faces of other similar particles (Figure 2.16b). In soils with significant positive charge, anion bridging may occur, particularly if multivalent anions are present (Figure 2.16c). In response to the various mechanisms of interparticle attraction, particles flocculate together to form larger units known as *domains*; these may be up to $5\,\mu m$ in diameter. In addition to electrostatic interparticle forces, interaggregate attraction can occur due to the binding effects of various organic compounds, fungal hyphae and plant roots. Organic polymers, such as polysaccharide gums, together with organic mucilages and fungal hyphae, facilitate the binding of domains to form microaggregates which may be up to 250 μm in diameter (e.g. Churchman and Tate 1987, Haynes and Swift 1990). At the larger scale, plant roots and fungal hyphae play an important role in the binding of microaggregates to form macroaggregates or *peds* which can be easily observed in the field. Inorganic cementing agents such as carbonate and iron compounds can perform a similar binding function in soils.

Because of the different scales of aggregation, it is often viewed in terms of a hierarchical model comprising building blocks which increase progressively in size (Figure 2.17) (e.g. Tisdall and Oades 1982, Miller and Jastrow 1990). The degree of persistence of aggregates varies between the different levels in the model. Electrostatic forces predominate at the domain level and are particularly resistant to change. Similarly, at the microaggregate level, binding forces are relatively persistent, although they vary to some extent according to the organic content of the soil (Piccolo and Mbagwu 1990). In contrast, at the macroaggregate level, binding forces are often transient, being closely influenced by variations in plant cover and the development of root networks.

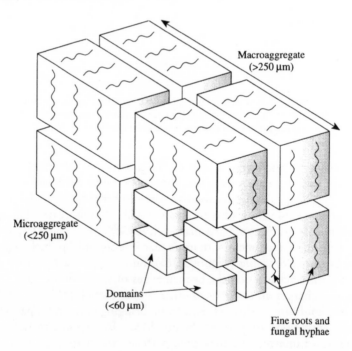

Figure 2.17 Diagrammatic representation of the hierarchical nature of soil aggregate formation (based on Greenland 1979)

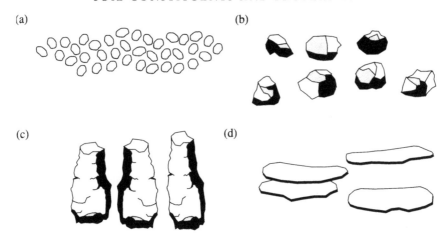

Figure 2.18 The main groups of ped morphology: (a) spheroidal, (b) blocky, (c) prismatic, (d) platy

Aggregates, or peds, which persist during wetting/drying and freezing/thawing cycles form the basis of soil *structure*. Soil structure is defined in terms of the size, shape and arrangement of particles, aggregates and pores. It is classified on the basis of ped morphology (shape), class (size) and grade (distinctiveness and durability). There are four main groups of ped morphology – spheroidal, blocky, prismatic and platy (Figure 2.18). Spheroidal peds are equidimensional in form and can be divided into granular (non-porous) and crumb (porous) types. Granular and crumb structures are most commonly found in the A horizons of soils, their development being facilitated by the presence of roots and other forms of organic material. Blocky peds are also equidimensional and may possess several curved or planar surfaces. Frequently they are observed in B horizons, particularly in soils with significant amounts of clay, where they develop in response to the effects of regular wetting and drying cycles. Prismatic and associated columnar peds are characterised by strong vertical and weak horizontal development. The tops of prismatic peds are poorly defined, whereas those of columnar peds are often rounded in form. Like blocky structures, prismatic and columnar structures often develop as a result of wetting and drying but are found in the B horizons of less clay-rich soils. Platy structure displays strong horizontal and weak vertical development, and often develops as a result of compaction. The structures are sometimes lenticular in form, with their centres being thicker than their edges, and these are often found in soils that are subjected to regular freezing and thawing cycles. Peds which comprise more than one structural type are often observed in soils and are known as *compound peds*. For example, large prismatic peds may be made

35

Table 2.4 Ped size classification

	Spheroidal	*Blocky*	*Prismatic*	*Platy*
Fine	< 2 mm	< 10 mm	< 20 mm	< 2 mm
Medium	2–5 mm	10–10 mm	20–50 mm	2–5 mm
Coarse	5–10 mm	20–50 mm	50–100 mm	5–10 mm
Very coarse	> 10 mm	> 50 mm	> 100 mm	> 10 mm

Source: Based on Hodgson 1976

up of smaller, blocky peds. It is also not uncommon to find that structures vary between the horizons within a soil profile, such that the A, B and C horizons could possess crumb, prismatic and blocky structures respectively.

Ped class or size is usually classified according to the system shown in Table 2.4. Grade, which describes the distinctiveness and durability of peds, may be expressed as structureless (apedal), or as weakly, moderately, well or strongly developed. An apedal soil is described as massive if it is coherent and as single grain if it is not. Usually, massive soils are fine-textured while single grain are coarse-textured. Weakly developed structure comprises poorly formed, indistinct and weakly coherent peds, most of which will be broken down if the soil is disturbed, while strongly developed structure comprises well formed, distinct and durable peds. The grade of structure depends on a number of soil characteristics, particularly moisture content and the quantity of aggregating colloids such as clay and humus; peds tend to be considerably more durable in dry soils and in those with high colloidal contents.

2.3.3 Pore space

Pore spaces vary dramatically in shape from spherical voids to tortuous, interconnecting cracks and channels. They also vary in size from large macropores of several cm in diameter to very fine micropores which may be < 1 μm in diameter. Generally, a diameter of 60 μm (0.06 mm) is taken as the size boundary between macro- and micropores. Pore space will influence both the *bulk density* and the *porosity* of a soil. Bulk density refers to the specific gravity of a bulk soil sample, usually collected as an undisturbed core. Its calculation is then derived from measurement of the mass and volume of the dried core. Values of bulk density are generally considerably lower than those of the particles which make up a soil, because they are based on both solid material and pore space rather than on the solid components alone. For example, the average density of soil particles is often assumed to be 2.65 g cm^{-3}, whereas bulk density values can range from around 2.0 g cm^{-3} in sandy soils and compacted clay layers to < 1.0 g cm^{-3} in organic soils. If the bulk density of a soil is not measured, then an average value of 1.33 g cm^{-3} is often assumed.

Porosity is a measure of the percentage volume of pore space, and can be determined indirectly from particle and bulk density as follows:

$$\text{Porosity (\%)} = 1 - \frac{\text{Bulk density}}{\text{Particle density}} \times 100$$

Consequently, a soil with a bulk density of 1.33 g cm^{-3} and a particle density of 2.65 g cm^{-3} will have a porosity of 50 percent. In some instances porosity is expressed as a ratio rather than a percentage, thus the pore space ratio (PSR) of a soil with 50 percent porosity is 0.5. Porosity can also be determined directly from an undisturbed core of known volume (V). Initially the core is saturated with water and the weight (Ws) recorded. It is then dried and re-weighed (Wd), and the difference between Ws and Wd represents the weight of water held by the core in its saturated state. This will be equivalent to the water volume, because the specific gravity of water is 1.0 g cm^{-3}. Hence, porosity is calculated as follows:

$$\text{Porosity (\%)} = \frac{\text{Ws} - \text{Wd}}{\text{V}} \times 100$$

Using undisturbed cores in their field moist state, it is possible to determine the proportions of water and air occupying the pore space. Water-filled porosity is equivalent to the volumetric water content (θv) of the soil. This is calculated from the weight of field moist soil (Wm), its weight after drying (Wd) and the volume of the core (V):

$$\theta v = \frac{\text{Wm} - \text{Wd}}{\text{V}} \times 100$$

As with total porosity, values of water-filled porosity can be expressed as a percentage or on a proportional basis. On average, values are about 25 percent or 0.25, although in a saturated soil the value will be equivalent to the total porosity, whereas in a dry soil it may be < 5 percent or 0.05. Air-filled porosity can be calculated by subtraction of water-filled porosity from total porosity. A more sophisticated method of porosity measurement is provided by image analysis, in which undisturbed sections of soils are photographed and the total area occupied by pores is quantified using an image analyser (e.g. Mackie-Dawson et al. 1988).

In addition to porosity, pore size distribution is an important soil characteristic. It is associated directly with water retention, drainage and aeration, and therefore has a major influence on plant growth. It is most often established from moisture characteristic or moisture retention curves (Figure 2.19) (e.g. Buchan and Grewal 1990). This involves determining the volumetric water content at various points over a range of tensions or suctions

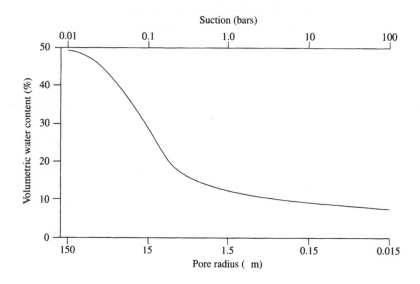

Figure 2.19 Soil moisture characteristic curve (based on White 1987)

(section 2.3.4) applied to an undisturbed soil core (Hall *et al.* 1977). As tension is increased, water is removed from progressively smaller pores and because tension is inversely proportional to pore radius, the volume of pores of a certain size can be determined from the amount of water extracted at the appropriate tension. In Figure 2.19, for example, the total porosity is 50 percent and, of this, 15 percent of pores are > 15 μm in diameter, 15 percent are 1.5–15 μm, 10 percent are 0.15–1.5 μm and 10 percent are < 0.15 μm.

Porosity and pore size distribution are influenced by a number of soil characteristics, particularly texture, degree of aggregation, bulk density, presence of swelling clays, and organic content. Closely packed sands, for example, could have a theoretical minimum porosity of about 10 percent. However, as soils usually possess a wide range of particle sizes and aggregates, and because soil particles are rarely spherical, values of porosity are usually much greater than this. Sandy and compacted clay soils may have porosities of < 40 percent, whereas a fine-textured A horizon can have values of > 60 percent.

2.3.4 Moisture

Soil water possesses free energy which is a measure of its potential for movement and change in the soil. In soils with a high moisture content, forces attracting the water to solid particles are weak and its free energy is high. As moisture content decreases, however, the attractive forces become progressively stronger and its free energy decreases. Soil moisture is affected by three types of force determined by soil properties which can either encourage or restrict water movement. First is *adsorption* whereby water molecules are attracted to the surfaces of colloids mainly by electrostatic forces. This is therefore an important process in soils with high clay or organic matter contents, in which such forces are high (section 2.4.2). Second is *capillarity* by which water is held in soil pores by adsorptive forces between the water and pore surfaces and surface tension forces at the water surface. The force by which the water is held increases with decreasing pore size, therefore water can drain out of large pores more easily than from smaller ones. The combined effect of adsorptive forces and capillarity is known as *matric suction*. The third force is *osmosis* which occurs between solutions of different ionic concentrations, water moving from lower to higher concentration solutions. This is important in saline soils in which high solute concentrations can form due to the relative ease of dissolution of salts.

These forces are usually expressed in units of atmospheres (atm), bars or kiloPascals (kPa), with 1 atm being approximately equal to 1 bar (1.02 atm = 1 bar), and 1 bar equal to 100 kPa. In the past, these forces were sometimes expressed as the logarithm of the *hydraulic head* (pF); this is the length of a column of water which is required to produce a given positive pressure, or which can be supported by a given negative pressure (tension). This concept is, however, rarely used today.

Soil water is held most strongly (30 to > 1,000 atm) when it is adsorbed onto colloidal particle surfaces in the form of thin films only a few molecules in thickness (hygroscopic water), or when it is bound up in mineral structures (structural water). Further away from particle surfaces, water is held in small micropores at tensions of 0.05–30 atm by capillarity; much of this water can be lost from the soil through evaporation and plant uptake. Water held at tensions below 0.05 atm occurs in the macropores, and this will usually drain rapidly from the soil, in two or three days, under the influence of gravity and is therefore known as *gravitational water*. Thus, on moving away from particle surfaces into progressively larger pores, tensional forces become progressively weaker, decreasing logarithmically with increasing pore size, until they are unable to counteract the effect of gravity (Figure 2.20). A soil which has lost all of its gravitational water through drainage contains the maximum level of plant-available water and is said to be at *field capacity*. If, however, a soil loses all of its available water through evaporation and plant uptake, without further wetting, then it is said to have reached *wilting point*. Soil water will be discussed in more detail in sections 6.2 and 6.5.2 in terms of the ways in which soil properties influence its movement, storage and availability for plant growth.

A variety of methods are available for the measurement of soil moisture. For example, it can be determined gravimetrically using bulk samples; these

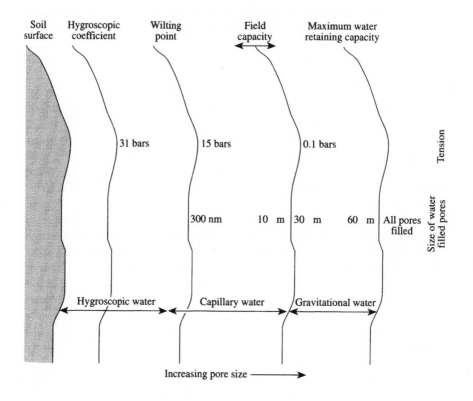

Figure 2.20 Soil–water relationships (based on FitzPatrick 1986)

are weighed in their field-moist and oven-dried states, the weight difference representing the moisture content, which is then expressed as a percentage of either field-moist or oven-dried soil. Alternatively, moisture content can be expressed on a volumetric basis using soil cores of known volume. Weight measurements are made as above and, given that the specific gravity of water is $1.0 \, g \, cm^{-3}$, the moisture content can then be expressed as a percentage of the soil volume. Changes in moisture content within a soil can be determined using a neutron probe. As the probe is lowered through aluminium tubes inserted into the soil, it emits neutrons which collide with the hydrogen nuclei contained in water molecules, and a sensor detects backscatter from the collisions, the intensity of which is in direct proportion to water content. Problems have, however, been experienced with the use of neutron probes, particularly in soils containing significant amounts of swelling clays, where cracking results from drying and associated shrinkage (e.g. Jarvis and Leeds-Harrison 1987).

The strength of tensional forces (matric potential) by which water is held in the soil can be determined using a tensiometer. This consists of a sealed plastic tube filled with water, with a ceramic cup at one end and a vacuum gauge at the other. The tube is set in the soil at the required depth and the gauge adjusted to zero. As water moves out into the soil through the cup, tension increases progressively until the water in the tensiometer is in equilibrium with that in the soil. At this point, movement ceases and the tension recorded on the gauge is a measure of the matric potential; tensiometers can also be connected to pressure transducers and logged electronically. The tensiometer can only be used at relatively low tensions, between about 0 and –1.0 atm, but this range is important within the context of plant growth; plants obtain water via their roots by exerting a tensional force of about 0.05–15 atm.

2.3.5 Temperature

Soil temperature is an extremely dynamic property, varying diurnally and seasonally, with the effects being most rapid and extreme towards the surface (Figure 2.21). On a diurnal time-scale, soils are heated during the day and the effect gradually extends downwards, perhaps taking several hours to reach a depth of 30 cm. At night soils cool rapidly at the surface and heat is transferred upwards from within the soil. Seasonal heating and cooling cycles operate in a similar manner, but they penetrate deeper into the soil than

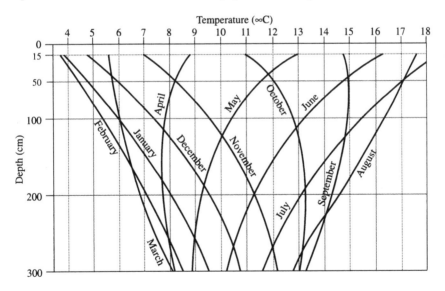

Figure 2.21 Seasonal variations in soil temperature in southern England (from FitzPatrick 1986)

diurnal cycles because the time-scale is much greater; diurnal cycles usually affect only the upper 30 cm or so, whereas seasonal cycles can penetrate to a depth of several metres. In order to detect their rapid variation, soil temperatures must be measured frequently and at a variety of depths, and are usually recorded electronically using a series of thermistors linked to an automatic recorder. Soil temperature is influenced by a number of soil properties, in particular texture, moisture and organic content; these will be discussed later in the context of soil influences on small-scale climatic characteristics (section 6.3.1).

2.3.6 Mechanics

Commonly encountered soil mechanical properties are strength, stability and consistence. The mechanical strength and stability of soils are derived from interparticle and interped forces responsible for the development of soil structure (section 2.3.2). Aggregate stability is a useful measure of the structural stability of soils, and is based on the ability of aggregates to survive wetting. During wetting, air is trapped in pores within the peds, and the pressure of the trapped air increases until it is sufficient to cause aggregate breakdown or *slaking*. In addition, water entering the aggregates interferes with electrostatic forces of interparticle attraction and may dissolve cementing agents, thus causing further weakening. Measurements of aggregate stability have included determining the proportion by weight of aggregates retained on a 2 mm sieve after wet-sieving or after ultrasonic dispersion, although such methods have been questioned because they subject the soil to artificial forces for relatively short periods of time (Matkin and Smart 1987). The strength and stability of a soil can also be assessed from its resistance to compression and shear, which provides a useful indication of the degree of cohesion. Resistance to compression can be determined using a penetrometer, whilst a shear vane can be used to measure resistance to shear. These characteristics can also be established from a triaxial shear test, in which a soil core is first subjected to compression in order to determine its load-bearing capacity, and then to a shear test (Selby 1993).

Soil strength is closely related to a number of soil properties, particularly texture, organic content, bulk density and moisture content. Coarse-textured soils frequently possess a relatively high degree of strength due to the often irregular nature of particle boundaries and the resulting large surface area of contact. Coarse-textured soils with smooth particle boundaries are considerably weaker, however, due to the smaller surface area of contact. Fine-textured soils are often strongly cohesive and well aggregated, and are particularly strong when dry. When wet, however, the cohesive forces tend to break down and soil strength decreases dramatically. The presence of organic matter improves the cohesive properties of soils and increases their strength. Similarly, soil strength increases with increasing bulk density, due largely to the reduction in total porosity.

In contrast to cohesion, dispersion may reduce the strength and stability of soils. This commonly occurs in soils whose colloidal fraction is dominated by adsorbed sodium, as often occurs in semi-arid environments. Sodium ions are only loosely held by colloidal particles, and are unable to neutralise fully the surface negative charge. This situation promotes interparticle repulsion rather than attraction, thus resulting in breakdown of soil structure. Structure may also be disrupted by repeated shrinkage and swelling, which can be particularly active in soils containing large amounts of swelling clays. The shrink–swell potential of a soil can be determined from the coefficient of linear extensibility (COLE), which describes the volume change of a length of soil on drying, and is calculated as follows:

$$COLE = \frac{Lm - Ld}{Ld}$$

where Lm and Ld are the moist and dry sample lengths respectively.

Soil consistence describes the change in state of soil, from solid through plastic to liquid, with increasing moisture content (Figure 2.22). It is

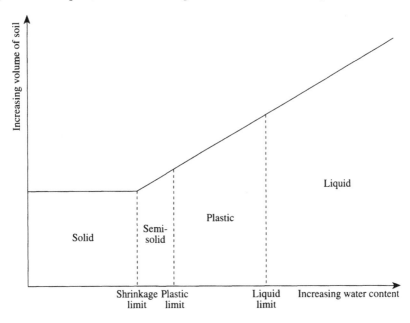

Figure 2.22 The consistency or Atterberg limits: relationship between soil volume and water content

43

established from the three Atterberg limits – shrinkage, plastic and liquid limits (Selby 1993). The shrinkage limit represents the minimum moisture content above which soil volume begins to increase with further wetting; below this limit soil volume remains constant irrespective of changes in moisture content. The plastic limit represents the minimum moisture content above which the soil becomes mouldable, and the liquid limit represents the minimum moisture content above which the soil flows under its own weight. The plastic and liquid limits in particular are closely associated with a number of soil properties, for example cation exchange capacity (section 2.4.2), clay content and specific surface area.

2.3.7 Colour

Soil colour is a characteristic which can easily be determined in the field, and which provides useful information regarding the presence or absence of certain soil constituents. For example, dark colours are usually indicative of high organic contents, while red colours are characteristic of soil rich in iron oxides, and blue-grey colours indicate the presence of iron in its reduced form. Such indications are, however, only general because, for example, manganese can also produce a dark coloration, while the intensity of colour is often related to moisture content. The identification of soil colour is also rather subjective, as different individuals interpret colour in different ways. In an attempt to reduce subjectivity, soil colour is most often determined using the Munsell system where soil samples are matched against standard colour charts. Munsell colour notation has three components – the *hue* which indicates the major colour(s) present, the *value* which is a measure of the degree of darkness or lightness of the colour, and the *chroma* which is a measure of colour intensity. Pages of the chart are shown according to hue, and each colour chip on the page has its own co-ordinate, expressed in terms of the value on the vertical axis and the chroma on the horizontal axis. A sample of soil is therefore represented by a notation according to the colour to which it most closely corresponds, and each notation has an equivalent description; for example, a soil with a hue of 10YR, a value of 3 and a chroma of 4 is represented as 10YR 3/4, which has the description yellowish brown. The objective determination of soil colour has been the subject of much discussion, and although the Munsell system is the most widely adopted, other methods and indices have been developed (e.g. Melville and Atkinson 1985, Barron and Torrent 1986).

2.4 SOIL CHEMICAL PROPERTIES

2.4.1 Elements and compounds

Elements and compounds in a soil occur in two principal forms – as the chemicals that make up the structure of the basic soil constituents, and as individual components which are held in the soil by interparticle attraction. The first of these are important in terms of soil properties only once the constituents start to break down, whereupon they are released into the soil, but the second of these are generally of more immediate importance because they are more readily available for interaction. The chemicals that make up the structure of mineral material are determined by total chemical analysis, for example by atomic absorption spectrometry following dissolution of the material in strong acids (Lim and Jackson 1982), or by the more rapid method of X-ray fluorescence spectrometry (Jones 1982). In terms of organic matter, the principal structural elements which are normally determined are carbon and nitrogen. Organic carbon content is usually determined by dichromate oxidation, and this can also be used to estimate the organic matter content of a soil by using a conversion factor of 1.7 (organic matter % = organic carbon % × 1.7). Organic matter content can also be determined by loss on ignition or by digestion with hydrogen peroxide, although these methods may not always be particularly accurate (Nelson and Sommers 1982). Nitrogen content is most often determined by the Kjeldahl digestion procedure (Rowell 1994).

Individual elements and compounds which are commonly examined as individual components include exchangeable bases, free iron and aluminium, carbonates and heavy metals. Exchangeable bases are held in the ion exchange complex of a soil (section 2.4.2) and are usually extracted using ammonium acetate, then analysed by flame emission or atomic absorption spectrometric methods (e.g. Rowell 1994). Exchangeable cation content is commonly expressed in units of milliequivalents (me) per 100 g of dry soil, which represents the amount of a cation that will replace or combine with 1 mg of hydrogen per 100 g of dry soil. Exchangeable base content can also be expressed in units of $cmol(+)kg^{-1}$, which is exactly the same as me per 100 g. Free iron and aluminium are present in soils in a number of forms, including sesquioxides (oxide and hydroxide compounds), poorly crystalline or amorphous allophanic and imogolitic materials (section 2.2.1), and organo-metallic complexes (section 3.2.2). These can be determined using a range of selective extractants (Bascomb 1968, Farmer et al. 1983). For example, total free iron and aluminium is often determined by extraction with sodium dithionite or by a dithionite-citrate-bicarbonate (DCB) extraction. Inorganic forms of iron and aluminium, which are poorly crystalline or amorphous, can be extracted by sodium oxalate, and organically bound forms by potassium pyrophosphate. The potency of these extractions decreases in the

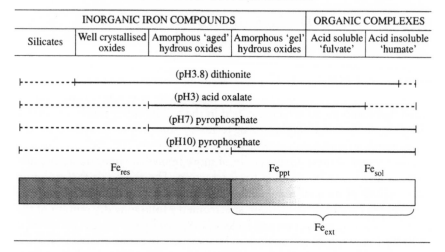

Figure 2.23 The variety of iron compounds in soils and their removal by different extractants (from Bascomb 1968)

order: DCB > oxalate > pyrophosphate, and there is thought to be relatively little overlap between them (Figure 2.23). It is therefore possible to determine levels of all these different forms by sequential extraction performed on the same sample. The most common carbonates in soils are compounds of calcium and magnesium, derived largely from carbonate-rich parent materials. Presence or absence of carbonates can be established simply by adding a few drops of dilute hydrochloric acid (HCl) to a small sample of soil; effervescence is observed if carbonates are present. More accurate measurement can be made in the laboratory, for example using a calcimeter (Baize 1993), which measures the calcium carbonate ($CaCO_3$) equivalent by determining the quantity of carbon dioxide (CO_2) evolved during treatment with hydrochloric acid:

$$2HCl + CaCO_3 \Rightarrow CaCl_2 + CO_2 + H_2O$$

Heavy metals can occur naturally in soils, but often occur at enhanced levels due to additions from motor vehicle emissions, sewage sludge applications, metal mining and smelting, and scrap metal processing (Thornton 1991). They can exist in a number of forms including numerous compounds, particularly oxides, sulphides and sulphates, metal cations such as lead, zinc and cadmium (Pb^{2+}, Zn^{2+} and Cd^{2+}), which may be adsorbed onto the surfaces of negatively charged colloidal particles, and organo-metallic complexes. Heavy metals are commonly extracted from soils by acid digestion, using a strong mineral acid such as nitric acid to determine total contents, and EDTA or a weak organic acid such as acetic acid to determine

46

'plant-available' levels (Rowell 1994). Contents are then measured by atomic absorption spectrometry.

2.4.2 Ion exchange

Ion exchange is a most important soil property in that it plays a key role in plant nutrition, and in a broader context, in the development of many chemical characteristics of soils. Central to ion exchange is the way in which ions are held on the surfaces of colloidal particles. Colloidal material consists mainly of clay and humus particles and is often referred to as the *clay-humus complex* or *exchange complex*. These particles have a high specific surface area (ratio of surface area to volume) and possess both high surface energy and significant surface charge. This charge is largely negative and can occur either as permanent charge or variable (pH dependent) charge. In the case of permanent charge, cations in mineral structures are replaced by cations of similar size but with lower valency (isomorphous substitution); such charge is independent of acidity. In the case of variable charge, hydrogen ions undergo reversible dissociation from surface groups such as $-OH$ and $-OH_2^+$ on the edges of clay minerals and oxides, and $-COOH$, $-OH$ and $-NH_2$ in organic material. As a result of their negative surface charge, colloidal particles behave like giant anions and are known as *micelles*. The micelles are able to attract cations onto their surfaces, a process known as *adsorption* (Figure 2.24). With increasing distance from the micelle surface, the concentration of cations decreases exponentially whilst the concentration of anions shows a reciprocal increase. The zone of adsorbed cations, together with the surface negative charge on the micelle, are often referred to as the *electrical double layer*.

The ease with which a cation is adsorbed depends on its valency and degree of hydration. Cations with a high valency have a high energy of

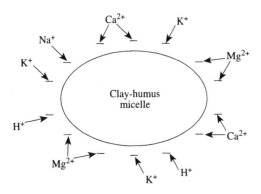

Figure 2.24 Clay-humus micelles and cation adsorption

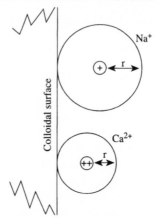

Figure 2.25 Factors influencing cation adsorption preference; r = radius of hydration (based on Foth 1990)

adsorption and are therefore adsorbed in preference to lower valency cations. For cations with equal valency, however, the one with the smallest *radius of hydration* (radius of the ion together with the associated water molecules surrounding it) is adsorbed preferentially because it can get closer to the micelle surface (Figure 2.25). The generally accepted sequence of preferential adsorption for base cations is $Ca^{2+} > Mg^{2+} > K^+ > Na^+$, and this is reflected in their usual proportions in the soil ($Ca^{2+} = 80\%$, $Mg^{2+} = 15\%$, $K^+ + Na^+ = 5\%$). Adsorbed cations can be exchanged for cations in the soil solution:

$$\text{Micelle-A} + B^+(aq) \Rightarrow \text{Micelle-B} + A^+(aq)$$

The sequence of preferential adsorption again applies, with cations of low valency or high radius of hydration being exchanged in preference for cations of higher valency or smaller radius of hydration, depending on their availability. Plants obtain many of their nutrients through this exchange mechanism, the nutrient cations being taken up through the root system in exchange for hydrogen ions.

Although cation adsorption and exchange tend to predominate in soils, in some circumstances anion adsorption and exchange are more common. While the former occur in soil containing significant amounts of humus, and clay minerals with a high specific surface area such as montmorillonite, vermiculite and illite, the latter occur in soil with limited humus content and with clay minerals with a low specific surface area such as kaolinite. Oxides of iron and aluminium are also significant in anion adsorption as they possess variable surface charge which is positive under acidic conditions (Figure 2.26). This positive charge results from the combination of hydrogen (H^+) ions with edge hydroxyl groups, for example:

48

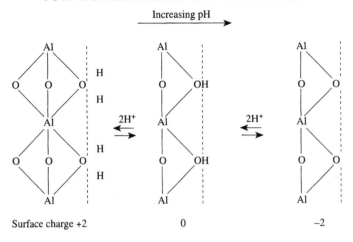

Figure 2.26 Influence of pH on surface charge in an aluminium oxide. The addition of H^+ under acid conditions produces positive charge, whereas the removal of H^+ at higher pH produces negative charge (from White 1987)

$$Al\text{-}OH + H^+ \Rightarrow Al\text{-}OH_2{}^+$$

As in the case of cations, anions with a high valency are able to get closer to surfaces than anions with a low valency. Thus sulphate ($SO_4{}^{2-}$) and phosphate ($PO_4{}^{3-}$) are often strongly held, whereas nitrate ($NO_3{}^-$) tends to be held rather weakly. Like cations, adsorbed anions can also be exchanged for anions in the soil solution.

Unlike Ca^{2+}, Mg^{2+} and K^+, which are important plant nutrients, Na^+ is toxic to many plant species, and it also has a deleterious effect on soil structure, promoting the dispersal of aggregates (section 2.3.2). The assessment of soil salinity can therefore be important in these instances, and various expressions are commonly used. The *sodium adsorption ratio* (SAR) represents the amount of Na^+ relative to Ca^{2+} and Mg^{2+}, and is determined as follows:

$$SAR = \frac{Na^+}{(Ca^{2+} + Mg^{2+})^{0.5}}$$

The SAR is approximately equal to the *exchangeable sodium percentage* (ESP) of the soil, where the Na^+ level, measured in me per 100 g (section 2.4.1), is expressed as a percentage of the total exchangeable base content. An ESP value of 15 percent is generally accepted as the threshold above which soils are considered to be sodium-affected (sodic). Soil salinity is usually determined from a saturation extract, where distilled water is added to the soil to produce a paste (Rowell 1994). Salinity is then established from the electrical conductivity of the saturation extract (EC_e).

The ability of soil to yield cations is measured by its *cation exchange capacity* or CEC. In effect it is a measure of the amount of negative surface charge, and hence of the potential for cation adsorption. CEC is often determined by ammonium acetate extraction, with cations on the soil exchange complex being displaced by ammonium ions, although this method can be unreliable due to the change in pH, induced by ammonium acetate treatment, producing a change in surface charge; alternative extraction methods have therefore been developed, but these have not yet become widely used (Baize 1993). As with exchangeable base content, CEC is expressed in units of me per 100 g or $cmol(+)kg^{-1}$. Values vary dramatically and tend to be highest in soils with high clay and organic contents. CEC values for organic matter may be 150–300 me per 100 g, virtually twice that of clay. Clay mineralogy also has a major influence on CEC, in terms of the charge density per unit area; CEC values range from 5–10 me per 100 g for kaolinite, through 30–40 for illite, to > 100 for vermiculite and montmorillonite.

From the content of exchangeable bases and CEC, it is possible to determine the proportion of exchangeable bases on the soil exchange complex, a property known as the *percentage base saturation*. This is calculated as follows:

$$\text{Base saturation (\%)} = \frac{(Ca^{2+} + Mg^{2+} + K^+ + Na^+)}{CEC} \times 100$$

The difference between CEC and total exchangeable base content provides an approximate measure of exchangeable hydrogen content, or *exchangeable acidity*:

$$\text{Exchangeable } H^+ = CEC - (Ca^{2+} + Mg^{2+} + K^+ + Na^+)$$

It is also possible to measure the *anion exchange capacity* (AEC) of a soil. Reliable methods of determination have been developed in relation to tropical soils, which often possess significant AEC values, unlike soils of temperate environments. Essentially these methods are based on the degree of adsorption of an index anion, usually Cl^- or NO_3^- (Rowell 1994).

2.4.3 Acidity and pH

Acids in aqueous solutions undergo dissociation to release their constituent ions, namely hydrogen (H^+) and an acid anion, for example:

$$H_2SO_4 \Rightarrow 2H^+ + SO_4^{2-}$$
$$\text{(sulphuric acid)} \qquad \text{(sulphate)}$$

Acidity is measured in terms of the H^+ ion concentration using the pH scale. The relationship between pH and H^+ ion concentration is inverse and logarithmic:

$$pH = -\log_{10} [H^+]$$

The pH scale ranges from 1.0 at the most acidic extreme to 14.0 at the alkaline extreme, with a value of 7.0 at neutrality. The pH of soils varies widely, from around 2.0 in acid sulphate soils to about 12.0 in alkaline sodic soils; good quality agricultural soils have a value around 6.0 to 7.0.

The measurement of soil pH is usually made in a standard suspension of 1:2.5 weight to volume (e.g. 10 g of soil in 25 ml distilled water), in order to ensure data comparability. Although distilled water is often used to make up the suspension, a suspension made with a dilute solution of calcium chloride (0.01M) is sometimes used in order to provide a more realistic value of H^+ concentration by minimising calcium release from the soil exchange complex. For this reason pH levels measured in calcium chloride suspension are generally lower than those recorded in a suspension made up with distilled water. The measurement itself can be made using electrometric (pH probe) or colorimetric techniques (Baize 1993).

In addition to H^+ ions, Al^{3+} ions play an important role in the generation of soil acidity, particularly in soils that are already acidic. Al^{3+} ions undergo hydrolysis during which H^+ ions are released into the soil solution:

$$Al^{3+} + H_2O \Rightarrow Al(OH)_2^+ + H^+$$
$$\text{(hydroxy-aluminium)}$$

Positively charged hydroxy-aluminium species can then occupy exchange sites, thus resulting in reduced CEC. Hydroxy-aluminium species may undergo further hydrolysis to produce yet more H^+ ions and stable aluminium hydroxide (gibbsite):

$$Al(OH)_2^+ + 2H_2O \Rightarrow Al(OH)_3 + 2H^+$$
$$\text{(gibbsite)}$$

Consequently, soil acidity promotes the development of further acidity through aluminium hydrolysis, and this becomes an important source of H^+ ions when soils become acidic. If soil pH falls below about 5.5, Al^{3+} ions themselves begin to occupy exchange sites. Because of their higher valency, Al^{3+} ions are adsorbed much more strongly than divalent and monovalent cations (section 2.4.2), therefore levels of exchangeable aluminium increase, and amounts of exchangeable bases decrease, as pH declines. Soil acidity is closely related to many other soil properties such as organic content, exchangeable base content and CEC (Table 2.5).

Table 2.5 Correlations (Pearson's *r*) between pH, total exchangeable base content, CEC and organic carbon content in morainic soils at Haugabreen, Norway (based on 43 samples from 11 soil profiles)

	pH	*Total exchangeable bases*	*CEC*
Total exchangeable bases	−0.37*		
CEC	−0.35*	+0.74***	
Organic C	−0.44**	+0.92***	+0.85***

Note: * = 95% significance level; ** = 99% significance level; *** = 99.9% significance level

2.4.4 Aeration

Soil aeration relates to the amount of oxygen present in the soil atmosphere. This can be assessed using both direct and indirect methods. The rate of oxygen diffusion through the soil can be measured directly using a platinum electrode (Rowell 1994), while the concentration of oxygen in a sample of soil air can be determined by gas–liquid chromatography (Smith and Arah 1991). A number of instruments are available for sampling the soil atmosphere and for monitoring changes in aeration, although problems have been experienced in their design, due largely to leakage (Moffat *et al.* 1990). Indirect assessment of soil aeration can be made from detection of fermentation and putrefaction products, either by quantitative analysis or simply by their odour. Limited aeration can also be seen by the appearance of reduced compounds of manganese and iron, which are characterised by the presence of black specks and pallid grey colours respectively.

A particularly useful indicator of the degree of soil aeration is the *redox potential* (Eh) or oxidation-reduction status. Redox or oxidation-reduction reactions are those in which a chemical species undergoes oxidation or reduction through the transfer of electrons (e^-). An example of such a reaction is the reduction of ferric (Fe III) hydroxide to ferrous (Fe II) hydroxide:

$$Fe(OH)_3 + e^- \Leftrightarrow Fe(OH)_2 + OH^-$$
(ferric hydroxide)   (ferrous hydroxide)

This reaction is reversible, with a tendency to change towards the left under oxidising conditions and towards the right under reducing conditions (see section 3.2.2 for further discussion of oxidation and reduction processes). The redox potential (Eh) is the difference in electrical potential between the two halves of this coupled reaction. It can be measured using a platinum electrode and an appropriate reference electrode; these are inserted into the soil and the electrical potential difference (V) is recorded, and referred to a standard pH value of 7.0. Eh values in aerobic soils are generally between 0.3 and 0.8 V, while in anaerobic soils they are often between 0.3 and −0.4 V.

Under aerobic conditions, electrons produced during respiration combine with oxygen. Under anaerobic conditions, however, oxygen is unavailable and other chemical species must act as electron receptors. Ferric iron compounds often take on this role, undergoing reduction to ferrous iron compounds (see previous equation). Each chemical species has a threshold redox potential below which it begins to act as an electron receptor and thus becomes unstable (Figure 2.27). Ferrous iron compounds are often present in the centre of peds where Eh may be 0.1 V lower than at the edges, whereas ferric compounds are more likely to occur between peds, in areas adjacent to root channels, and in patches where texture is relatively coarse. Such micro-scale variation in distribution of the different iron compounds can often give the soil a mottled appearance, where blue-grey colours of ferrous compounds are interspersed with orange-brown colours of ferric compounds. The degree of soil aeration is therefore closely associated with porosity and pore size distribution, and is inversely related to soil water content. It is also related to texture and the degree of structural development. Optimum levels

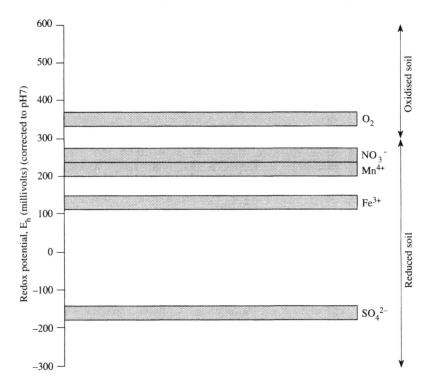

Figure 2.27 Threshold redox potential below which selected oxidised species become unstable (based on Patrick and Mahapatra 1968)

of aeration are most often found in well structured soils with a fine loamy texture.

2.5 SUMMARY

In this chapter we have examined the basic constituents of soil, namely mineral and organic material, water and air, and have looked at some of the most important properties that are relevant to the study of soil in an environmental context. Mineral material is derived largely from the parent material and consists primarily of silicate minerals. The dominant elements in these minerals are silicon, oxygen, aluminium, iron and base cations, and these are arranged in a variety of ways to produce a wide range of structural groups, the main types of which are the framework, chain and sheet silicates. The sheet silicate group in particular, which includes the clay minerals, has a major influence on many physical and chemical soil characteristics.

The soil organic fraction is derived from vegetation and organisms. In terms of its chemical composition organic matter consists predominantly of carbon, nitrogen, oxygen and hydrogen which are arranged in a variety of ways to produce a wide range of compounds, some of which have particularly complex structures. The main groups of compounds include carbohydrates, proteins and amino acids, lignin and humus, the latter representing the end-product of organic matter decomposition. A wide variety of soil organisms have evolved and these are responsible for breakdown of organic material. They are often classified according to their size; micro-organisms consist of bacteria and actinomycetes, fungi, algae and protozoa, mesofauna include nematodes, arthropods and annelids, and macrofauna include larger molluscs, beetles, insect larvae and larger vertebrates. These organisms vary widely in their tolerance of different soil conditions and some play a highly specialised role in the decomposition process.

Water and air occupy the soil pore space and act as the transport and circulation system of the soil. These constituents are particularly dynamic in terms of their spatial and temporal variability, with relative proportions changing rapidly in response to the timing of rainfall events. Their chemical composition is also dynamic in its variability, as these constituents interact closely with the surfaces of mineral and organic particles through ion exchange processes. Soil water is derived largely from precipitation and ground-water sources. It contains a variety of dissolved solids and gases, including base cations, acidic components, iron and aluminium, and in some instances pollutants such as heavy metals. Movement of water through the soil depends largely on pore size distribution; in macropores it is weakly held and moves rapidly under the influence of gravity, while in micropores it is tightly held and moves slowly by capillary action. Soil air represents a mixture of gases derived from the atmosphere and the respiration of soil organisms and comprises mainly nitrogen, oxygen, carbon dioxide and water vapour. In comparison with the atmosphere, soil air contains less oxygen and more carbon dioxide. The main process by which gaseous exchange occurs between the soil and the atmosphere is diffusion, whose rate increases with increasing porosity. Diffusion can only occur effectively if pore continuity is adequate, and is restricted in compacted soils or those with surface crusting.

Soil properties have been examined under physical and chemical headings. Physical properties have included those relating to mineral particles, aggregates, pores, moisture, temperature, mechanics and colour. A fundamental property of mineral particles is texture, which describes the relative proportions of different size classes of material. Particle sorting, shape, surface texture and mineralogy are additional properties of importance in the environmental study of soils. Soil aggregates are described in terms of their shape, size and durability, which together define soil structure. Aggregate development can be viewed in terms of a hierarchical model in which ion bridging processes lead to the development of domains; these form microaggregates through the agencies of fungal hyphae and related organic compounds, which in turn form macroaggregates or peds, promoted by plant roots. Pore space is described in terms of porosity and pore size distribution. These properties play a key role in the balance between soil moisture content and the degree of aeration in soils. Bulk density is also influenced by pore space characteristics, being inversely related to porosity.

Soil moisture characteristics include not only the water content, but also the different components of water – hygroscopic, capillary and gravitational – and the nature and strength of the tensional forces through which water is retained in the soil against gravity. Thermal characteristics of soils are influenced in particular by texture, moisture and organic content. Near the surface, soil temperature changes rapidly in response to changes in air temperature, but with increasing depth temperature changes become smaller and less rapid. The mechanical properties of soils cover a wide range of measures including compressive and shear strength, swelling and dispersal characteristics, and consistence. Consistence describes the change in physical behaviour of the soil with increasing water content, and is expressed in terms of the three Atterberg limits – shrinkage, plastic and liquid limit. Soil colour is most often described using the Munsell system, and relates closely to the presence of constituents such as organic matter, which produces dark colours, and iron compounds, which can produce orange or red colours under aerobic conditions or blue-grey colours when reduced.

In terms of chemical properties, elements and compounds commonly examined in soil-environmental studies include exchangeable bases, organically bound and inorganic forms of free iron and aluminium, carbonates and heavy metals, plus carbon and nitrogen associated with soil organic matter. Ion exchange plays a central role in many soil chemical characteristics, particularly exchangeable base content, CEC and base saturation. Colloidal particles carry negative charges which attract cations, and these adsorbed cations may be exchanged for other cations, depending on their concentration, valency and radius of hydration. In a similar way, anions may also undergo adsorption and exchange. Soil acidity, measured in terms of pH, is a particularly important soil property and relates to the content of exchangeable hydrogen ions. Under acidic conditions, aluminium hydrolysis may lead to enhanced acidity. Soil aeration is another important soil characteristic, affecting the oxidation and reduction of various chemical species, and is often measured in terms of the redox potential (Eh).

3

SOIL FORMATION – PROCESSES AND PROFILES

3.1 INTRODUCTION

The processes by which soil formation, or *pedogenesis*, occurs, are known collectively as *pedogenic processes*, of which there are four main groups – additions, transformations, transfers and losses (Figure 1.2). Additions involve both organic and mineral material, and can occur at the surface or within the soil itself. These materials are then transformed by the processes of organic matter decomposition, mineral weathering and clay mineral formation. Soil components can also undergo transfer from one part of the soil to another by a variety of processes which involve either transport in water or mechanical mixing. Some of the material will be lost from the soil via a number of processes, either as individual components or in combination with others. The pedogenic processes produce soil horizons, which combine to form soil profiles, different horizon combinations giving rise to different soil types. This chapter will first examine each group of pedogenic processes in turn, although it is important to recognise that these processes rarely occur in isolation; soil profiles are products of the combined operation of a number of processes. This will be considered later in the chapter, by examining the formation of soil horizons and a series of pedogenic pathways which produce basic types of soil profile. Finally, the classification of soil profiles will be discussed, looking at different methods of classification and their relative merits.

3.2 PEDOGENIC PROCESSES

3.2.1 Additions

In the context of soil formation, the material added to a soil can take two main forms – organic matter and mineral material – and addition can occur either at the surface or within the soil itself. Organic matter derived from above the ground includes material of local origin, supplied by vegetation growing in the soil or animals living at the surface, and also material derived from some distance away, being transported to the site by a variety of

57

mechanisms. Locally derived material usually takes the form of plant litter and animal droppings, or larger but less frequent inputs when a whole organism dies, while transported material can include, for example, wind-blown leaves, stems carried by running water, a dead tree falling down a steep slope, or manure added to a soil to assist cultivation. Mineral material can also be added in a similar way, being blown or washed onto a soil, moved downslope under gravity or added by human activity which could include, for example, agrochemical application, refuse disposal or the movement of soil material during construction. Mineral material can also be carried by water in a dissolved form and therefore added via precipitation or water moving downslope. Other sources of mineral material include volcanic ash and glacial or marine sediments, although the volume of such materials may cause complete burial of the soil rather than additions to the profile (section 5.1).

Like above-ground inputs, subsurface additions can be locally derived or transported from elsewhere. Organic inputs of local derivation comprise plant roots, soil fauna and micro-organisms, while more distant material can be carried in solution or as small particles in water draining laterally through the soil. Mineral material can also be carried from some distance away by lateral drainage. The movement of soil materials is discussed further in sections 3.2.4, 6.4.2 and 6.4.3.

3.2.2 Transformations

Once organic material has been added to a soil, it will undergo decomposition, unless this is prevented by environmental circumstances or the material is already fully decomposed. Most organic matter decomposition relates to enzymic oxidation by organisms living in the soil or at its surface (Brady 1990). As carbon and hydrogen make up the bulk of dry organic matter, the oxidation process can be shown as:

$$-(C, 4H) + 2O_2 \quad \Rightarrow \quad CO_2 + 2H_2O + energy$$
$$\text{enzymic oxidation}$$

The breakdown of complex organic structures also leads to the formation of a variety of more simple, inorganic products. This process is known as *mineralisation*, and is an important source of plant nutrients such as nitrogen, sulphur and phosphorus; cations such as Ca^{2+}, Mg^{2+} and K^+ are also released.

During organic matter decomposition, an additional set of processes occurs known as *humification*. These involve the complex reaction of various decomposition products to produce large, complex molecular chains, or *polymers*. These reactions result in the formation of a stable end-product known as humus (section 2.2.2), comprising small, colloidal particles, $< 2 \mu m$ in size which, like clays, have a net negative charge, and are important in soil

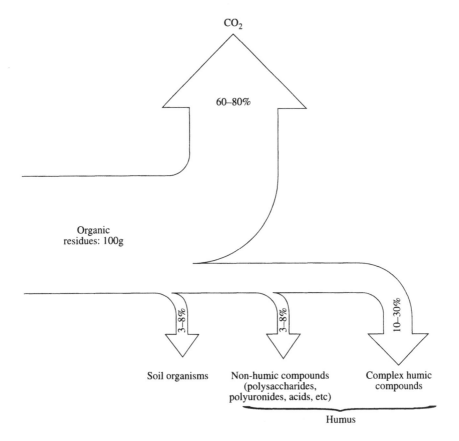

Figure 3.1 The fate of 100 g of organic residues 1 year after incorporation into the soil (from Brady 1990)

aggregation and cation exchange reactions (sections 2.3.2 and 2.4.2). During one year of organic matter decomposition, approximately 60–80 percent of the organic residues are oxidised and returned to the atmosphere as carbon dioxide, the rest remaining in the soil either as humus or in the bodies of soil organisms (Figure 3.1).

Two groups of compounds can be recognised in humus chemistry – humic and non-humic groups (Brady 1990) (Figure 3.1). Humic substances account for around 60–80 percent of the soil organic matter and are characterised by aromatic, ring-type structures, which include polyphenols and polyquinones. These substances form by a variety of biological and chemical processes, the details and relative importance of which remain unclear. The non-humic group represents around 20–30 percent of the soil organic matter and is generally less complex and less resistant to breakdown by micro-

59

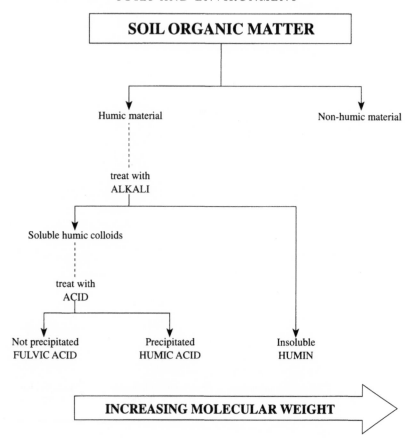

Figure 3.2 Fractionation of soil organic matter (from Ross 1989)

organisms. It includes polysaccharides and polyuronides, along with organic acids and some protein-like substances. The components of humic material can be separated by chemical fractionation into three main groups – fulvic acid, humic acid and humin (Figure 3.2). These differ in their molecular weight, and the extent and strength of polymer bonding increases with molecular weight.

Mineral transformation can occur either by weathering or by the formation of clay minerals. Weathering is often considered to be of either a mechanical (physical) or chemical nature. White *et al.* (1992) have, however, argued that this distinction is invalid according to the principles of physics, and have suggested the use of two alternative categories based on the scale at which processes occur – *brittle fracture*, which operates at the scale of the rock mass, rock fragment or mineral crystal, and *crystal lattice breakdown*, which operates at the molecular and sub-molecular scales. Processes causing brittle fracture involve the application of mechanical stress or the release of strain, which can occur either as a single event or on a cyclical basis. The principal mechanisms which have been recognised are thermal expansion and contraction, freeze/thaw, salt crystal growth, root wedging and strain release. Although not always operative within soils themselves, these processes can be particularly important in the initial stages of parent material breakdown and in preparing this material for crystal lattice breakdown.

The effectiveness of thermal expansion and contraction, which occurs by diurnal heating and cooling, has been questioned, although there is general agreement that wetting and drying may increase its effectiveness in situations where the two processes are occurring simultaneously (Ollier 1984, Goudie 1989). Freeze–thaw, which operates as a result of water increasing its volume by about 9 percent on freezing, produces three basic types of process – frost scaling, in which thin layers of ice form parallel to the rock surface; frost splitting, which occurs in massive rock; and frost wedging, which is probably dominant in most cases, in which cracks within a rock are exploited (Lautridou 1988). Again, its effectiveness as a weathering mechanism has been questioned (White 1976, Fahey 1983), although this remains open to debate because the processes involved are incompletely understood. Salt crystal growth can occur for three main reasons – evaporation of the salt solution, decreased solubility due to falling temperature, or mixing of two different salt solutions with the same major cation (Goudie 1989). Although some salts cause little breakdown, others, such as calcium chloride and magnesium sulphate can result in significant breakdown (Goudie *et al.* 1970). Root wedging is the process whereby plant roots extend into rock crevices and exert pressures on the rock as they extend and widen (Ollier 1984), although its effectiveness is probably confined to weakly cohesive materials. In contrast, strain release is a process capable of producing major rock fracture. This occurs when an overburden is removed from a rock by erosion, causing it to rebound elastically, fracturing perpendicular to the direction of unloading. As this often results in sheets of fractured rock lying roughly parallel to the ground surface, the process is also known as *onion-skin* weathering or *exfoliation*.

Although a wide range of complex processes are involved in crystal lattice breakdown, these are often assigned, albeit simplistically, to a number of basic categories – solution, carbonation, hydration, hydrolysis, oxidation/reduction and organic complexing (Ross 1989). Solution involves the breakdown of ionic solids in water to produce a stable ionic solution, as can be seen, for example, in the case of sodium chloride (NaCl):

$$NaCl + H_2O \Leftrightarrow Na^+ + OH^- + H^+ + Cl^-$$

The arrows point in both directions, indicating that the reaction can reverse if the water is removed, for example, by evaporation, which causes the sodium chloride to precipitate out of solution. Carbonation can occur following the solution of carbon dioxide (CO_2) in water, to produce carbonic acid:

$$H_2O + CO_2 \Leftrightarrow H_2CO_3$$

A common carbonation reaction in soils is the dissolution of calcium carbonate ($CaCO_3$) to produce soluble calcium bicarbonate:

$$CaCO_3 + H_2CO_3 \Leftrightarrow Ca^{2+} + 2HCO_3^-$$

Hydration is the process whereby compounds absorb water, often causing them to expand and thus to create pressures which can result in mineral breakdown. In soils the compounds most often affected by this type of weathering are usually those of iron, for example the hydration of hematite (Fe_2O_3) to goethite:

$$Fe_2O_3 + H_2O \Leftrightarrow 2FeOOH$$

The reverse reaction is of course known as dehydration, which occurs as a result of water being removed from compounds. The pressure changes associated with the repeated alternation between hydration and dehydration can be particularly effective in causing mineral decomposition (Figure 3.3).

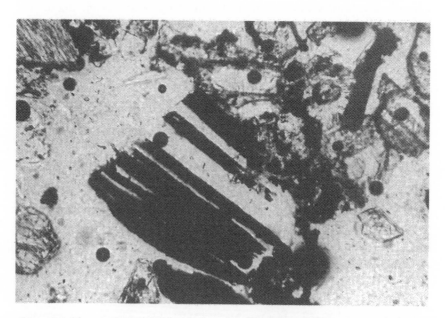

Figure 3.3 Photomicrograph of a biotite grain opening up along its lattice planes due to weathering by hydration (frame width = 2 mm)

Hydrolysis occurs when metal cations within mineral structures are replaced by hydrogen ions in soil waters. Hydrogen possesses two important characteristics which allow this to occur. First, because it is very small relative to other ions, it can more properly be considered as a subatomic particle (i.e. proton) rather than as an ion, and these always associate with other ions or molecules rather than occurring in isolation; second, it is unique in being able to form bridging bonds between negatively charged atoms such as oxygen (Curtis 1976a). It can therefore migrate in chemical systems involving oxygen, and does so in such a way as to allow metal cations to be liberated into solution. An example of such a reaction is the weathering of plagioclase feldspar to produce kaolinite and silica along with sodium ions:

$$2NaAlSi_3O_8 + 2H^+ + H_2O \Leftrightarrow Al_2Si_2O_5(OH)_4 + 4SiO_2 + 2Na^+$$
(plagioclase)   (kaolinite)   (silica)

Oxidation/reduction operates in soils which experience fluctuating levels of oxygen within them. Oxidation and reduction are concerned with the respective removal and addition of electrons. This is because oxygen will accept electrons when present in a soil system, but if oxygen is unavailable or restricted, other elements must act as electron receptors (section 2.4.4). In the context of soil weathering, the two elements most commonly involved in the oxidation/reduction process are iron and manganese, both of which are divalent in their reduced state, but increase to a valency of three and four respectively with the addition of oxygen and consequent loss of an electron, for example:

$$4Fe^{2+} + O_2 + 4H^+ \Leftrightarrow 4Fe^{3+} + 2H_2O$$

Oxidation/reduction reactions can also be shown in terms of electron (e^-) transfer, for example:

$$MnO_2 + 4H^+ + 2e^- \Leftrightarrow Mn^{2+} + 2H_2O$$

Iron and manganese in their reduced forms are known as *ferrous* and *manganous* respectively, whereas in their oxidised forms they are known as *ferric* and *manganic* respectively. Oxidation/reduction reactions are important in weathering because in their reduced forms iron and manganese are soluble and can therefore be removed from the site of weathering (section 3.2.3). Organic complexing, or *chelation*, occurs when metal atoms form covalent-type bonds with organic molecules which can render the metal soluble. This is a particularly important process in the case of iron and aluminium, which are otherwise usually insoluble. Organic complexes are thought to be formed from materials in leaf leachates and also from humic and fulvic acids (Ross 1989).

Although the six major chemical weathering reactions in soils have been considered individually above, it is important to recognise that these reactions rarely operate in isolation within a soil; one type of reaction may

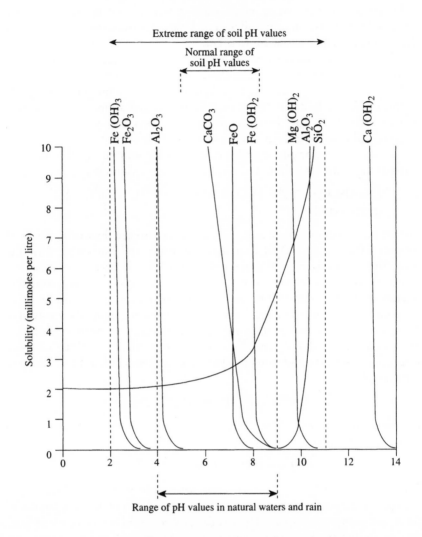

Figure 3.4 Solubility in relation to pH (from Ross 1989)

be dominant in a particular soil or for a given mineral, but the reactions usually operate simultaneously, both within the soil as a whole and often upon any one mineral. The weatherability of soil minerals varies, however, depending on their chemical characteristics (e.g. Curtis 1976a, b), pH conditions (Figure 3.4) and the weathering regime as determined by environmental factors (chapter 4).

An important additional aspect of mineral transformation involves the

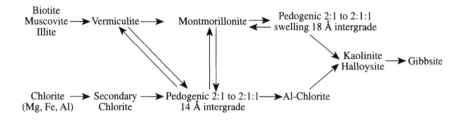

Figure 3.5 Pathways of clay mineral transformation (from Jackson 1964)

formation of clay minerals. This can occur in one of two ways – *transformation*, which involves a change from one type of mineral to another, and *synthesis* (also known as *neoformation*), in which a new clay mineral is formed from weathering products. Transformation usually involves changes in the interlayers (section 2.2.1), by ion exchange, the addition of hydroxyl ions or the removal of siloxane sheets or silica from the structure. Reactions can occur along a layer, or move inwards from the edge of a crystal. The ease of alteration depends on the extent to which a crystal structure has to undergo change. For example, changes within and between the 2:1 and 2:1:1 groups of clay minerals can occur relatively easily and are generally reversible because the basic crystal structure is retained. Ion exchange involves the replacement of interlayer ions by ions in the soil solution, as in the case of the transformation of illite to vermiculite by the replacement of K^+ by Mg^{2+} and Ca^{2+}. Changes from 2:1 to 1:1 clay minerals are more difficult to achieve because they involve a change to the basic crystal structure, and are not easily reversed (Birkeland 1984a). A series of clay mineral alteration pathways are generally recognised, with both reversible and non-reversible components (Figure 3.5).

Synthesis involves the recombination of silica, alumina and cations, released during weathering, to produce new clay minerals. The type of clays formed depends primarily on the type and concentration of cations, and pH. For example, illite formation occurs in acid conditions with high concentrations of K^+ relative to other cations, while smectite forms under neutral or alkaline conditions with high concentrations of Ca^{2+}, Mg^{2+} and Na^+. The solubility of silica and alumina in relation to pH is also important; aluminium becomes insoluble above pH 5, so if values rise above this, it becomes precipitated, and can then adsorb silica. The amount of adsorption in acid conditions is therefore limited, which results in kaolinite clay formation in which silica contents are relatively low. In higher pH environments, however, more silica is adsorbed, which produces clays such as smectites (Duchaufour 1982).

3.2.3 Transfers

Water can transport material within a soil either in solution or suspension. This process is known as *translocation*, although the term *leaching* is also commonly used to refer to the movement of material in solution. While transport usually occurs in a downwards direction, it can in some cases operate laterally or upwards. Material carried in soil water can be derived from material which forms direct additions to the soil, or which is the product of transformation. Whether it is carried in solution or suspension will depend on its size and solubility (Birkeland 1984a), and these factors also determine whether it is redeposited in the soil or lost from it.

Transfer in solution can occur in the form of soluble material produced during transformation processes (section 3.2.2). Solution can also occur by the presence of H^+ and Al^{3+} ions in the soil drainage water, which displace base cations from the soil exchange complex (sections 2.4.2 and 2.4.3). The passage of mobile anions through the soil can also attract base cations into solution. For example, HCO_3^- anions result from the dissociation of carbonic acid:

$$H_2CO_3 \Leftrightarrow H^+ + HCO_3^-$$
(carbonic acid)

and these can therefore stimulate the leaching process. Unless all the base cations are redeposited elsewhere in the soil, they will gradually be lost via the ground drainage waters and the soil will therefore become progressively more acidic. Transfer in solution can also occur as a result of the uptake of nutrients by plant roots; these can be stored in the plant or returned to the soil via the plant litter, therefore taking the form of a *nutrient cycle* (Stevenson 1986).

Redeposition of material carried in solution will occur when a change in soil conditions renders it insoluble. The most obvious cause is removal of the water supply as a soil dries out, but other changes can also be important, such as pH, temperature, exchange sites or, in the case of elements made soluble by reduction, oxygen availability. For example, increased pH values, which often occur lower down a soil profile, may render certain iron and aluminium compounds insoluble (Figure 3.4), while a temperature decrease with depth may reduce solubility. Materials can also be redeposited if they carry an electric charge which enables them to be attracted to the exchange sites of a clay or organic colloid (section 2.4.2). Elements such as iron and manganese, which can be reduced and translocated under the anaerobic conditions associated with very wet soils, may become insoluble on moving to a drier, more aerated part of a soil, or into one in which more oxygen is available due to the oxidising effect of certain micro-organisms, or decreased mesofaunal respiration. In the case of materials carried in the form of soluble organic complexes (section 3.2.2), redeposition can occur due to reduced solubility at higher pH levels, biodegradation of the organic components or polymerisation, which is also enhanced at higher pH (Ross 1989).

Translocation in suspension will occur in the case of insoluble particles which are sufficiently light to be moved by soil water and sufficiently small to be able to pass through the soil pores. The most commonly transported materials in this context are clay and organic colloids, whose movement is aided by a lack of flocculating or cementing agents (McKeague 1983). In acid soils soluble organic matter can also enhance clay mobility by forming a hydrophilic envelope around the particles, protecting them from flocculating cations (Duchaufour 1982). The size and charge of colloidal particles will also influence their transport potential. For example, smectites are generally the smallest size of clay particles and are therefore most easily transported, while kaolinites are generally larger and have a relatively low charge which can be easily neutralised and are therefore amongst the least mobile types (Duchaufour 1982). The deposition of clays often takes the form of coatings, known as *clay skins, illuviation cutans* or *argillans* (Brewer 1976), which can be seen lining the walls of soil pores or surrounding larger mineral grains or aggregates (Figure 2.13d, Plate 1). In soils where water movement is rapid and pores are large, material up to a few millimetres in size can be carried in suspension. This includes silt, sand, faecal pellets

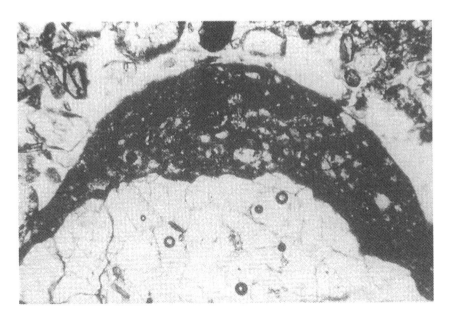

Figure 3.6 Photomicrograph of a deposit of translocated silt and fine sand on the upper surface of a larger particle (frame width = 2 mm)

67

Figure 3.7 A termite mound (2 m in height) in west Africa

and other organic fragments, and in extreme conditions gravel. Deposition of this material occurs due to the evaporation of water or a reduction in pore size, and since the particles lack the electrostatic bonding properties of colloidal material, their deposits are generally confined to the upper surfaces of particles (Figure 3.6). Of course, materials transported by water are not always redeposited in the soil. Some are often lost as outputs to the ground-water drainage system and reach rivers, ultimately to be deposited in lakes or oceans.

Mechanical transfer of soil materials can occur by a variety of processes, both biological and inorganic, referred to collectively as *pedoturbation*. Of the biological processes, also known as *bioturbation*, the most important are generally soil mixing by burrowing animals and by human activity. For example, it has been estimated that earthworms can produce over 20 kg m^{-2} a^{-1} of cast material (Lee 1985), derived usually from within a metre of the surface, while termites can bring mound-building material to the surface from depths of over 10 m (Figure 3.7); erosion of the mounds can add up to an estimated 0.2 mm a^{-1} of soil to the surface (Lee 1983). Earthworm burrowing can also concentrate larger material into a residual, subsurface layer or *stone line*, above which the cast material accumulates. In some instances bioturbation can produce complex layering in soils by the accumulation and burial of material moved by soil fauna (Johnson 1990). Cultivation practices, particularly ploughing, can typically mix soil to a depth of 20–50 cm, although larger implements can cause mixing to greater depths, as can excavation, such as that associated with quarrying, road construction, pipeline installation and urbanisation. Other processes may also be operative such as root decay, causing channel collapse, displacement during root growth, and root movement during strong winds, while mixing by tree-throw may also be important locally (Finlayson 1985).

Inorganic mechanical transfer processes involve primarily alternating expansion and contraction of the soil as a result of wetting and drying or freezing and thawing, although freeze/thaw can also be responsible for the separation of different sizes of soil particles as well as soil mixing. Alternate wetting and drying is particularly effective where large contents of expanding lattice clay minerals are present. On drying, the clays contract and vertical cracks form in the soil, into which material can fall. When the soil becomes wet again, the clays expand, and the new infill becomes incorporated into the surrounding material. On a smaller scale, clays can become reorientated within soil aggregates by the stresses produced during expansion. This often occurs such that the clay particles become orientated preferentially with their optical axes aligned perpendicular to the direction of stress. When viewed microscopically (in crossed polarised light), this gives an appearance (known as *birefringence*) similar to that of argillans formed by the redeposition of clay carried in suspension. The features are known as *pressure cutans* or *stress cutans* (Brewer 1976) (Plate 2), and can usually be distinguished from argillans in that the latter have a sharp boundary while the boundary of stress cutans is diffuse as the reorientation gradually becomes less strong away from the point of maximum stress at which contact with other aggregates occurs.

The transfer of material resulting from freezing of the soil involves a variety of processes, some of which remain unclear. One such process is frost sorting, whereby mineral material is sorted into different sizes within the soil, either vertically or laterally. Vertical sorting can occur when soil is freezing downwards or upwards. In the former case, larger particles in the soil will allow a more rapid penetration of the freezing isotherm through them than will the surrounding soil, because of their higher thermal conductivity, therefore ice can form beneath these particles before the surrounding soil becomes completely frozen. If the particles are near the surface, the pressures created by the freezing water beneath them may be sufficient to cause upward displacement. Alternatively, if these processes are restricted to the overlying frozen soil, they may displace the unfrozen soil adjacent to the larger particles laterally or downwards (Williams and Smith 1989). Stones can also be pushed upwards into unfrozen soil above a freezing isotherm which is advancing towards the surface (Mackay 1984). Due to lateral variations in soil thermal properties, freezing rates may vary horizontally, which can also contribute to sorting. By a combination of these processes, material can therefore become sorted into different sizes, usually

Figure 3.8 Patterned ground (sorted stone polygons) in northwest England

with larger particles becoming concentrated towards the surface or forming various types of pattern at the surface (Figure 3.8). Such forms, known as *patterned ground*, have been extensively reviewed by Washburn (1979).

Another transfer process related to soil freezing is the mixing of soil by alternate freezing and thawing, a process known as *cryoturbation*. Soil displacements have traditionally been explained in terms of *cryostatic pressure*, which results from differential freezing rates such that wet, unfrozen pockets of soil are subjected to pressures from adjacent areas where freezing is occurring. However, there is little convincing evidence to support this process – it is unlikely that unfrozen soil could be intruded under pressure into frozen material, and it is likely that unfrozen areas will be ones of water removal (as water moves towards the freezing isotherm) where pore water is therefore under tension, rather than compression (French 1988, Williams and Smith 1989). An alternative to the concept of cryostatic pressure is the cell-like movement of soil caused by different extents of ice lens formation in the initially unfrozen, or *active*, layer (French 1988). It appears that soils rich in silt are most susceptible to disturbance by frost processes (Figure 3.9). These can develop large ice contents because their associated pore sizes allow water to move through the soil towards the freezing layer more effectively than larger pores associated with coarser-textured soils or the small pores of clays (Williams and Smith 1989).

A process which can result in a similar pattern of mixing is that of density

Figure 3.9 A silty soil in northern Norway showing the effects of frost disturbance in the upper part of the profile

loading, where overlying, denser material sinks under gravity into underlying, less dense material, which in turn rises into the denser layer. For this to occur the soil must have an excess water supply so that intergranular contacts are lost and the soil therefore becomes liquefied (Vandenberghe 1988). This can sometimes occur during soil melting, but can also occur in unfrozen soils if the hydrological conditions allow. The effect of density loading is to produce involutions, similar to those associated with frost disturbance. In more advanced cases, droplets of denser material become detached and sink through the underlying less dense material.

Other inorganic mixing processes include soil displacement during the growth or solution of crystals and, less commonly, by water upwelling towards the surface or during earthquakes (Hole 1961, Previtali 1992). Mixing can also occur by differential movement of soil on a slope. Movement can occur in a variety of directions due to a combination of the processes described above, operating along with gravity, and at different rates at different depths beneath the surface (Finlayson 1985, section 6.4.3).

3.2.4 Losses

Material can be lost from soils in four main forms – gases, solutes, particulate material and via vegetation removal. As in the case of additions, the processes involved can usefully be divided into surface and subsurface categories. Surface losses include gases which are produced during organic matter decomposition and lost to the atmosphere, solutes which are taken up as nutrients by vegetation and then lost when the vegetation is removed, for example by harvesting of crops or removal of trees, particulate material which is lost by water or wind erosion, and the upper parts of profiles which may be removed *en masse* by erosion or human activity. In the case of gases and solutes, the significance of losses via the surface will depend on the extent to which they are dissolved and lost by subsurface drainage, which in turn will depend on climate and land use (Hornung 1990).

Removal of particulate material by wind will be most effective in the case

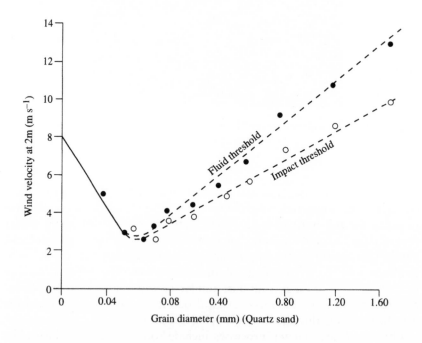

Figure 3.10 The relationship between grain size and wind velocity (from Warren 1979). Fluid and impact thresholds are the velocities required to initiate and maintain movement respectively

of soils with a high silt or fine sand content, as this size of material is more easily entrained than larger, heavier particles or small clay particles which resist entrainment due to their greater aggregation (Figure 3.10). Organic matter is also prone to wind erosion, and its lower density relative to mineral material means that larger particles can more easily be carried. Low moisture content, poor aggregation and sparse vegetation covers will also enhance susceptibility to erosion. Particle detachment occurs in response to the creation of eddies at the ground surface, and as a result of impacts from particles which are already in motion. Small particles (< 0.05 mm in diameter) are transported by aerial dispersion and may be carried to elevations of several thousand metres. Particles of intermediate size (0.05–0.5 mm) are transported within a metre or so of the ground surface by the process of saltation. In contrast, large particles (> 0.5 mm) are moved along the ground, largely as a result of impacts from saltating particles, by the process of creep. Saltation is the most important process of wind erosion in terms of the quantity of material moved, and it is estimated that 55–72 percent of wind-eroded particles are transported in this way (Morgan 1986). In the case of erosion by water, soils with weak aggregation and low vegetation covers are particularly susceptible because the aggregates can be easily broken down by direct raindrop impact. This can also result in surface compaction and sealing to form a crust which may be several mm in thickness (e.g. Valentin 1991), which impedes infiltration, therefore enhancing the loss of material by surface runoff (Le Bissonnais and Singer 1993) (section 6.4.2).

Erosion losses can also occur *en masse*, for example by glacier ice, if there is a major change in environmental conditions (section 5.1). Losses of material by human activity can occur in many ways, but the most common are associated with the removal of topsoil for use as a resource or during construction. For example, organic-rich soil may be removed for fuel or horticultural purposes, while mineral soil may be extracted to improve the cultivation capabilities of a soil elsewhere. Removal of soil is also often associated with the construction of roads and buildings, or with landscaping programmes.

Subsurface losses can occur in solute or solid form and can involve any of the products of addition and transformation, along with those materials undergoing transfer where conditions for total redeposition within the soil do not occur. The extent of solute outputs will clearly depend on the solubility of the materials involved, along with factors such as temperature, which will control the rate at which reactions occur, and the speed of water movement, which will determine the time available for reactions to occur (Crabtree 1986). Material lost in a solid, particulate form will only occur with any significance in soils containing large pores or other forms of passageway. In coarse-textured or uncompacted soils, pore spaces may be sufficiently large to allow material to move out of the soil and into drainage channels, but particulate losses are more usually associated with the development of soil pipes, whose diameter can range from a few centimetres up to several metres. These are found in soils which experience cracking, due to the occurrence of either highly expandable clay minerals such as smectite or high organic contents, and which also contain a subsurface layer of restricted permeability (Jones 1981, McCaig 1985). Soil pipes allow large volumes of water to be moved rapidly through them, resulting in the removal of both solid particles and dissolved material.

3.3 SOIL PROFILES

Having examined the various processes responsible for soil formation on an individual basis, we shall now see how these processes operate in combination to produce soil profiles. In order to do so, we shall start by looking at the main characteristics of soil profiles – their formation as weathering products of the parent material and the development of horizons within them. Reference to a particular soil horizon is usually made by assigning a letter, which can then be refined by adding letters or numbers as prefixes or suffixes. A number of systems have been devised for this purpose, but two in particular have received widespread usage – the FAO-Unesco (1974, 1989) and the Soil Survey Staff (1975, 1992) schemes. Both have many characteristics in common, and indeed have many similarities with various other systems, and they will therefore form the basis of horizon designation in this chapter. After examining horizon differentiation, we shall then look at the main trends or pathways along which soil formation can proceed and the principal types of profile that result.

3.3.1 Horizon differentiation

The material from which a soil develops is known as the parent material; this may be bedrock or a geologically recent, superficial deposit (section 1.1). At the start of soil formation the parent material is unaltered. In the case of solid bedrock, this is referred to as the *R layer*. However, as weathering begins to operate, the material starts to undergo alteration and in this condition is referred to as the *C horizon*. This term is also given to unconsolidated parent material, which comprises a superficial deposit overlying bedrock, before weathering has commenced. In some instances where a superficial deposit is only thin, the weathering processes may extend down through this material into the underlying bedrock, in which case a composite C horizon may result.

With continued weathering, the soil material becomes increasingly transformed so that the original structure of the parent material is lost, and this is known as a *B horizon*, although B horizons are also recognised on the basis of transfer processes. The boundary between C and weathered B horizons is often transitional, reflecting the gradual increase in weathering from one to the other. This increase can be recognised in a number of ways. For example, there is often a decrease in the number and size of stones up through a profile as a result of an increasing intensity of weathering, and this may be accompanied by an increase in fine particles and in the amount of iron staining as iron compounds are released. The formation of clay and release of oxides during weathering may encourage aggregation and therefore allow structure to develop in the B horizon (section 2.3.2). Colour changes can also occur within a profile as iron compounds are transformed, as in the oxidation of ferric hydroxide to hematite which results in a colour change from yellow to red. Changes in texture and colour are not, however, reliable indicators of weathering if based solely on field observation, because they can also result from various transfer processes.

Progressive weathering may also change the shape and surface characteristics of mineral particles. For example, particles may become more rounded with the progression of solution weathering, or may become etched (Locke 1986) (Figure 2.13b and c). During weathering, surface and near-surface zones of mineral particles may become altered. This can take two forms – *weathering rinds* or *grus* (Birkeland 1984a). Weathering rinds occur as alteration zones around a particle which remains essentially intact, for example a zone of staining resulting from iron oxidation, while grus is a zone of physical disintegration around the margin of a rock as a result of weathering, usually of igneous rocks. The thickness of weathering rinds and grus will therefore usually show an increase from the C to the B horizon.

Differences in weathering can also be seen by the use of mineral or chemical ratios. This is based on the assumption that certain minerals are broken down, or chemicals released, more easily than others, and ratios between the two will therefore vary according to the extent of weathering (Brewer 1976, Marshall 1977, Santos *et al.* 1986). For example, it is generally considered that minerals such as quartz, garnet and zircon are resistant to weathering while those such as feldspars, amphiboles and pyroxenes are less resistant. As weathering proceeds, the ratio of resistant to less resistant minerals within any particular size fraction will increase, and the magnitude of the ratio will therefore increase from lower to upper horizons. This is demonstrated in Table 3.1, where site 1 shows the smallest difference in weathering intensity between the upper and lower horizons while site 3 shows the greatest difference; site 4 is almost devoid of feldspars at both depths. There are, however, various problems associated with the use of weathering ratios. For example, there is some uncertainty about the relative resistance of certain minerals to weathering, and also a problem concerning the statistical validity of occurrence estimates of the minerals used in ratio calculations in view of the number of grains counted (Brewer 1976). A further problem concerns the mineralogical and chemical variability of the parent material; unless this material is homogeneous, variations in mineral or chemical ratios with depth cannot be considered with

Table 3.1 Quartz:feldspar ratios of the 250–500 μm fraction of some eastern Canadian soils

Soil	Horizon	Quartz:feldspar	Soil	Horizon	Quartz:feldspar
1	Upper	6.2	3	Upper	100+
	Lower	3.7		Lower	19
2	Upper	13	4	Upper	100+
	Lower	6.0		Lower	100+

Source: McKeague and Brydon 1970

any certainty to be a function solely of weathering.

A widely used alternative to weathering ratios in the observation of parent material alteration is the mineralogical investigation of clay transformation and synthesis. X-ray diffraction analysis of soil clays allows the constituent minerals to be identified (Brindley and Brown 1980) (section 2.3.1), and the nature of alteration can be inferred from variations in clay mineral type and proportion with depth. For example, both soils in Table 3.2 show decreases in chlorite and mica, and increases in smectite up through the profiles, indicating clay transformation and synthesis. As in the case of weathering ratios, for clay mineralogy to provide useful information about mineral alteration, the soil parent material must be mineralogically homogeneous.

Soils which receive organic additions often possess a surface horizon which differs from the underlying horizons on account of its higher content of organic matter or the form in which this material occurs; this is referred to as an O or A *horizon*. The nature of the surface horizon will depend to a large extent on the balance between the processes of organic matter addition and its subsequent transformation, transfer and loss, and this balance will be determined by environmental conditions (section 4.2). In cases where the

Table 3.2 Clay mineral contents of two Canadian Spodosols

Horizon	Smec.	Verm.	Chlo.	Mica	Kaol.	Chlo.-verm.	Mica-verm.	Mica-smec.
A	++++	–	–	tr	+	–	–	+
B	–	+	+	+	+	++	–	–
C	–	+	++	++	+	+	+	–
A	++	++	–	+	tr	–	+	tr
B	tr	++	+	+	tr	+	+	tr
C	–	+	++	++	tr	+	+	tr

Source: Ross 1980
Note: Smec. = smectite; Verm. = vermiculite; Chlo. = chlorite; Kaol. = kaolinite. tr < 10%, trace; + 10–25%, minor; ++ 25–50%, moderate; ++++ 75–100%, dominant

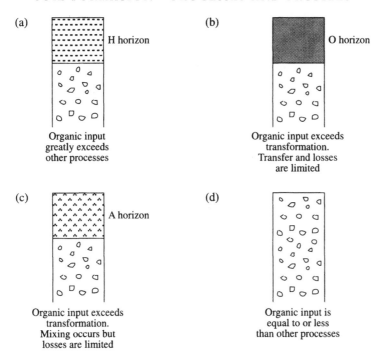

Figure 3.11 Models of surface horizon formation

organic input greatly exceeds these other processes, there will be a large accumulation of organic matter at the soil surface (Figure 3.11a). Such an accumulation is referred to in the FAO-Unesco system as an *H horizon* and is usually associated with poorly drained, anaerobic conditions in which little biological activity is occurring either in the form of organic matter decomposition or bioturbation.

When organic addition exceeds transformation to a lesser extent, but where transfer in the form of mixing remains limited, along with losses, an organic-rich surface horizon results as in the previous case, but different extents of organic matter decomposition can be recognised within it (Figure 3.11b). This is referred to as an *O horizon*. Since most organic additions are via the surface, the longer material remains in the soil without bioturbation, the deeper it will become, and because organic matter decomposition will also proceed through time, the degree of decomposition will therefore increase with depth. The most recently added material, at the surface, is known as the *litter layer*, and this is usually easily recognised because much of the original structure of the material remains intact. Below this layer progressive decomposition is apparent. The original organic components are less easily identified as they become physically comminuted by the biting action of mesofauna and discoloured by the chemical attack of micro-organisms (Babel 1975). The walls of stems and roots and the veins of leaves become fragmented and the internal cellular structures begin to disappear, often becoming replaced by the faecal material of the mesofauna responsible for their ingestion. With increasing decomposition at greater depth the organic matter comprises humus and faecal material whose original components can no longer be easily recognised (Plate 3). O horizons are usually associated with acid soil conditions (pH < 5.5), which are unfavourable for burrowing mesofauna and therefore have a sharp boundary with the underlying mineral soil. Such conditions are also associated with an acid-tolerant flora, whose organic material is not very nutritious and is often therefore not very highly decomposed. Organic matter of this type is traditionally referred to as *mor*.

In soils where organic addition exceeds transformation, but in which transfer of organic matter in the form of mixing also occurs, although losses remain limited, the resulting surface horizon will contain a mixture of organic and mineral material (Figure 3.11c). This is known as an *A horizon*, and contrasts with the previous types of horizon on account of its lower organic content and its merging boundary with the underlying material, due to mixing. Suffixes commonly applied to A horizons are *p* to denote disturbance by ploughing or other forms of tillage and, in the case of the FAO-Unesco system, *h* to denote organic accumulation where no disturbance has occurred. Although the litter layer is usually present, there is no obvious progression of increasing decomposition with depth; organic matter is derived from material which is relatively easily broken down compared to that of O horizons and therefore appears in an advanced state of decomposition at all depths, mixed with mineral material. This usually produces a crumb or granular structure (section 2.3.2, Figure 3.12). This can be made up of loosely bound aggregates containing colloidal organic and mineral material along with larger mineral and organic fragments. Soils in which mixing occurs are usually not very acid (pH > 5.5) and their flora generally produce more nutritious organic matter than in the previous case. This is known traditionally as *mull*. Surface horizon organic matter whose acidity and extent of decomposition and mixing are intermediate between those of mor and mull is known as *moder*.

Soils in which the addition of organic matter is equal to or less than its transformation, transfer or loss will contain little or no organic matter in their surface horizons, since the organic material will be removed from the soil surface as quickly as it is added (Figure 3.11d). It is important to stress that the precise balance between organic matter addition, transformation, transfer and loss will often result in conditions which are transitional between the categories described above. This balance will also determine the thickness of a surface horizon; in general, the greater the addition of organic

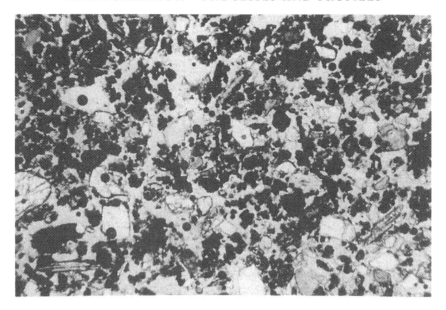

Figure 3.12 Photomicrograph of granular structure in a surface organic-rich horizon, showing mineral material mixed with decomposed organic matter mainly in the form of faecal pellets (frame width = 4 mm)

matter relative to the other three processes, the thicker the horizon.

In addition to the C and B horizons produced by weathering, subsurface horizons can also form by transfer processes. These can be of two basic types – those from which material is removed and those into which material is transferred. Although we have seen that there are many processes which can cause transfer (section 3.2.3), in terms of soil horizon differentiation these processes relate mainly to the transfer of material by water, rather than by mechanical means. Horizons from which material is washed are known as *eluvial* or *E horizons*, and those in which material is redeposited are termed *illuvial* or *B horizons*. In the strict sense, E horizons are defined on the basis of removal of silicate clay, iron or aluminium, either individually or in some combination. The characteristics of the horizons will obviously depend on the type of material undergoing transfer. In the case of eluvial horizons, material removed in solution will change their chemical content, and in some cases their colour and structure also. For example, if iron is removed the eluvial horizon will become lighter in colour and may also lose its structure if iron was acting as an aggregating agent. Where material is removed in suspension, a change in texture can result if mineral particles are involved. For example, the eluviation of clay and silt will cause the horizon to become coarser-textured as the fines are removed.

Illuvial horizons can be characterised on the basis of their texture, structure, colour or chemistry, depending on which soil components have been deposited, and a variety of different B horizon types can therefore be recognised. Horizons containing deposits of illuvial clay are designated *Bt horizons* and can be recognised by their finer texture compared to adjacent horizons, although this is not conclusive evidence because fine textures can also result from weathering. A more certain indication is the presence of argillans when viewed microscopically in thin section (Plate 1). The argillans may also give ped faces a shiny appearance, although this is not always easily visible. The presence of Bt horizons is often associated with a well developed structure, due to the aggregating effects of the illuvial clay.

Horizons which receive material in solution, for example calcium carbonate, gypsum, sesquioxides, sodium and silica may be distinguished chemically, and are given the suffixes *k*, *y*, *s*, *n* and *q* respectively. Some materials may also produce physical characteristics such as cementation, where accumulations are high and the soil pores become filled with deposited material, or nodules may form where localised concentrations of illuvial material accumulate. Cementation and concretion are denoted by the suffixes *m* and *c* respectively; for example, a horizon cemented by illuvial carbonates would be denoted *Bmk* while a zone of sesquioxide concretion would be denoted *Bcs*. In some cases, material carried in solution may reach the C horizon, and here the same suffixes are applied. Illuvial accumulations can also be recognised on the basis of colour as, for example, in the case of organic matter and iron which will often be darker in colour than an overlying E horizon or underlying C horizon, although again this may be the result of weathering rather than translocation in the case of iron. The deposition of these materials is seen particularly well if the soil is viewed microscopically in thin section, where the deposits often occur as coatings around mineral particles or along the walls of pores (Plate 4). Coatings of organic matter and iron have been termed *organans* and *ferrans* respectively by Brewer (1976). Depositional coatings of other types of illuvial material have also been given similar terms, for example *manganans* (manganese) and *silans* (silica).

Organic matter and sesquioxides are often transported in association with one another as organo-metallic complexes (section 3.2.2), and various chemical extraction procedures have been developed to identify organically bound sesquioxides (section 2.4.1). Bs horizons often possess a granular structure, although its origin is open to debate; it has been considered by some to result from the deposition of organo-metallic complexes, although others have argued that it represents residues from organic matter decomposed by soil mesofauna (De Coninck and Righi 1983).

In addition to the receipt of illuvial material, B horizons can form by the residual concentration of sesquioxides after other materials have been removed. These are associated with low latitude soils (section 4.3.3) and are characterised by strong weathering which results in a low silt content, a sand fraction in which no weatherable minerals remain, and a clay fraction comprising typically kaolinite and iron and aluminium oxihydrates. They also have a granular structure whose origin it is thought can relate to a combination of weathering, eluviation and bioturbation (Stoops 1983).

Soils may contain horizons which are of low porosity, known as *pans*, which restrict water movement and root growth. They can be formed by a variety of processes, and are known by a variety of terms, for example, *calcrete, plinthite, silcrete, fragipan, petrogypsic horizons* and *placic horizons*. Calcrete is a hard, indurated deposit of calcium carbonate, also known as a *petrocalcic horizon* (Soil Survey Staff 1975). Gile *et al.* (1966) recognised a number of stages in its formation (Figure 3.13). Carbonate deposits form initially when water droplets evaporate from the underside of stones, then gradually the pores become filled with illuvial carbonate. When this is complete, any further downward water movement is impeded, which causes water to pond up above this zone and subsequent evaporation episodes leave deposits which occur as a series of horizontal layers. Carbonate may also become concentrated by biological accumulation, or added via rainwater, atmospheric dust or sea spray (Klappa 1983). Calcrete can also form by a number of non-pedological processes, for example, the evaporation of water in shallow, ephemeral lakes or when artesian water rises to the surface (Goudie 1983, Milnes 1992).

Plinthite is an iron-rich material formed by the concentration of iron moved

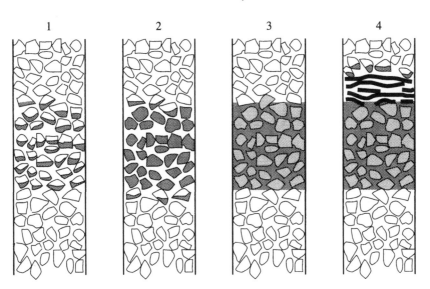

Figure 3.13 A schematic diagram showing the stages of illuvial carbonate accumulation in a gravelly soil (from Gile *et al.* 1966). Note how the carbonate is deposited in horizontal layers in stage 4

in solution from elsewhere, and occurs in strongly weathered soils in which clay synthesis and translocation of silica in solution also occur. On repeated wetting and drying or exposure to solar heating it hardens irreversibly to ironstone. Silcrete, also known as *duripan* (Soil Survey Staff 1975), is an indurated material cemented by various forms of secondary silica. Various origins have been proposed, including the residual concentration of silica following intense leaching of aluminium and other major cations under acidic conditions, and the leaching of silica in ground water (Milnes and Thiry 1992).

Fragipans have a high bulk density and are very hard when dry, forming vertical cracks and a prismatic type of structure (section 2.3.2). They restrict water movement and are therefore often mottled due to gleying. Their origin is unclear but may relate to compaction by ground ice, with aggregation by clay minerals, iron and aluminium compounds or silica (Soil Survey Staff 1975, Birkeland 1984a, Payton 1992). Petrogypsic horizons are indurated layers of gypsum-rich material, the gypsum content usually exceeding 60 percent (Soil Survey Staff 1992), which again can form by translocation or by ground-water concentration (Watson 1992). Placic horizons consist of a dark coloured pan cemented by iron, iron and manganese or an iron-organic complex, and normally do not exceed 10 mm in thickness.

Finally subsurface horizons can possess various characteristics which are not related to transfer. For example, in soils with periodic or permanent saturation, the reduction of iron resulting from the associated anaerobic conditions can produce a blue-grey coloration known as *gleying*. In zones of permanent saturation this colour occurs throughout, while in zones of periodic saturation a mottled appearance results, with anaerobic areas of reduced iron being interspersed with orange-brown, aerobic areas of oxidised iron (section 2.4.4). Variations in the degree of aeration occur because of different rates of wetting and drying due to small-scale variations in soil hydrological properties, particularly porosity. Gleyed horizons are given the suffix g, which can be applied to both B and C horizons, for example Btg and Cg.

In some cases, additions to a soil profile may be so great that the material does not form a horizon by the processes described above, but buries any existing horizons that have already developed. This can occur, for example, if a soil is inundated by a river, the sea or an ice sheet, which covers the soil with thick deposits of sediment . A similar situation could occur in the case of volcanic activity, mass movement, aeolian transport or human activity (Figure 3.14). Any type of horizon can become buried in this way, and is denoted by the suffix *b*. Buried horizons can be of great importance in the study of environmental history, and are considered further in Chapter 5.

Mechanical transfer processes can also alter profiles, for example in the case of cryoturbation which can cause horizon boundaries to become contorted (Figure 3.9) or, in more extreme cases, portions of horizons may become detached and incorporated into adjacent horizons. Disturbance by

Figure 3.14 A buried soil beneath colluvium

bioturbation or downslope movement can inhibit horizon development, causing the soil to become homogenised. Soil losses can cause disturbance either to individual horizons or to complete profiles. For example, a surface mineral-organic horizon may become partially or completely removed by excavation, while erosion may result in the loss of both A and B horizons. Profiles truncated in this way can, like buried soils, be useful in examining environmental history (see Chapter 5).

3.3.2 Pedogenic pathways

The combination of horizons within a soil forms a soil profile. Profiles will vary according to the nature of the horizons they contain, which will be determined by the way in which the soil-forming processes operate, and this in turn is influenced by environmental conditions. Because these conditions vary widely over the earth's surface, an almost infinite variety of soil profiles results. However, it is possible to recognise a number of basic pathways along which soil formation can proceed, which results in a number of basic categories of soil type. These will be discussed in this section in the context of traditional terms which have been used to categorise soils for many years and which are still in common usage. Broad environmental associations will also be given here, but pedogenic–environmental relationships will be discussed in more detail in the next chapter.

In environments where weathering is not excessively restricted by environmental factors, a weathered B horizon will develop above the C horizon, and this can be recognised in terms of various physical, chemical and mineralogical characteristics (section 3.3.1). The most strongly weathered profiles have been categorised into three types by Duchaufour (1982) – *fersiallitic, ferruginous* and *ferrallitic*. Fersiallitic soils are characterised by clay mineral synthesis, and the dehydration of iron oxides and subsequent crystallisation to hematite, a process known as *rubification*. This gives the profile a reddened appearance, as in the case of reddened soils occurring over limestone, known as *terra fusca* where rubification is slight or *terra rossa* where it is more marked. Ferruginous soils show greater clay synthesis and also the partial solution of silica, although rubification may not occur. Ferrallitic soils, which are also often rubified, represent the most advanced stage of weathering, with greater silica mobility and the formation of kaolinite.

Many soils, however, can be grouped according to the nature of transfer processes. For example, in dry environments where there is a net upward movement of water, alkaline components, particularly salts, which would normally be easily leached downwards or out of the profile, are moved upwards towards the surface by capillary action, where they can accumulate as a thin surface crust or as a horizon. Where salts are involved, the resulting profiles are known as *solonchak* (Plate 5). They are light coloured, and usually have a low organic content and alkaline pH values. Two soil types related to solonchak are the *solonetz* and *solod*. These result from leaching, which initially causes dispersion of the organic matter and hydrolysis of salts, producing a very high pH (up to a value of 10); this is the solonetz phase. With continued leaching the organic matter is moved further down the profile and the pH in the upper part of the profile decreases as the salts are also leached downwards, giving a solod (Fanning and Fanning 1989). Where calcium is the major element involved in transfer, either by leaching or upward capillary movement, the process is known as *calcification*, and in extreme cases can lead to the formation of calcrete (section 3.3.1). In cases of more moderate calcification, *chernozems* can form (Plate 6); these are associated with grassland areas and have a mull surface horizon and neutral or slightly alkaline pH values.

If the extent to which alkaline components are leached exceeds the rate at which they are replaced by weathering, the soil will gradually become more acid in its upper part, as the exchange sites on the clay-humus complex become occupied by hydrogen ions and the alkaline components, particularly sodium and potassium, are lost from the profile via ground drainage waters. Soils which show a limited degree of acidification are often associated with broadleaf forest and are known simply as *brown earths* (Plate 7). They have a mull or moder surface horizon, and in the more alkaline varieties of profile, the main transfer processes are the mixing of organic matter by burrowing organisms, and only slight leaching of basic cations. In the more

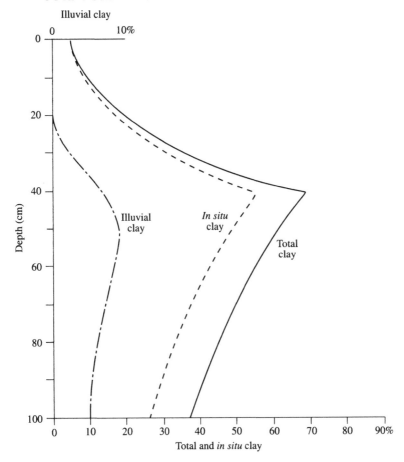

Figure 3.15 Depth functions of illuvial, *in situ* and total clay (from Brewer 1968)

acid variety, leaching of basic cations can be greater, resulting in less aggregation in the upper part of the profile. This can allow clay to become washed downwards, a process known as *lessivage*, which may therefore form a Bt horizon showing characteristic argillans (Plate 1). These profiles are normally slightly to moderately acid, with mull or moder types of organic matter and pH values in the order of 4.5–6.5 (McKeague 1983). Such profiles are sometimes referred to as *argillic brown earths*, the term argillic being used to distinguish these types from ordinary brown earths in which clay movement is absent or limited. Although these are characterised by the presence of clay translocation, the amount of illuvial clay in the Bt horizon is often much lower than that of the clay produced by *in situ* weathering (Figure 3.15). Clay translocation and Bt horizon formation also occurs in weathered, fersiallitic soils, but decreases in ferruginous and ferrallitic soils because the kaolinite which is formed is resistant to movement by water (Duchaufour 1982).

Soils in which sesquioxides are transferred are usually moderately to highly acid, either because they are developed from an acidic parent material or because most of the alkaline components have already been leached out of the profile. Surface horizons are generally of the mor type, with pH values below 4.5. The movement of sesquioxides is traditionally associated with organic complexing, which is enhanced under acidic conditions (section 3.2.3), and the resulting profiles develop a light-coloured E horizon from which the iron staining has been removed, overlying a zone of illuviation (Plate 8). These profiles are known as *podzols* and can be divided into two main types, those in which the illuvial zone comprises a single horizon of the Bs type, known as *ferric* podzols, and those which have an additional organic-rich illuvial horizon between the E and Bs, known as *humo-ferric* podzols. The formation of Bs horizons has been discussed in the previous section, but in some cases the illuvial zone may contain a layer of strong iron cementation, often associated with restricted drainage, which is thought to form where illuviation is strong in relation to bioturbation (De Coninck and Righi 1983). The importance of the organic complexing process in the formation of podzols has become questioned; it has been argued that sesquioxides are concentrated in B horizons largely by inorganic processes involving translocation in solution, with the additional organic complexing with aluminium occurring within the B horizon following the attack of the clay minerals imogolite and proto-imogolite (Farmer 1982, Ross 1989). It may, however, be the case that the extent to which organic and inorganic processes are involved in podzol formation is determined by the amount of organic matter present, with organic complexing being more important in soils with a higher organic matter content (Wang *et al.* 1986).

Highly leached and weathered profiles are found mainly in low latitude regions and are known traditionally as *laterites*, being characterised by the process of ferrallitisation (Duchaufour 1982). Their horizon characteristics vary greatly, depending on the materials undergoing transfer (Tardy 1992), but two basic sets of processes can be distinguished. The first type, referred to as *latosolisation*, is associated with mafic rocks high in weatherable minerals, for example basalt, in which silica and bases are leached out of the upper part of the profile, leaving a residual accumulation of sesquioxides (Plate 9). The second type, known as *lateritisation*, involves the formation of plinthite (section 3.3.1), which, because the concentration of the iron is often not even, commonly takes a nodular form (Fanning and Fanning 1989). Laterite profiles are often much deeper than the other types discussed above due to intense weathering operating over long periods of time, with depths of 100 m or more being recorded (Gerrard 1988); profiles of the other soil types generally do not exceed a few metres in depth and it is not unusual for them to be less than 1 metre deep.

As will be seen in the next chapter, environmental conditions can cause soils to be poorly developed. In such profiles all the pedogenic processes may be restricted, and some may be absent entirely. Although a soil, by definition, must show evidence of some weathering or organic matter accumulation, it does not necessarily need to experience significant transformations or transfers of material, and a poorly developed profile can therefore possess only a weathered C horizon and/or a surface organic horizon, with no B horizon. Soils which contain only an A and C horizon are often referred to simply as A/C profiles (Plate 10). They may also be known as *lithosols* if developed from weakly weathered bedrock, *regosols* if formed in regolith, or *rankers* if occupying steep slopes. A special category of poorly developed profile occurs in the case of soils developed on limestone parent material, where the absence of a B horizon does not necessarily result from limited weathering but from the fact that all the weathering products are removed from the profile in solution; this type of soil is known as a *rendzina* (Plate 11). Another special category of soil is the *andosol* or *andisol* which develops from parent materials of airborne volcanic deposits and is distinguished from other soils by its low bulk density and high content of amorphous or low crystallinity weathering products (Shoji *et al.* 1993).

The profiles described above form in freely draining conditions, but there are many soils whose profiles develop as a result of restricted drainage. These can be divided into organic and mineral varieties. Where transformations, transfers and losses are restricted by poor drainage, organic additions will undergo little decomposition (section 3.3.1), resulting in the formation of a *peat*; this process is sometimes known as *paludification* (Shotyk 1992). Peats may be of varying thickness and pH, but all usually show the water table close to the surface. Within mineral soils, restricted drainage causes gleying (section 3.3.1) and in cases of permanent saturation, transfer processes are also inhibited due mainly to lack of water movement and bioturbation. Profiles of this type are known as *gleys* (Plate 12), although gleying can also occur within other soil types in which saturation is periodic. Two basic types of gley can be recognised – where saturation occurs due to the local water table extending up into the profile, gleying increases with depth and these profiles are termed *ground-water gleys*. In contrast, profiles which have a perched water table in their upper part, caused for example by compaction or clay illuviation, and become progressively well drained with depth are called *surface-water gleys*.

It is important to note that the pedogenic pathways outlined above relate to processes which characterise particular types of soil, but that these may not be the only processes operating in these soils. For example, clay translocation can operate in rubified soils and gleying can occur in brown earths and podzols if the drainage is impeded.

3.3.3 Soil classification

The profiles discussed above allow a basic distinction between soil types, but this is obviously generalised and qualitative. There have, however, been numerous attempts throughout the history of soil science to classify soil profiles in a more detailed and systematic way. Soils may be classified for a variety of reasons, but here we will consider only genetic classification systems. Most of these use a hierarchical approach, whereby a series of major groups are identified and then each is subdivided at a number of levels of increasing detail. A similar system is of course used in this book – the ten chapters are each divided into a number of headings, some of which are further subdivided.

As in the case of horizon designation, probably the best known and most widely used systems of classification are those developed by the FAO-Unesco (1989) and Soil Survey Staff (1975, 1992), and these will be focused on here. The FAO-Unesco system, based on an earlier scheme (FAO-Unesco 1974), uses 28 *major soil groupings*, which are in total subdivided into 153 *units*. The names of the units are derived from a variety of linguistic sources, some of the terms simply being adopted from existing, traditional terms and others being newly devised (Table 3.3). Many of the characteristics used for classification are morphological, although occasionally refer to processes (Luvisols, Vertisols) or chemistry (Acrisols, Ferralsols). Division of the units ranges from two in the case of Greyzems, to nine in the case of Cambisols. These divisions are made according to terms derived mainly from Greek and Latin roots, most of the terms being morphological, and some being the same as those used to identify the soil units themselves. Such a system, while being fairly comprehensive, is only qualitative, so in order to make it less subjective, various sets of quantitatively defined diagnostic horizons and properties have been devised for allocating soils to particular units and sub-units. Five types of surface horizon and five types of B horizon are recognised, along with six other horizon types, and 26 diagnostic properties are defined, mainly on the basis of morphological and chemical properties.

The Soil Survey Staff (1975) system distinguished ten major soil groups, but an eleventh group – Andisols – has recently been devised (Shoji *et al.* 1993) and appears in the 1992 scheme. The groups are known as *orders* and, as in the case of the FAO-Unesco system, their names are derived from a variety of linguistic roots (Table 3.4). Some refer to morphological character-istics (Histosol, Mollisol, Spodosol), while others relate to processes (Incep-tisol, Vertisol, Ultisol) or chemical properties (Alfisol, Oxisol). The orders are divided into 47 *sub-orders* using terms again derived mainly from Greek and Latin sources and based on a variety of characteristics. The number of sub-orders ranges from two to seven per order, with most having four or five. Again, a number of diagnostic horizons and properties are also recognised. The sub-orders are divided into *great groups* using a variety of terms based

Table 3.3 Formative elements used for naming FAO-Unesco (1989) major soil groupings

Major soil grouping	Formative element	Connotation
Acrisols	L. *acer, acetum* (strong acid)	Low base saturation
Alisols	L. *alumen*	High aluminium content
Andosols	Jap. *an* (dark), *do* (soil)	Dark surface horizon (rich in volcanic glass)
Anthrosols	Gr. *anthropos* (man)	Resulting from human activities
Arenosols	Gr. *arena* (sand)	Weakly developed, coarse-textured soils
Calcisols	L. *calx* (lime)	Calcium carbonate accumulation
Cambisols	L. *cambiare* (to change)	Changes in colour, structure and consistence
Chernozems	Rus. *chern* (black), *zemlja* (earth, land)	Black, rich in organic matter
Ferralsols	L. *ferrum, alumen*	High sesquioxide content
Fluvisols	L. *fluvius* (river)	Alluvial deposits
Gleysols	Rus. *gley* (mucky soil mass)	Excess water
Greyzems	AS *grey*, Rus. *zemlja* (earth, land)	Uncoated silt and quartz grains within organic-rich layers
Gypsisols	L. *gypsum*	Calcium sulphate accumulation
Histosols	Gr. *histos* (tissue)	Fresh or partly decomposed organic matter
Kastanozems	L. *castanea* (chestnut), Rus. *zemlja* (earth, land)	Organic-rich, brown colour
Leptosols	Gr. *leptos* (thin)	Weakly developed, shallow soils
Lixisols	L. *lixivia* (washing)	Clay accumulation and strong weathering
Luvisols	L. *lueve* (to wash)	Clay accumulation
Nitisols	L. *nitidus* (shiny)	Shiny ped faces
Phaeozems	Gr. *phaios* (dusky), Rus. *zemlja* (earth, land)	Organic-rich, dark colour
Planosols	L. *planus* (flat, level)	Seasonal surface waterlogging on level or depressed relief
Plinthosols	Gr. *plinthos* (brick)	Mottled clayey materials which harden on exposure
Podzols	Rus. *pod* (under), *zola* (ash)	Strongly bleached horizon
Podzoluvisols	Podzols and Luvisols	
Regosols	Gr. *rhegos* (blanket)	Loose mantle of material
Solonchaks	Rus. *sol* (salt)	Salty area
Solonetz	Rus. *sol* (salt), *etz* (strongly)	
Vertisols	L. *vertere* (to turn)	Turnover of surface soil

Note: AS = Anglo Saxon; Gr. = Greek; Jap. = Japanese; L. = Latin; Rus. = Russian.

Table 3.4 Formative elements used for naming the Soil Survey Staff (1975, 1992) soil orders

Order	Formative element	Connotation
Alfisols	M. *Alf* (from Pedalfer)	Clay translocation
Andisols	M. *And* (from Andosol)	Volcanic soil (see Table 3.3)
Aridisols	L. *aridus* (dry)	Arid soil
Entisols	M. *Ent* (from Recent)	Early stages of soil formation
Histosols	Gr. *histos* (tissue)	Limited organic matter decomposition
Inceptisols	L. *inceptum* (beginning)	Limited soil development
Mollisols	Gr. *mollis* (soft)	Dark brown/black surface horizon, soft when dry
Oxisols	Fr. *oxide* (oxide)	Oxic horizon or plinthite
Spodosols	Gr. *spodos* (wood ash)	Bleached eluvial horizon
Ultisols	L. *ultimus* (last)	Highly developed soil
Vertisols	L. *verto* (turn)	Turnover of soil

Note: Fr. = French; Gr. = Greek; L. = Latin; M. = meaningless syllable.

on the diagnostic properties, which gives around 225 categories. Further division is made into *sub-groups* by adding various adjectives to the great group names, and still further subdivision can be made into *families*, primarily on the basis of particle size, mineralogy and temperature regime. A final level of subdivision, the *series*, is also possible, but this is usually derived from the name of the locality in which that type of soil was first recognised, and therefore has no real value in providing information about soil characteristics or genesis; series are used in soil mapping (section 9.2.1). Although this scheme provides the most comprehensive soil classification available, it recognises that not all soils will be accommodated within it, and additional categories have been proposed, particularly for soils influenced by

Table 3.5 Approximate relationships between the Soil Survey Staff (1975, 1992) soil orders and FAO-Unesco (1989) major soil groupings

Soil Survey Staff soil orders	FAO-Unesco major soil groupings
Alfisols	Luvisols
Andisols	Andosols
Aridisols	Calcisols, Gypsisols, Solonchaks, Solonetz
Entisols	Arenosols, Fluvisols, Leptosols, Regosols
Histosols	Histosols
Inceptisols	Cambisols
Mollisols	Chernozems, Greyzems, Kastanozems, Phaeozems
Oxisols	Alisols, Ferralsols, Nitosols, Plinthosols
Spodosols	Podzols
Ultisols	Acrisols, Lixisols
Vertisols	Vertisols

human activity such as waste disposal, construction and dredging (Fanning and Fanning 1989). Approximate relationships between the Soil Survey Staff soil orders and FAO-Unesco major soil groupings are shown in Table 3.5.

Hierarchical classification systems have also been developed elsewhere for use within a national context, for example in the British Isles (Avery 1990), Canada (Canada Soil Survey Committee 1978) and Australia (Moore *et al.* 1983). The number of possible designations is less than in the Soil Survey Staff system, but still allows a fairly comprehensive classification. For example, the British Isles system has three levels within the hierarchy, comprising six major soil groups (lithomorphic soils, brown soils, podzols, gley soils, man-made soils and peat soils), 23 soil groups and 94 sub-groups.

Despite the widespread use of hierarchical classification systems, various problems are associated with them. An obvious problem for many non-specialist users is the complex and unfamiliar terminology, which makes the classification and recognition of soil types rather laborious. Another problem is that the criteria used to distinguish soils at any one level within a hierarchy are not all defined using the same characteristics; for example, a mixture of morphological, genetic and chemical characteristics is used in defining the first level of the hierarchy in the systems discussed above. The use of quantitatively defined criteria for differentiation of soils means that certain soil types which are very similar in many respects can be allocated to different categories if they narrowly fail to meet the criteria being used. This occurs, for example, in the case of the Spodosol category of the Soil Survey Staff system, where podzols cannot be classified as such because their B horizons do not comply with the criteria used for defining a spodic horizon (Avery *et al.* 1977, McKeague *et al.* 1983, Lietzke and McGuire 1987). The opposite situation could also occur in the case of soils which are in many respects different being placed in the same category if they happen to meet the differentiating criteria which have been selected. A further problem is the difference in the number of subdivisions of the various categories at any one level within the hierarchy. For example, using the FAO-Unesco system, Cambisols can be allocated to one of nine possible categories whereas only two subdivisions of Greyzems are possible. Finally, it is clear that selection of the differentiating criteria is subjective, with only a limited number of numerous possible soil characteristics being chosen and value judgements being made about the relative importance of criteria both within and between hierarchical levels.

Some of the problems associated with traditional classification systems have been overcome by the development of numerical methods. A variety of methods are available, but the two most commonly used are ordination and the construction of dendrograms (Webster and Oliver 1990). Ordination methods involve the use of principal-components analysis and can be considered as follows. If the relationships between a range of soil properties for a number of soils are considered to be represented by a series of points

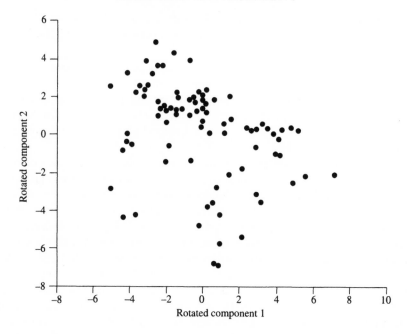

Figure 3.16 Scatter of 85 sampling sites rotated in the plane of the first two principal components (from Webster and Oliver 1990)

in a multidimensional space, these can be reduced mathematically to one or a few principal axes which account for as much of the total variance as possible and in this form the differentiation between the soils can become relatively easy. For example, 15 properties measured at two depths for 85 sites were subjected to principal-components analysis by Cuanalo and Webster (1970), who found that the first two latent roots, out of 30, accounted for nearly half the variance in the sample. Projection of the sites on to the plane defined by the first two principal axes therefore gave the most informative single display of relationships between the soils. Rotation of the factor axes allows each of the original variates to contribute strongly to one of the factors and much less so to the others, making the plot more simple to interpret (Figure 3.16). As a result, the first (horizontal) axis strongly represented properties related to the moisture of the soil, while the second (vertical) axis related strongly to textural properties. The soils are therefore grouped in Figure 3.16 such that the driest soils are on the right and the wettest on the left, while the heaviest-textured soils are at the top and the lightest at the bottom. The small number of sites located in the bottom left-hand quarter of the plot therefore indicates that there are relatively few wet, light-textured soils.

The construction of dendrograms is concerned with the distances between the sites plotted as a result of principal-components analysis, and is an alternative hierarchical approach to the systems discussed earlier. The most closely grouped sites form the lowest level of the hierarchy and increasingly distant groups are linked at higher levels (Figure 3.17). However, although

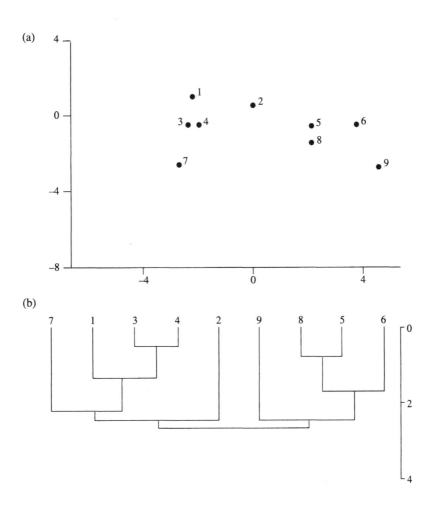

Figure 3.17 Numerical hierarchical classification: (a) a scatter of nine points in the plane of the first two principal components, (b) a single-linkage dendrogram for the nine points (from Webster and Oliver 1990)

hierarchical numerical classifications are often applied to soils, they are designed for populations which show nested clustering, which is rarely found in the case of soils; consequently non-hierarchical numerical methods such as canonical variate analysis may be used as an alternative (Webster and Oliver 1990).

The advantage of numerical methods is that they allow a large number of soil characteristics to be taken into account without any preconception of their relative importance. They do, however, suffer from the problem that no single attribute is either sufficient or necessary to confer class membership, and it is therefore very difficult to construct an identification key (Avery 1990). They also require a large quantity of data for each soil being classified, and to date their use has therefore been confined to the local, rather than national, scale.

3.4 SUMMARY

In this chapter we have examined two main aspects of soil formation – the basic processes involved and the profiles which result from the operation of these processes. The processes have been identified under four main headings – additions, transformations, transfers and losses. Additions occur either via the surface or subsurface and can involve both organic and mineral material; this material is often locally derived, but can also include material transported from some distance away, by wind, water, gravity, animals or humans. Transformations also involve both organic and mineral material. Organic matter undergoes decomposition by a variety of organisms, which produces humus along with a number of by-products, while mineral material experiences weathering by brittle fracture or crystal lattice breakdown. Mineral alteration can also occur by transformation or synthesis of clay minerals. Pedogenic transfers can operate in two main ways – by water or mechanically. Water transfers material in solution or suspension, while mechanical transfer can occur by many processes, including bioturbation, alternating expansion and contraction, freezing, density loading, crystal growth, gravity movement, upwelling water and earthquakes. Losses from the soil, like additions to it, can occur via either the surface or subsurface. Surface losses may be by natural erosion processes or human activity, while subsurface losses involve solute leaching or material removed in suspension through large pores or pipes.

The combined operation of the soil-forming processes leads to profile differentiation. Weathering causes disintegration of the parent material, known in its early stages as the C horizon, and changes up through the profile due to differing intensities of weathering are usually expressed in terms of a change in colour or texture observable in the field, or a change in mineralogy or chemistry identifiable by laboratory analysis. Surface horizon formation often involves the accumulation of organic matter, which can occur above and independently of the mineral soil, as in the case of peat, or can be characterised by the mixing of organic and mineral material to produce an A horizon. Subsurface horizons are often characterised by features relating to transfer processes, particularly those involving movement of material by water. Horizons from which material has been removed in this way are known as eluvial or E horizons, while those into which material has been transferred are known as illuvial or B horizons. Some soil horizons may have a very low porosity and these are known as pans; all are characterised by the concentration of chemical compounds, except fragipan whose origin is uncertain but may relate to compaction by ground ice. Poorly drained soils are usually characterised by gleying, and this can affect any soil horizon in which iron compounds are available for reduction.

A number of profile types can be recognised on the basis of the way in which the soil-forming processes operate in combination; these sequences are referred to as pedogenic pathways. Profiles can be grouped according to strong weathering characteristics (fersiallitic, ferruginous and ferrallitic soils), the transfer of alkaline components (solonchak, solonetz, solod, chernozems and brown earths), clay (argillic brown earths) or sesquioxides and silica (podzols and laterites), features associated with poor drainage (peat and gleys), and restricted development (lithosols, regosols, rankers and rendzinas). The classification of soil types is considered from two perspectives – traditional and numerical. Traditional classifications are usually hierarchical, and although these allow soils to be differentiated in great detail, they are problematic in that the use of a limited number of rigid defining criteria can lead to the artificial separation of similar soils. Numerical methods help to address this problem by considering a large range of soil properties. These are subjected to principal-components analysis, which allows soils to be grouped by ordination or the use of dendrograms. These methods, however, require large data sets, and to date their use has therefore been confined to specific, local cases rather than to a general, national level.

4

SOIL FORMATION AND ENVIRONMENT

4.1 INTRODUCTION

It has long been recognised that the operation of pedogenic processes is determined by a number of environmental factors operating through the medium of time and that these therefore determine the type of soil produced in any particular location. These ideas were originally formulated in late nineteenth-century Russia when V.V. Dokuchaev identified the principal factors as climate, biota, parent material, relief and landscape age, and N.M. Sibertsev produced a system of soil classification based on bioclimatic zones. This system recognised that soils varied with latitude according to climatic and vegetation conditions, and gave rise to the concept of *zonal* soils (Table 4.1). It was also recognised, however, that these were not the only two factors controlling soil type distribution and that in some instances soils were more closely associated with parent material or drainage conditions. Such soils were termed *intrazonal*. A third category of soils was also recognised – *azonal* – to which could be assigned soils that were poorly developed, due mainly to lack of time, such as those occurring on recent alluvial deposits.

The influence of environmental factors on soil formation was considered in a more systematic manner following the work of Jenny (1941, 1946). Using the equation:

$$s = f\,(cl, o, r, p, t)$$

where s = soil, cl = climate, o = organisms, r = relief, p = parent material and t = time, it was suggested that if any four of these factors could be held constant, the equation could be solved for the fifth factor. This led to the concept of *sequences* and *functions*; sequences occurred in particular environmental situations where the variation in soil conditions could be attributed principally to variations in only one of the factors, the other four factors remaining effectively constant, while functions were the equations which described the way in which a soil property was influenced by any one factor. Five types of sequence and function were thus proposed: climo-, bio-, topo-, litho- and chrono-. It is clear, however, that in reality all the factors are interdependent and any single factor cannot therefore vary without at

Table 4.1 The classification of world soils according to N.M. Sibertsev

Zone	Soil type
Normal or zonal soils	
Boreal	Tundra (dark brown) soils
Taiga	Light grey podzolised soils
Forest-steppe	Grey and dark grey soils
Steppe	Chernozem
Desert-steppe	Chestnut and brown soils
Desert zone	Aerial soils, yellow soils, white soils
Sub-tropical and tropical forest	Laterite or red soils
Transitional or intrazonal soils	
Dry land moor soils or moor-meadow soils	
Carbonate-containing soils (rendzinas)	
Secondary alkaline soils	
Abnormal or azonal soils	
Moor soils	
Alluvial soils	
Aeolian soils	

Source: Fenwick and Knapp 1982

least some variation occurring in one or more of the other factors, and consequently the value of this approach has been questioned (Johnson and Watson-Stegner 1987). Nevertheless, one factor may often be the dominant, if not the only, variable controlling soil formation (Richardson and Edmonds 1987), and in these instances Jenny's concept still forms a useful starting point for the study of environmental influences on soil formation.

The first part of this chapter examines in turn the influence of the four principal environmental factors, along with the time factor, on the processes of soil formation – additions, transformations, transfers and losses. This is followed by an examination of the ways in which these factors operate in combination in various environments in order to provide an overview of soil formation at the global scale. The important influence of human activity on soils is often considered separately from the biotic factor, as is the case here; this will be examined within a historical context in the next chapter and in the context of contemporary land use in Chapters 7 and 8.

4.2 CONTROLS ON SOIL FORMATION

4.2.1 Climate

Climatic influences on soil surface additions occur via aeolian and water transport, and mass movement. Aeolian transport is greatest in arid areas, where the stabilising effects of soil moisture and vegetation cover are low, and also in areas of strong winds, whereas water transport and mass movement are greatest in areas of high rainfall or rapid inputs of water, for example by seasonal melting of snow and ice. Climate also influences soil surface organic additions indirectly due to its control on vegetation types and their growth rates. Generally, growth rates increase with temperature and/or humidity and therefore so does the quantity of organic material added to the soil; subsurface additions are also controlled indirectly by the effect of climate on biota (section 4.2.2).

Climate influences the transformation of organic components by its effect on rates of organic matter decay; maximum rates of decay usually take place in the temperature range 25–35°C (Ross 1989). At a global scale, the rate of decomposition of organic components therefore generally increases from

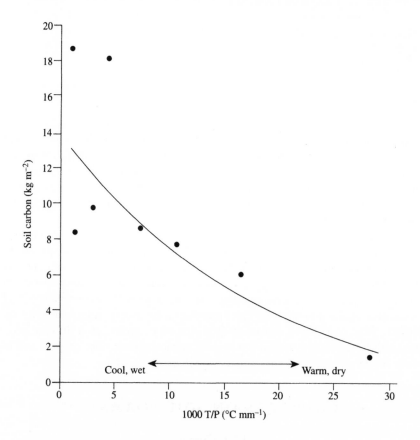

Figure 4.1 Relationship between soil organic carbon content and climate in New Zealand (from Tate 1992). On the horizontal axis T = mean annual temperature and P = annual precipitation

high to low latitude (section 4.3). The accumulation of organic matter in a soil is determined by the balance between the rate at which material is added and the rate at which it is decomposed (section 3.3.1). The influence of climate on these processes has been shown in New Zealand, where soil carbon content was found to increase from warm, dry to cool, wet environments due to decreasing decomposition relative to accumulation (Tate 1992) (Figure 4.1). A similar situation has been reported in the case of grassland soils in the USA, where organic carbon content increases with precipitation (Burke et al. 1989).

The rate of mineral transformation via crystal lattice breakdown increases with temperature; in general the rate of chemical reactions doubles for every 10°C rise in temperature (Curtis 1976a). However, for certain weathering processes it is the fluctuation in temperature, rather than the absolute value, which is important, as in the case of thermal expansion–contraction and freeze–thaw (section 3.2.2). Moisture also affects weathering, since it is a vital ingredient in many weathering processes and its availability is determined by precipitation and evaporation, the latter in turn increasing with temperature and windspeed. Temperature controls water availability via frozen ground conditions, as water is unavailable for weathering when it is locked in the ground as ice, and long durations of snow cover may delay ground thawing and hence further inhibit weathering. Fluctuations in moisture, often temperature-controlled, are also important in the case of weathering mechanisms such as salt crystal growth and hydration–dehydration.

The transfer of soil materials is also highly dependent on temperature and moisture. These will directly influence the processes involving water movement and mechanical transfer and also have an indirect influence via their control on the type and quantity of vegetation and on the organisms associated with bioturbation. Since translocation is moisture-dependent, its potential for downward movement will be greatest in humid environments, particularly where temperatures and windspeeds, and therefore evapotranspiration losses, are low, although in the case of upward movement of salts the reverse obviously applies. As in the case of weathering, where temperatures are sufficiently low to cause frozen ground conditions, translocation will be inhibited due to water being unavailable. In terms of mechanical transfer, fluctuations in temperature and moisture are important in order for frost processes and wetting and drying to be effective. While temperature and rainfall clearly have an important effect on translocation, wind can also affect soil moisture, and hence translocation potential, via evapotranspiration.

The influence of climate on weathering and translocation has been demonstrated in a number of studies at different scales. For example, Strakhov (1967) has shown how, at the global scale, weathering processes and the distribution of weathering products relate to precipitation and temperature along a transect from high to low latitudes, giving rise to particular

99

soil types within various latitudinal zones (Figure 4.2); this will be discussed further in section 4.3. A medium-scale illustration is the study by Arkley (1967), which showed that soil orders in the western USA could be grouped on the basis of three climatic parameters – calculated actual evapotranspiration, leaching index and mean annual temperature (Figure 4.3). This showed that Ultisols and Spodosols are associated with high leaching and Aridisols with low leaching, while Mollisols and Alfisols occupy an intermediate position which varies according to temperature. It is important to recognise, however, that these trends should only be accepted at a general level because other environmental factors will also have an important influence on these soils, which probably explains the extent of overlap of the groups. A smaller-scale study is that of Rabenhorst and Wilding (1986), which showed that in the drier, western part of the Edwards Plateau, Texas, the soils were characterised by calcium carbonate accumulation while in the more humid, eastern area the soils experienced greater translocation, being generally non-calcareous and often possessing argillic horizons. Another smaller-scale example is the study by Schaetzl (1991a), which examined the distribution of podzols (Spodosols) in southern Michigan. This considered that areas of most strongly developed podzols were related to those where thicker snow

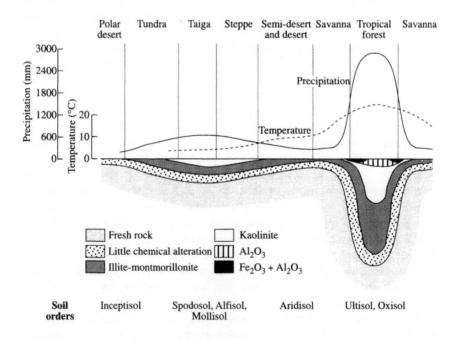

Figure 4.2 Latitudinal variation in weathering in response to climate (based on Strakhov 1967)

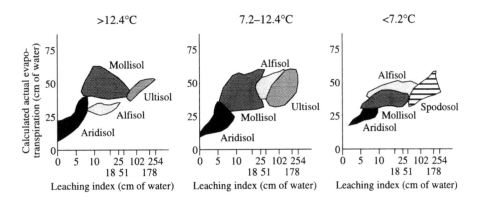

Figure 4.3 Relationship of soil orders to actual evapotranspiration and leaching index (based on Arkley 1967). Temperatures are mean annual values

covers and wetter autumns restricted soil freezing (section 6.3.1) and therefore allowed greater translocation of organo-metallic complexes.

Soil losses are controlled via climatic influences on water and wind transport, and also indirectly by influences on biota. Losses from wind erosion are favoured by arid, windy climates in which low soil moisture contents and vegetation covers make for easy removal of particles. Losses of particulate material by water erosion are, like additions, greatest in areas of high rainfall intensity or rapid influxes of water, and are also enhanced by low vegetation covers (Young and Saunders 1986, Selby 1993). In the case of losses by solution, there is in general an increase with mean annual temperature and precipitation, although again it is important to recognise that this relationship can be complicated by other factors, particularly parent material and drainage conditions (Crabtree 1986).

4.2.2 Biota

Soils and biota occur in intimate association, and the importance of the effect of one on the other cannot be overestimated. Biota are also influenced to a great extent by climate; additions of organic matter will depend largely on the type and density of flora and fauna, but these will in turn be determined by climatic conditions. Thus litter production relates broadly to bioclimatic regions, the greatest amounts occurring in tropical forests where plant growth is rapid and lowest amounts in tundra regions where growth is restricted by temperature (Table 4.2). Additions of organic matter derived from some distance away will be controlled by the size and density of the material, by its aerodynamic or hydrodynamic characteristics and by the ease with which it can become attached to animals. For example, humus, seeds,

Table 4.2 Litter input rates for different ecosystems

Ecosystem	Litter input (t ha^{-1} a^{-1})
Tundra	1.5
Boreal forest	7.5
Temperate deciduous forest	11.15
Temperate grassland	7.5
Savanna	9.5
Tropical forest	30

Source: Swift et al. 1979

leaves and other small organic components will be more easily transported by wind or water than large stems or branches, while material with irregular or adhesive surfaces will be easily attached to the fur of an animal and can therefore be transported as the animal moves from one area to another.

Transformation of organic matter via the processes of decomposition will be determined by the type of vegetation and the proportion of organisms of different functions. For example, vegetation with a low content of nutrients and high levels of phenol, waxes or lignin will be less rapidly and extensively decomposed than more nutritious vegetation in which these constituents occur at lower levels (Figure 4.4). Thus soils which support needleleaf trees often have a mor type of humus (section 3.3.1) and high C:N ratios, while those supporting broadleaf trees have a moder or mull type of humus and lower C:N ratios (Klemmedson 1987, Van Wesemael and Veer 1992, Bernier et al. 1993). Organisms can be classified according to their function as follows: decomposers, which live in the surface organic horizon and ingest only organic matter; burrowers, which inhabit the subsurface, mineral part of the soil and may ingest both organic and mineral material; grazers, which live at the soil surface and eat vegetation growing in the soil; and predators, which live either at the surface or within the soil and prey on these other groups of organisms. Decomposers, burrowers and grazers transform organic matter directly, while predators ingest these organisms before returning them to the soil via excretion. Biota can also influence transformations in terms of mineral weathering. For example, organic acids can cause hydrolysis, and organo-metallic complexing can take normally stable iron and aluminium compounds into solution (Sohet et al. 1988, Lundström 1993).

Biota can influence the transfer of both organic and mineral material. For example, the ease with which organic matter is carried in solution will depend on its molecular weight and its rate of polymerisation; materials of high molecular weight or those which polymerise quickly will be least mobile. The ease of organic matter transport in suspension will depend mainly on its size; small humus particles will be more easily carried than faecal pellets or other larger organic fragments, although the high surface charge of humus

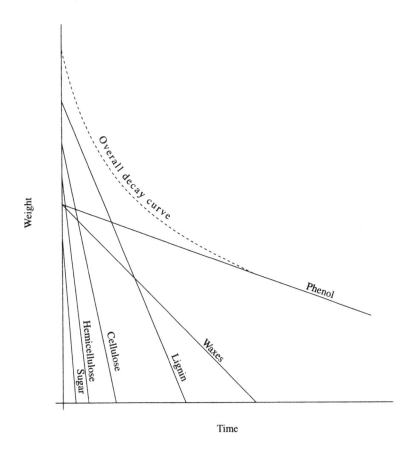

Figure 4.4 Decay curves for the principal constituents of plant litter (from Minderman 1968)

(section 2.4.2) may make its initial movement more difficult. Both organic and mineral material can be transferred by burrowing, the extent of this process being determined by the type and number of organisms present (Hopkins *et al.* 1988). Burrowing earthworms can consume up to 90 t of temperate grassland soil ha^{-1} a^{-1} (Edwards and Lofty 1977), mixing both organic and mineral material, while in the tropics termites can carry large quantities of fine material from depths of over 10 m to construct termitaria at the surface (Lee 1983) (Figure 3.7); this material is eventually added back to the soil when these constructions are abandoned, which can cause layers of material to develop (Johnson 1990). While burrowing is therefore an important means of soil transfer, it can also inhibit the formation of horizons by the continual disturbance of material (Hole 1981).

103

Vegetation can influence transfer processes in terms of the addition of organic acids which can lower pH values in the upper part of the soil (Binkley and Sollins 1990, Dahlgren *et al.* 1991) (Figure 4.5). The influence of biota on organic complexing can also be important in the transfer of metals in solution, particularly iron and aluminium (sections 3.2.3 and 3.3.2) which laboratory experiments have shown can be complexed by solutions derived from a variety of tree and heathland types (e.g. Fisher and Yan 1984, Ross 1989). The effectiveness of this process can be seen in the field by examining the soil at particular locations relative to individual trees. For example, deep podzols occur beneath individual large, and therefore probably old, kauri trees in New Zealand while adjacent soils are not podzolised (Buol *et al.* 1980). In general, needleleaf trees are associated with soil acidification and podzolisation, broadleaf trees have more limited acidifying effects and herbaceous species tend not to cause these effects (Miles 1985).

Organisms will determine soil losses in terms of gaseous outputs, and surface and subsurface lateral removal. Gaseous losses are related to organic matter decomposition and will depend on the type of organic matter. Surface and subsurface losses will be determined by the ease with which organic components are moved, as discussed already in the case of additions and transfers, but in addition will be controlled by the effect of organisms on slope stability. For example, their aggregating effects will help to stabilise soil material on slopes, while large burrowing fauna may destabilise a soil, either

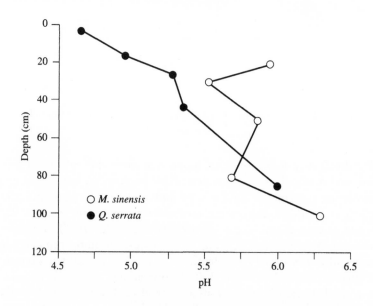

Figure 4.5 The pH of soil solutions beneath Japanese pampas grass (*M. sinensis*) and Japanese oak (*Q. serrata*) (from Dahlgren *et al.* 1991)

mechanically or by their burrows acting as major routeways for subsurface erosion. Vegetation will also help to protect the soil surface from erosion by wind and water (Selby 1993).

Because of the close association between biota and climate, it is often not easy to distinguish biosequences from climosequences. The studies referred to in the previous section, for example, will all involve biotic variations along with the climatic variations. The influence of biota on soil formation is therefore best seen at a small scale, in particular by reference to soil characteristics of profiles occurring beneath different vegetation types within a limited area as, for example, in the case of the study by Boettcher and Kalisz (1990). This compared soil conditions beneath individual trees of two types, showing that conditions were less acid and organic decomposition greater under *Liriodendron tulipifera* than under *Tsuga canadensis*; at this scale a mosaic of profiles therefore occurs, reflecting the distribution of individual species.

4.2.3 Parent material

The major influence of parent material on soil formation is expressed principally in terms of weathering, as this determines the amount of mineral material available to a soil profile and also the subsequent weathering characteristics of the soil itself. The nature and extent of weathering will be determined by the resistance of the minerals which make up the parent material and also by various physical characteristics of the parent material itself. Many studies have sought by a variety of means to determine the relative resistance of minerals and, although the results vary in detail, there is broad agreement over a number of the commonly occurring minerals in soils (Brewer 1976). For example, quartz and zircon are generally considered to be amongst the most resistant minerals, while biotite and staurolite are of intermediate resistance and olivine and augite are easily weathered. Differences in mineral resistance relate to the nature of the bonding between the atoms of their constituent elements (Curtis 1976a, b).

The physical properties of a rock will also determine its resistance to weathering, such as cleavage, porosity, water absorption, coefficient of expansion and thermal conductivity (Brunsden 1979, Gerrard 1988). Such factors will control the ease with which water will enter and reside in a rock to cause chemical weathering, and also the extent to which a rock is subjected to stresses during physical weathering. However, the relationship is not straightforward. For example, a porous rock will allow relatively large volumes of water to enter it, but will often be made up of larger mineral grains with a smaller total surface area than a lower-porosity, finer-grained rock, and the latter will therefore often weather more easily (Schaetzl 1991b). Like mineral stability, various studies have examined the relative stability of rock types, although the results differ according to the weathering process involved, and generalisations are therefore difficult. For example, Birkeland (1984a) has proposed the following sequence for temperate North America: quartzite, chert > granite, basalt > sandstone, siltstone > dolomite, limestone, while for chemical weathering of igneous rocks Gerrard (1988) has proposed the sequence: granite > syenite > diorite > gabbro > basalt. For weathering by salt crystal growth, Brunsden (1979) has reported the sequence: diorite > dolerite > granite > gneiss > shale > sandstone > limestone > chalk. The position of a rock within a sequence can also be altered by environmental conditions, particularly climate. For example, in humid regions granite is more resistant to weathering than limestone while in arid regions the reverse is the case (Crabtree 1986).

The nature of weathering as determined by parent material is fundamental to many soil characteristics. For example, the thickness of a soil profile is dependent in part on the susceptibility of its parent material to weathering, and in cases of extreme resistance shallow lithosols will often result (section 3.3.2). The texture of a soil may also relate to the resistance of the parent material to weathering, more resistant materials giving stonier or coarser-textured soils. Texture will also be influenced by the grain size of the parent material, with shales weathering to give fine-textured soils and glacial meltwater deposits producing coarse-textured soils. The textural characteristics will in turn influence other properties such as aggregation and porosity. The parent material will also affect to a large extent the chemical characteristics of a soil via the type and quantity of elements and minerals released during weathering (Curi and Franzmeier 1987). Two well-known examples of soil types whose characteristics are strongly influenced by their parent material are rendzinas and andosols. Rendzinas form over limestone parent materials which weather by carbonation (section 3.2.2) and therefore lack the development of a B horizon, while andosols develop in airborne volcanic deposits and are distinctive for their low bulk densities and high contents of amorphous or low crystallinity weathering products (section 3.3.2).

Parent material also influences the operation of transfers, particularly those relating to transport by water. This will be determined by the components available for transfer, which will depend on the mineralogical and weathering characteristics of the material, and the ease with which they can be transported, as determined by the permeability, and therefore drainage, of the material. For example, parent materials weathering to produce acidic soils will provide few bases for translocation, while clay translocation will be inhibited in a parent material which is impermeable (Levine et al. 1989). Similarly, translocation can be restricted in very coarse-textured soils due to the excessively rapid drainage of water (Schaetzl 1991b). The chemical content of a parent material will also influence translocation; a soil weathered from an alkaline parent material can experience a large transfer of bases if the climate is sufficiently humid, whereas few bases will be available for translocation in a soil derived from an acid parent material, but the acid conditions may encourage sesquioxide translocation via organic complexing (sections 3.2.3 and 3.3.2). The permeability of a rock depends on various factors such as the size, density and orientation of joints and bedding planes and the size and density of pores. A sequence of relative permeability reported by Brunsden (1979) is as follows: gravel > sand > sandstone > limestone > clay > shale > igneous. Hence a soil developed in a gravel or sand will have a high translocation potential and may therefore become acid as bases are removed, while one developed from clay, shale or igneous parent material will have a lower potential and may develop gleying characteristics as a result of restricted drainage. Effects of parent material on soil losses also relate largely to its influence on drainage conditions. For example, a permeable parent material will allow free drainage and therefore losses will tend to be downwards, whereas a less permeable material will restrict downward water movement and therefore drainage, if the slope permits, will be lateral. Drainage effects will be discussed further in the following section.

4.2.4 Relief

Relief influences soil formation via the movement of material on slopes, and also through its control on drainage. Two important factors can be distinguished – slope angle and position on a slope. The potential for gravity-induced movement increases with slope angle, so additions are often greatest in footslope soils which receive material moved down steep slopes, and in extreme cases this can lead to burial of the soil (section 3.3.1). High slope angles will also encourage free drainage and consequently transformation processes. Conversely, low slope angles often produce poor drainage which will limit transformation processes, often leading to poorly decomposed organic matter in surface horizons and gleying in subsurface horizons (section 3.3.1). Such conditions are therefore often associated with soils on valley floors or in enclosed hollows.

The relationship between slope and the transfer of material in soils is not straightforward. For example, high-angle slopes will encourage rapid drainage which may be effective in transferring material in suspension, but may inhibit solutional transfer because of a reduced reaction time, that is the time available for water passing through the soil to cause dissolution. Slope angle will also determine translocation potential in terms of the amount of precipitation received per unit area of surface, the amount decreasing as the angle increases (Figure 4.6). Also in the case of mechanical mixing caused by downslope movement of soil *en masse*, Finlayson (1985) has argued that there is no clear relationship between slope angle and rate of movement, with movement occurring upslope as well as downslope in some studies of soil creep. Poor drainage on low-angled slopes inhibits burrowing organisms and therefore pedoturbation. Soil losses via the surface can be influenced by slope angle, material being more easily removed downslope under gravity or by

(a)

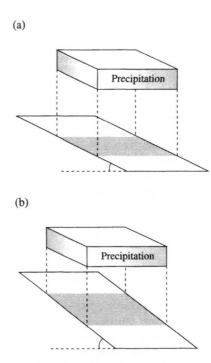

(b)

Figure 4.6 The influence of slope angle on precipitation received at a surface:
(a) gentle slope, (b) steep slope – the same volume of precipitation as for the
gentle slope is distributed over a larger surface area

rainsplash or surface-flowing water at steeper angles (Thornes 1979, Selby
1993), although losses may be compensated by material received from
upslope. Subsurface outputs by water will have the same response to slope
angle as those of transfer discussed above. Carter and Ciolkosz (1991)
examined a number of pedogenic characteristics of soils occurring on a range
of slopes in the eastern USA. Solum depth, B horizon thickness and indices
of development based on clay, Fe and Al contents were found to decrease
with increasing slope angle. This was attributed to greater erosion and lateral
leaching, and less effective precipitation per unit area (Figure 4.6) on steeper
slopes. Erosion of soil on steep slopes not only inhibits soil development by
thinning profiles, but also by reducing the time available for pedogenesis
under stable conditions (Rebertus and Buol 1985, Marron and Popenoe
1986).

In addition to slope angle, the position of a soil on a slope will influence
the operation of pedogenic processes. For example, gravity-induced losses
are more likely to occur from soils in upslope positions while those in

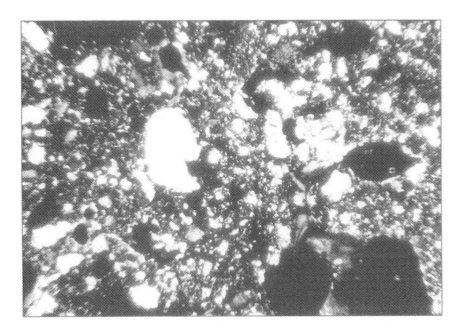

Plate 1 Photomicrograph of illuviation cutans (argillans), appearing as orange-yellow concentrations lining the dark pores (specimen viewed in crossed polarised light; frame width = 4 mm). Courtesy of C.J. Chartres

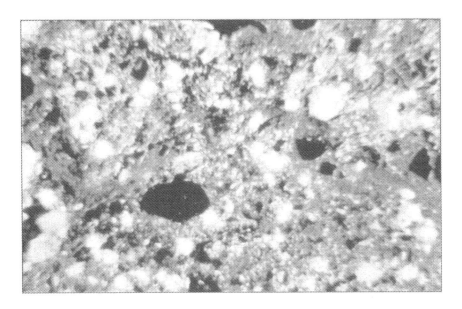

Plate 2 Photomicrograph of stress cutans. The orientated clay appears as orange-yellow zones within the matrix (specimen viewed in crossed polarised light; frame width = 4 mm). Courtesy of C.J. Chartres

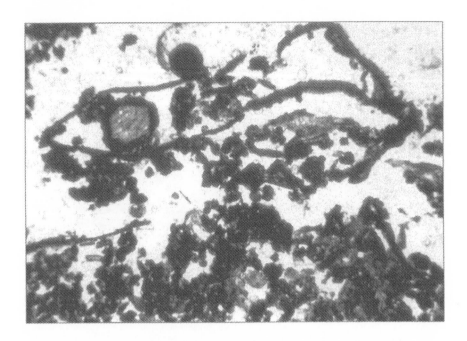

Plate 3 Photomicrograph of decomposing organic matter. The upper part of the photograph shows plant fragments still partially intact, while the lower part shows more highly decomposed organic matter, much of which occurs in the form of faecal pellets (specimen viewed in plain light; frame width = 4 mm)

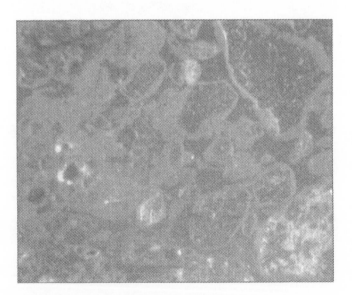

Plate 4 Photomicrograph of orange iron deposits around dark mineral grains (specimen viewed in incident light; frame width = 4 mm)

Plate 5 A solonchak. The photograph shows the top 0.5 m of the profile

Plate 6 A chernozem, characterised by a well developed mull surface horizon 0.5 m thick

Plate 7 A brown earth, showing a mull surface horizon over a weathered B horizon. The photograph shows the top 1 m of the profile

Plate 8 A podzol profile. The characteristic bleached eluvial horizon overlies a darker illuvial horizon

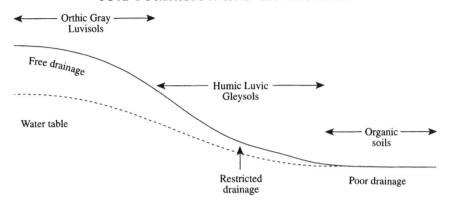

Figure 4.7 An example of a catena (based on Donald *et al.* 1993)

downslope positions are more likely to receive additions (Swanson 1985, Graham and Buol 1990, Karathanasis and Golrick 1991, Webb and Burgham 1994). The operation of transformation and transfer processes will depend to a large extent on drainage conditions, as discussed above, and these can also be determined by slope position. Freely drained soils are most likely to occur in the upper portions of slopes which will therefore usually have a higher potential for vertical translocation than those further downslope, although soils in footslope positions can receive constituents transferred laterally from upslope (Donald *et al.* 1993, Förster 1993). Drainage can also influence weathering processes, as in the case of the formation of iron minerals in Oxisols in Brazil (Macedo and Bryant 1987).

A sequence of soils whose developmental differences can be attributed largely to their position on a slope is often referred to as a *catena*. An example is given in Figure 4.7, from the study of Donald *et al.* (1993) in south-central Canada, in which Organic Soils were found on low slope angles where moisture contents were higher and therefore rates of decomposition were lower; gleys (Humic Luvic Gleysols) occurred in the footslope area where the water table was at a shallow depth, and the more acid, weathered and leached soils (Orthic Gray Luvisols) were located further upslope, where leaching could proceed unhindered.

Relief can therefore both enhance and inhibit soil formation. Enhancement occurs on slopes which allow the optimum amount of additions and transfers, with minimum losses. Inhibition occurs particularly in the case of high-angle slopes, where profile development is limited due to movement of material *en masse*, which restricts horizon development and generally results in thin profiles. Such conditions are therefore characterised by rankers (section 3.3.2). With large additions, soils may become buried by material from upslope, while large losses may remove profiles completely.

4.2.5 Time

Time is the medium through which all the environmental factors operate, and its role in soil formation is therefore integral to the understanding of soil-forming environments. In a general sense, soils will become increasingly developed through time, as long as they are not subject to rapid and excessive additions or losses; these can restrict soil development by adding large quantities of new material on which the soil-forming processes must start to operate, or by removing portions of a soil as it develops. In extreme cases, additions may bury a soil to a depth which prevents it undergoing any further development, while losses may lead to the complete removal of a profile.

Many chronosequence studies have examined profiles developed on surfaces of differing age and compared their extents of development. In order to quantify these studies, the age of the surfaces must be known. It is also important that the environmental conditions are as uniform as possible over the study area so that differences in the degree of development can be attributed to time differences rather than to those of environmental factors. Datable surfaces on which soil formation has occurred can be produced by a variety of natural and artificial means and commonly used examples include sequences of glacial moraine ridges, river terraces, slope deposits, aeolian dunes, volcanic deposits and mine spoil material. Many of these do not exceed a few decades, centuries or millennia in age, although some may be tens or hundreds of thousands of years old. Dating of the more recent surfaces can sometimes be achieved using documentary records or historical maps, while older surfaces usually require some form of incremental dating such as dendrochronology, lichenometry or radiometric methods. The dating of surfaces up to several millennia in age can be reasonably accurate, but that of older surfaces is usually less secure. An additional problem with very long chronosequences is that environmental factors, particularly climate and biota, have varied through time (Birkeland 1992).

The extent of soil development in chronosequences can be assessed on the basis of individual properties, or on groups of properties combined to produce a *profile development index* (PDI). Various indices have been devised; for example Harden (1982) used eight field properties which were quantified on a points system, allowing horizon and parent material values to be compared. Differences between the two were normalised, giving a scale from 0 to 1, with 1 representing maximum development.

Many chronosequences have been studied in a wide range of environments and although details differ, a number of general pedogenic trends can be recognised starting from time zero, when soil formation commences on a new surface. For example, organic matter content usually shows a progressive increase as vegetation colonises the surface and becomes increasingly dense, although after a time the rate of input of organic matter may

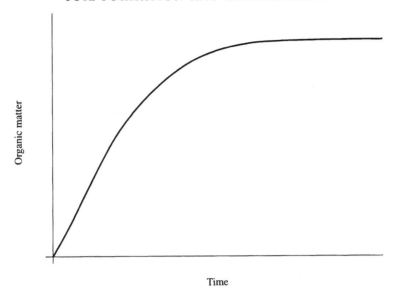

Figure 4.8 Accumulation of organic matter through time

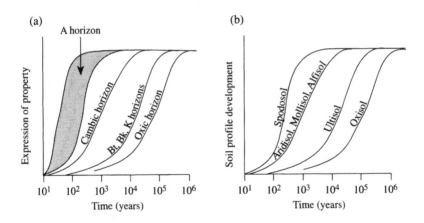

Figure 4.9 The time required to form (a) certain types of horizon and (b) various soil types (from Birkeland 1984a)

become balanced by losses during decomposition, and an equilibrium may therefore be attained such that no further increase occurs (Figure 4.8). Similarly, as weathering proceeds the solum usually increases in thickness, and the operation of transfer processes increases the content of illuvial

111

components, such as clay or sesquioxides, in B horizons. In contrast, certain constituents may decrease as development progresses, notably alkaline components which can be easily leached out of a profile in humid climates, producing a decrease in pH through time.

The time required for various soil properties and soil types to reach equilibrium has been discussed by Birkeland (1984a), again using data from previous chronosequence studies, from which it appears that great variations occur. For example, A horizons may reach equilibrium in as little as a few hundred years, whereas oxic horizons may require a million years (Figure 4.9a). Similarly, Spodosols may reach equilibrium within a thousand years, and visually recognisable podzolic characteristics can form in an even shorter time (e.g. Ellis 1980), while Oxisols may require more than a million years (Figure 4.9b).

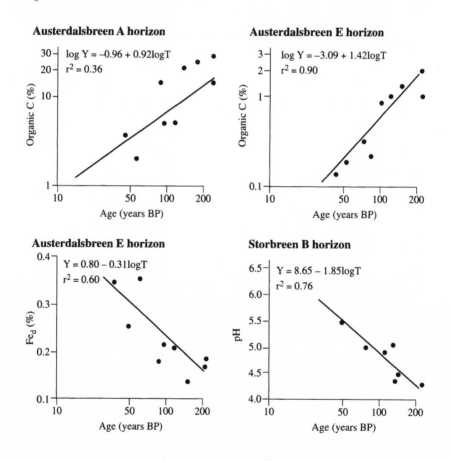

Figure 4.10 Examples of chronofunctions for soils developed on moraine ridge sequences in southern Norway (from Mellor 1985)

Chronofunctions can be described mathematically using a variety of models. Bockheim (1980) tested various models using chronosequence data from a range of different environments and concluded that the single-logarithmic model:

$$Y = a + b \log X$$

provided the highest degree of explanation when regressing a soil property (Y) against time (X) using linear regression techniques. This has been corroborated by studies in specific environments (e.g. Birkeland 1984b, Mellor 1985) (Figure 4.10), although the use of other models has also been advocated (e.g. Schaetzl et al. 1994).

Bockheim (1980) concluded that climate exerts an important control on the rate of soil development, with the rates of increase in solum thickness, oxidation depth, salt accumulation and B horizon clay content all being positively correlated with mean annual temperature, while total nitrogen showed a negative correlation. Rates of increase in B horizon clay content and solum thickness were also positively correlated with parent material clay content. In contrast, pH and base saturation were found to decrease with time independently of climate, biota and parent material. However, Birkeland (1992) synthesised the data from a wide range of chronosequences and found that in the case of clay and iron accumulation and a field-based PDI, the relationship with climate was unclear.

It is important to note that studies of soil development through time do not involve chronosequences in the strict sense, because all the other soil-forming factors do not remain constant; although parent materials and relief may be similar throughout a temporal sequence, vegetation can change markedly over time due to successional effects (Miles 1985). Over long time periods climate can also change, influencing vegetation as it does so, and human activity can bring about additional environmental changes; these aspects will be examined in the following chapter. It is also important to note that pedogenesis does not always progress through time; processes such as pedoturbation and erosion can cause soil development to be regressive (Johnson and Watson-Stegner 1987).

4.3 PEDOGENIC ENVIRONMENTS

It is not possible to rank the soil-forming factors in order of importance – they are all equally important, although within any one environment certain factors may have a more dominant influence on soil formation than others. At the local scale, this dominance will depend on the nature of the environment; for example, if parent material varies widely, this will probably account for many of the differences in soil formation within that area, whereas in an area with great variations in relief, this factor may dominate. At the regional and global scale, however, climate is the dominant factor in causing differences in soil formation, both by its direct influence on the operation of the soil-forming processes and by its indirect effect on biota. The remaining factors will then account for variations in soil formation within a particular climatic zone. In this section we will examine soil formation in different climatic environments in order to provide a global overview.

The definition of macro-scale climatic zones is not easy, because climate is a continuum with no rigid boundaries. However, for the purposes of this discussion, it is proposed to recognise four broad divisions – high, mid- and low latitudes, plus mountain regions whose climate varies altitudinally rather than latitudinally.

4.3.1 High latitudes

These areas can be loosely defined as those poleward of 60° latitude, and therefore comprise the northern parts of North America and Eurasia, along with Antarctica. Temperatures and precipitation both vary widely, with mean winter temperatures in the range 0 to –40°C and summer values in the range 0–15°C, although these remain below zero in much of Antarctica. Mean annual precipitation ranges from < 250 mm in the polar deserts but can reach 1,000–1,500 mm towards the more temperate margins. Vegetation may be absent or largely limited to thin lichen crusts in the polar deserts, but elsewhere is dominated by tundra shrubs and herbs. At high latitudes the soils may be permanently frozen below a certain depth, thawing only in their uppermost part during the summer months. Ground which is permanently frozen is known as *permafrost*, and can extend to depths ranging from a few metres to over 1,000 m depending on climatic severity (Harry 1988, Williams and Smith 1989). Precise correlation of permafrost distribution with ground temperatures is difficult, but as a general statement permafrost is fairly continuous where mean ground temperatures are several degrees below zero, and becomes discontinuous in areas with temperatures around 0°C (Williams and Smith 1989). Above the permafrost is the zone of seasonal thawing, known as the *active layer*; this increases in thickness as the permafrost becomes thinner and more sporadic.

Organic additions are generally low in comparison with warmer environments, due to low litter production rates (Table 4.2). In contrast, mineral additions can be high. For example, in areas of sparse vegetation and dry soil, wind erosion can occur on a large scale (Washburn 1979). Aeolian additions to soils can range from a few millimetres to several metres in thickness, and may therefore result in the complete burial of profiles. Additions can also occur due to movement of material from upslope, either by surface wash or mass movement (Lewkowicz 1988), and these can also cause profile burial if operating on a large scale.

Organic matter transformation is slow due to low soil temperatures,

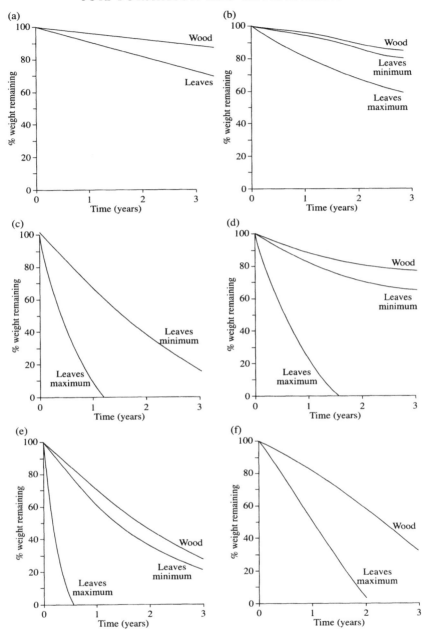

Figure 4.11 Decomposition of wood and leaves as related to different climates and vegetation types: (a) tundra, (b) taiga, (c) mid-latitude grassland, (d) mid-latitude broadleaf forest, (e) equatorial forest, (f) savanna grassland (based on Swift *et al.* 1979)

which inhibit the activity of mesofauna and micro-organisms (Figure 4.11a), and many soils are frozen for several months of the year (Williams and Smith 1989) which further restricts decomposition. However, where soils are well drained, their surface horizons are not excessively thick because although decomposition is slow, it can generally keep pace with the slow rate of organic input. At poorly drained sites, anaerobic conditions further limit decomposition, and poorly decomposed, peaty organic horizons result, referred to in arctic areas as *bog soils* (Tedrow 1977).

Although mineral additions via the surface can be high, material produced from parent material weathering is generally less than in warmer environments. This is because of the temperature-dependent nature of chemical weathering reactions (section 4.2.1) and also because water, which is fundamental to the operation of many weathering processes, is locked up as ice for much of the year or limited in high polar areas due to aridity (Ugolini 1986, Campbell and Claridge 1992). Even the effectiveness of freeze–thaw weathering (section 3.2.2) has been questioned, and other processes, such as hydration, may be at least as important (White 1976, Fahey 1983). In areas recently exposed by retreating glaciers or ice sheets, strain release can be important in weakening bedrock in preparation for the operation of other weathering processes.

Like weathering, the transfer of material will depend to a large extent on the availability of water. Where soils thaw for a sufficiently long period each year, translocation of bases can result in acid profiles. Translocation of sesquioxides can also occur because lichens and tundra heathland plants can cause organo-metallic complexing (Ugolini 1986). Because of restricted water availability, however, soils often do not become fully developed podzols, but occur as a genetically related but less well-developed form known as *arctic brown soil* (Tedrow 1977, Campbell and Claridge 1992). This lacks the characteristic bleached eluvial horizon of podzols, although it possesses an illuvial horizon of sesquioxide accumulation (Plate 13). In arid, polar desert areas there may be a net movement of water up towards the soil surface, which can lead to the upward transfer of salts in solution, where they are deposited on the underside of stones or as a crust at the surface.

Although the period of water availability in cold environments may be limited, large volumes of water can be released during this time if the melting of snow and ice is rapid. This can lead to the translocation in suspension of material of sand size or even of gravel, if the soil pores are sufficiently large, as is often the case for a number of reasons. First, when soil freezes the water within it forms small lenses of ice, known as *segregation ice*, which produce quite large pores on melting. Second, soils in high latitudes frequently occur on parent materials of glacial sediments recently exposed from the ice and therefore have not had time to become highly compacted, and third, glacial sediment deposited by meltwater comprises mainly coarse, highly porous material. The deposition of translocated material occurs on the upper surfaces of peds and individual larger particles, due to its lack of adhesion as compared to clay, and is best seen in thin section (Figure 3.6) (Ellis 1983). The peds are often lenticular in shape, resulting from the compaction associated with segregation ice lens growth (Figure 4.12).

Mechanical transfer by organic agencies – bioturbation and human disturbance – is generally low due to low populations on account of cold or frozen ground conditions, but transfer by frost action and mass movement can be high. As long as there is sufficient moisture, freezing and thawing will cause cryoturbation and heaving (Pollard 1988, Schunke and Zoltai 1988) (Figure 4.13), particularly in silt-rich soils, and also sorting (section 3.2.3). Transfers can range from localised reorientation of particles to the total intermixing of a profile. The former can take the form of vertical orientation of stones by freezing at stable sites, or the rotation and downslope orientation of particles at sites experiencing mass movement, as in the case of *gelifluction* (sometimes also referred to as *solifluction*), in which an upper, thawed layer of soil moves downslope over a frozen, subsoil layer (Washburn 1979, Nelson 1985). Such rotation can result in illuvial coatings, originally deposited on the upper surface of stones, appearing on their sides or undersides as the stones become reorientated during soil movement (Harris and Ellis 1980). Gelifluction can also cause the inversion of soil profiles if it

Figure 4.12 Photomicrograph of lenticular structure in an arctic brown soil (frame width = 4 mm)

Figure 4.13 Earth hummocks resulting from frost disturbance

Figure 4.14 Inversion of soil horizons at the front of a gelifluction lobe. The thin, inverted bleached eluvial horizon is seen overlying the inverted surface organic-rich horizon

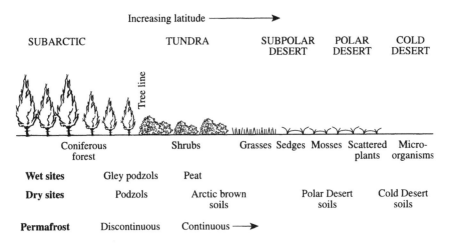

Figure 4.15 Latitudinal transect of soils in cold climate regions (from Campbell and Claridge 1992)

is associated with lobes; as the soil reaches the front of a lobe it gradually overturns in a movement analogous to that of a caterpillar track (Figure 4.14). Complete intermixing of profiles is associated with higher energy forms of mass movement such as mudflows or debris avalanches (section 6.4.3).

Gaseous losses from cold environment soils are limited due to the low levels of biological activity (Swift *et al.* 1979). However, other forms of soil loss can often be high due to the active aeolian sediment transport and surface wash processes already mentioned in the context of additions. Losses by wind erosion are greatest in areas of dry soil and low vegetation cover, while those by water transport are greatest on steep slopes, where water infiltration is impeded by subsurface frozen ground conditions and where large influxes of water occur due to rapid melting (Lewkowicz 1988).

According to the soil classification systems discussed in the previous chapter (section 3.3.3), high latitude soils fall largely into the Inceptisol, Entisol and Spodosol categories of the Soil Survey Staff (1975, 1992) system, along with Histosols in poorly drained areas, and into the Regosol, Leptosol, Cambisol and Podzol categories of the FAO-Unesco (1974, 1989) system, with Gleysols and Histosols in areas of poor drainage. It is possible to identify a latitudinal pedogenic gradient, for soils at freely drained sites, which takes the form of decreasing downward transfer and increasing upward transfer as moisture diminishes polewards. Hence there is a general progression from podzols through arctic brown soils to polar desert and cold desert soils (Figure 4.15).

4.3.2 Mid-latitudes

Mid-latitude areas can be considered as those lying between 60° and 30° north and south of the equator, and therefore include much of North America, Europe and Asia along with the northernmost part of Africa and the southernmost parts of South America, Africa, Australia and New Zealand. Mean summer temperatures lie within the approximate range 15–30°C and wide variations in winter temperatures occur from 15°C down to –30°C in the continental interiors. In contrast to high latitudes, however, temperatures do not seriously impede the operation of the soil-forming processes, except towards the poleward margins. Soil formation does vary markedly, however, in response to variations in precipitation; mean annual values can range from < 250 mm in continental interiors to 2,000 mm in more oceanic areas. Moisture restriction due to evapotranspiration can also be important in the warmer areas. The natural vegetation of the more humid environments is characterised by needleleaf forest (or taiga) in the cooler areas and broadleaf forest in the warmer areas. In the drier regions forest gives way to grassland or 'Mediterranean' shrubland, while in the driest areas desert forms predominate. However, in many mid-latitude areas the natural vegetation has been modified by human activity, which has had a variety of effects on soils (section 5.2.2).

Organic additions will depend on the type of vegetation cover – for example, inputs of litter from grassland and needleleaf forest are generally lower than those from broadleaf forest (Table 4.2). Additions will also decrease where vegetation covers become incomplete due to restricted moisture availability, as in southern Eurasia, north Africa and southwest USA. Additions of mineral material are largely moisture-controlled. For example, aeolian additions only assume significance in semi-arid areas and additions by surface sediment transport will be most marked in areas having a flashy precipitation regime, again mainly in semi-arid regions (Selby 1993).

Transformation of organic material operates to a greater extent than in cold environments due to higher temperatures and increased levels of biological activity. Variations do occur within mid-latitude regions, however, due to environmental differences; for example, rates of organic matter decomposition are lower under needleleaf forest (*taiga*) than under grassland and broadleaf forest (Figure 4.11b–d). Mid-latitude surface organic horizons are not thinner than those of high latitudes, however, because the greater levels of decomposition are matched by greater additions. Indeed, at well drained sites these horizons are often thicker than at those of cold environments, due also to more extensive bioturbation. At poorly drained sites, however, decomposition will be inhibited regardless of temperature, and poorly decomposed, more peaty organic horizons will develop.

On the basis of a doubling in the rate of weathering for every 10°C increase in temperature (section 4.2.1), rates of transformation by mineral weathering are two to three times greater than at high latitudes. Freeze–thaw and strain release generally play a more minor role than in colder environments, and weathering is often dominated by hydration, oxidation and hydrolysis (section 3.2.2). Solution is also important in base-rich soils, and organic complexing is important where vegetation is acid-tolerant, as in areas of heathland and needleleaf forest. In warmer areas with a dry season, fersiallitisation occurs (Duchaufour 1982), resulting in rubification (section 3.3.2), which has a poleward limit between 30° and 40° latitude (Tardy 1992). Rubification of soils can occasionally occur above 40° latitude, but this is usually considered to be related to past environmental conditions (section 5.2.1). In the case of terra rossas developed over limestone, the origin of the reddening has been debated; it may be the result of *in situ* weathering or may be inherited from volcanic or aeolian deposits (Macleod 1980, Olson *et al.* 1980, Pye 1992).

The extent of transfer processes depends largely on moisture availability. With limited precipitation, as in central Asia, north Africa and the southwest USA, Aridisols will form. Where salts are concentrated towards the surface, solonchaks result. With slightly higher precipitation, downward translocation will occur but will be confined largely to bases, as in the continental interiors of North America and Eurasia. However, overall base loss may be limited by upward capillary action, which helps to retain bases near the surface. Grassland dominates such areas, and the resulting soils are of the chernozem (Mollisol) type (section 3.3.2). These soils are also characterised by extensive bioturbation due to the warm summer temperatures, neutral to alkaline soil conditions and relatively easily decomposed organic matter. In areas of moderate precipitation, characterised by broadleaf woodland, such as eastern USA and much of Europe, leaching of bases will be greater and the soils are therefore generally less alkaline, giving brown earths (Alfisols) (section 3.3.2). In colder, high precipitation areas such as northeast North America and northern Europe, leaching of bases is more extensive and acidic soils are therefore common. Associated acid-tolerant vegetation produces effective chelating agents, causing organo-metallic complexing, so many of the soils show podzolic characteristics (section 3.3.2). As in the case of the arctic brown soil of high latitudes, less well developed forms of podzolic soils also form in mid-latitudes where environmental conditions restrict full development. These occur, for example, in the British Isles, where they are known as *podzolic brown soils* (Avery 1990). Although climate and vegetation are normally the restrictive factors, in some cases iron-rich parent materials can also limit podzolisation (Ellis 1988).

Mechanical transfers by bioturbation are greater at mid- than at high latitudes. Soil temperatures are higher, freezing is less severe and there is a greater availability of organic matter as a food source. Bioturbation is greatest in neutral to slightly alkaline soils, where mixing of organic and mineral material can occur to depths in excess of 1 metre. Bioturbation decreases in intensity and depth in the more acid soil types. The effectiveness of other forms of mechanical transfer depends on temperature and moisture conditions. For example, mixing by wetting and drying will be greatest in soils which receive frequent but periodic precipitation, while mixing by frost-related processes will be greatest in soils towards the poleward margins. Mixing by human activity, such as cultivation or construction, is generally greater than at high latitudes, but will obviously vary with soil type and population density.

Differences in losses between soils relate mainly to differences in organisms and temperature. Gaseous losses are generally higher from mid-latitude soils than from those of high latitudes due to greater biological activity (Swift et al. 1979). Solute losses are also higher due to longer reaction times resulting from longer durations of unfrozen ground. Human activity often causes greater losses, particularly where soils are under cultivation and susceptible to surface erosion (section 8.2.1). Losses from piping can also be higher due to more limited frozen ground conditions, while losses via aeolian activity and surface wash are generally lower because of higher surface vegetation covers and fewer rapid water inputs. In drier areas, however, such losses will increase (Young and Saunders 1986).

The principal soil types in mid-latitudes clearly relate to bioclimatic gradients. Using the Soil Survey Staff (1975, 1992) terminology (section 3.3.3), Spodosols dominate in the cooler, wetter, needleleaf forest areas, with Alfisols in the warmer, broadleaf forest regions, Mollisols in the drier grassland areas and Aridisols in the driest zones. The terminology for these groups according to the FAO-Unesco (1989) classification scheme is shown in Table 3.5.

4.3.3 Low latitudes

Low latitudes can be taken as occupying areas below about 30° latitude, and therefore comprise Central America, much of South America and Australia, and most of Africa and southern Asia. Temperatures average around 25–40°C in summer and large variations occur in winter, from 25–30°C near the equator to 10°C in the Asian interior. Precipitation varies enormously from < 250 mm per year in north Africa, southwest Asia and central Australia to over 2,000 mm in equatorial regions, and in extreme cases over 5,000 mm. Precipitation is the major control on vegetation, which comprises rainforest in the wettest areas, seasonal forest and scrub in areas with seasonal rainfall contrasts, grassland in the drier areas and desert forms in the driest regions.

In humid areas vegetation growth rates are extremely rapid and large quantities of litter are added to the soil (Table 4.2). Where surface wash occurs, mineral additions can be high relative to humid mid-latitude areas, but will depend on factors such as the density of vegetation cover and rainfall intensity (Selby 1993). In drier low latitude areas, such as the savanna grasslands, organic additions are lower (Table 4.2). In the driest areas organic inputs are minimal, but sparse vegetation covers, high wind speeds and heavy but infrequent precipitation can cause high inputs of mineral material, particularly in the silt and clay size fractions (Watson 1992).

In humid regions rates of organic matter decomposition are very rapid, particularly in equatorial forests, due to temperatures which allow optimum levels of biotic activity (Figure 4.11e). Despite high organic inputs, surface organic horizons are often not very thick because decomposition keeps pace with them. In drier areas, such as savanna grasslands, decomposition is slower because of lower biotic activity (Figure 4.11f), but the lower levels of litter addition, due to less dense vegetation covers, prevent thick surface horizons from forming. Mineral weathering in humid regions is extremely rapid due to the rate at which chemical reactions occur at high temperatures; they may be two to three times quicker than in mid-latitudes. As a result, profiles can attain thicknesses of several tens of metres, as compared to a maximum of only a few metres in more temperate soils. Another reason for this is the much greater period of time which is often available for weathering at low latitudes due to greater stability of the weathered mantle, resulting from the absence of glacial and/or periglacial erosion during colder periods in the past (section 5.2.1). Extensive weathering results in ferruginous and ferrallitic soils (section 3.3.2), often accompanied by high levels of iron dehydration and crystallisation, producing rubified, hematite-rich soils. In more arid areas, the relative extent and importance of the various weathering processes remain unclear, although it is likely that salt plays an important role in the breakdown of rock material (Goudie 1989, Watson 1992).

In humid areas, transfer of soil constituents can be extensive, in association with ferrallitisation. Because of the dominance of kaolinite clays formed by weathering, clay translocation is restricted (Duchaufour 1982), but sesquioxides and silica are mobile, along with bases. At a basic level, two main types of transfer-related process can be recognised – latosolisation and lateritisation (section 3.3.2). The former results in sesquioxide-rich residual oxic horizons from which silica and bases have been leached; in climates with a long dry season these can produce iron crusts or ferricretes where iron accumulation is intense (Nahon 1986). The latter produces illuvial horizons containing kaolinite and translocated silica, often along with plinthite (section 3.3.1). Many variations have, however, been recognised on these themes, according to the type and distribution of weathering products within the profile (Mohr et al. 1972, Duchaufour 1982, Tardy 1992). For example, Tardy (1992) recognised eight types of lateritic soil, ranging from profiles rich in quartz and low in kaolinite, which resemble podzols, through Oxisols in which kaolinite, hematite and gibbsite have formed and which may also have developed hematite nodules, to highly weathered and leached Ultisols. Profiles may show a dominant concentration of iron or aluminium, which Tardy (1992) has associated with climate; iron accumulation occurs in areas

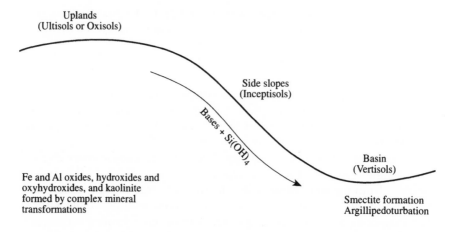

Uplands
(Ultisols or Oxisols)

Side slopes
(Inceptisols)

Bases + Si(OH)₄

Basin
(Vertisols)

Fe and Al oxides, hydroxides and
oxyhydroxides, and kaolinite
formed by complex mineral
transformations

Smectite formation
Argillipedoturbation

Figure 4.16 Idealised landscape cross-section for wet–dry tropical regions
(from Fanning and Fanning 1989)

with rainfall below about 1,700 mm per year, temperatures of around 28°C and average relative humidity around 60 percent, while aluminium concentration, which can produce bauxite where leaching is strong (Schellmann 1994), occurs preferentially in areas of higher rainfall, lower temperature and higher relative humidity.

In areas which receive bases and silica transported in solution from upslope, smectite clays can form (Figure 4.16). These are highly expansive and produce Vertisols, which are characterised by the vertical mixing of material due to cyclical shrinking and swelling (section 3.2.3). Vertisols occur most extensively in the eastern half of Australia, where they often produce a hummocky micro-relief known as *gilgai* (Hubble *et al*.1983), and to a more limited extent in west-central India. Silcrete is also found in low latitude soils, for example in Australia and southern Africa, indicating either the residual accumulation of silica or its concentration in solution by ground-water movement. Its formation is, however, generally considered to require periods of time longer than those of the lifespan of individual soils and therefore to be associated with past environmental conditions (Milnes and Thiry 1992) (section 5.2.1).

In drier regions, translocation is restricted to the most easily dissolved alkaline components. These can accumulate in illuvial horizons in large quantities because, in contrast to more humid environments, they are not lost from the profile by progressive leaching outputs. Calcrete can form in this way (section 3.3.1). In areas where there is insufficient downward leaching, salts may be transferred up towards the surface by capillary rise, thus forming solonchaks (section 3.3.2). Salt can also be concentrated by evaporation of shallow lakes, ephemeral streams or artesian water, by *in situ* weathering of salt-rich minerals or by atmospheric additions via rainwater or sea spray (Isbell *et al.* 1983, Watson 1992). Crusting can occur where salts become concentrated, the main types of crust being of either a halite (sodium) or gypsum (calcium) type, and in the latter case indurated petrogypsic horizons (section 3.3.1) may form; where both salts are present, two-tiered crusts can develop, with the more mobile halite material forming a crust beneath the less mobile gypsum crust (Watson 1992). Of the mechanical transfer processes, bioturbation can be particularly effective in the more humid areas due to high populations of burrowing organisms, especially termites (section 3.2.3), whereas in seasonally dry regions, transfers by wetting and drying can be important, especially in Vertisols.

Losses of gases are high in humid areas due to high organic matter decomposition rates (Swift *et al.* 1979). In contrast, outputs via aeolian activity are low, although losses via surface wash and solution can be high (Young and Saunders 1986, Selby 1993). In drier areas gaseous losses are low due to the limited quantity of organic matter, while losses via aeolian transport and surface wash can be high.

The distribution of soil types at low latitudes relates closely to moisture gradients. According to the Soil Survey Staff (1975, 1992) terminology (section 3.3.3), Oxisols and Ultisols dominate in the more humid regions, Alfisols and Vertisols in seasonally dry areas and Aridisols and Entisols in the driest areas. The terminology for these groups according to the FAO-Unesco (1989) classification scheme is shown in Table 3.5.

4.3.4 Mountain regions

Because climate and vegetation vary with altitude, mountain regions can possess a wide range of pedogenic conditions over short horizontal distances and for this reason are best considered separately from latitudinally defined regions. The effect of increasing altitude is to decrease temperature and often to increase precipitation, therefore an altitudinal sequence of soils can often be recognised which, in extreme cases, can range from low latitude conditions at low altitude to high latitude conditions at high altitude, such as occurs in the Andes, east Africa and the Himalaya.

Additions of organic material are generally low at high altitude due to restricted temperatures and often incomplete ground covers of vegetation, although in drier areas the reverse may apply due to increased rainfall at higher altitude and therefore denser vegetation covers (Amundson *et al.* 1989). Mineral inputs can be high, particularly at footslope locations, due to steep slopes with highly active downslope movement. Atmospheric inputs may also be important in the form of volcanic ash and aeolian material. For example, additions of volcanic ash to soils in the Cascade Range of North America are often rich in iron and aluminium which can increase the contents of these elements significantly, while aeolian sources include material from dry flood plains and from desert regions; calcite in dust added to alpine soils in Colorado, USA, maintains the pH of surface horizons close to neutrality (Litaor 1987), while the Alps of France and Switzerland are regularly supplied with Saharan dust which can similarly raise the pH of soil surface horizons (Legros 1992).

As in the case of high latitudes, organic matter decomposition in high altitude soils is often restricted due to reduced temperatures, as reported, for example, in Nepal (Righi and Lorphelin 1987) and Costa Rica (Grieve *et al.* 1990). This can lead to increased organic matter accumulation and increased C:N ratios, often also associated with the changes in vegetation type which accompany those in altitude (Alexander *et al.* 1993). Weathering may also be limited for the same reason, as in Papua New Guinea (Chartres and Pain 1984) and eastern Nepal (Baumler and Zech 1994). At high altitudes, meltwater from snow or ice can also cause effective solution weathering due to the relatively large amounts of CO_2 which can be dissolved in water at low temperature. It has also been suggested that the high silt contents which are often found in alpine soils are due to freeze–thaw weathering, while low clay contents indicate restricted chemical weathering, although high silt contents may also relate to aeolian inputs as noted above (Legros 1992).

The effectiveness of transfer processes will often vary with altitude. For example, translocation usually increases with altitude, in response to increased precipitation and decreased evapotranspiration, but may decrease at the highest altitudes if inhibited by frozen ground conditions. For example, in the French Alps a sequence of soils was reported by Legros (1992) in which Dystric Cambisols (acid brown earths) occurred on the valley footslopes at around 1,000 m altitude, while more highly leached podzols occurred further upslope, but these gave way to shallow, weakly developed rankers at above 2,300 m (Figure 4.17). Where conditions are too severe for podzols to develop at high altitude, *alpine brown soils* can result; these are similar to the arctic brown soils of high latitudes (section 4.3.1), although they have higher organic contents in their A horizons (Bockheim 1978). Transfers by organisms usually decrease with increasing altitude, but mixing in general may increase due to greater freeze–thaw activity and slope instability. Soil losses in mountain regions can be particularly high due to steep slope angles and often incomplete vegetation covers (Young and Saunders 1986), giving shallow, poorly developed profiles. Mountain soils can be of many different types, but generally show a decrease in development with increasing altitude, due to climatic, biotic and topographic restrictions.

4.4 SUMMARY

This chapter has examined soil formation from two viewpoints – the individual factors that influence the operation of the soil-forming processes,

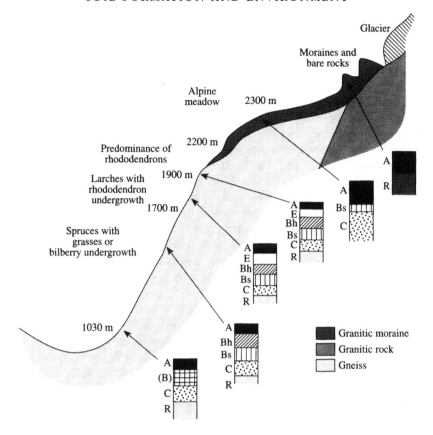

Figure 4.17 Altitudinal distribution of soils in the northern French Alps
(from Legros and Cabidoche 1977)

and the way in which these factors combine to produce soils characteristic of
major regions of the world. The individual factors are identified as climate,
biota, parent material, relief and time. Climate is of particular importance in
terms of determining the temperature, and therefore rate, at which pedogenic
processes operate, and also the availability of water, which is fundamental to
many of these processes. Biota are influenced to a large extent by climate, but
they also have a direct effect on soil formation via rates of organic matter
addition and decomposition, and the type of material involved. They can also
influence weathering and the transfer of soil components in solution and by
mechanical mixing, and human activity can exert a major control on soil
additions and losses, along with mixing via cultivation. The principal effect
of parent material on soil formation is seen in terms of mineral transformation, which is determined by the resistance and reaction of the material to

weathering processes. It can, however, also influence transfers and losses in terms of the materials available for participation in these processes, and the ease of drainage, and therefore of removal, of these materials. Relief can have an important influence on the transfer of material on slopes, and also on drainage, which can affect transformation, transfers and losses; high slope angles will encourage downslope movement and rapid drainage, whereas low angles will limit these processes and can therefore give rise to processes such as peat accumulation and gleying.

Unlike the previous four factors, time does not exert a direct control on soil formation, but is the medium through which the other factors operate. Soils generally become more developed through time with the progressive operation of pedogenic processes, and many chronosequence studies have demonstrated this principle, which in some instances has also been expressed quantitatively in terms of chronofunctions. After a certain period of time, however, soils may reach an equilibrium condition, when they can be said to be fully developed. The time required to reach this condition differs between soil horizons and types, ranging from as little as a few hundred years to as much as a million years.

Although the soil-forming factors operate in combination, it is useful to clarify their role in pedogenesis by considering them separately. However, their combined effect can be seen by considering soil formation in different latitudinal zones. At this scale, climate and biota will have a dominant influence on soil formation, and soil types can therefore be related in general terms to bioclimatic zones (Figure 4.18, Table 4.3), although other factors – parent material, relief and time – can have an important influence at the local scale.

High latitudes are dominated by Inceptisols, Entisols and Spodosols. They are characterised by low additions of organic matter, although those of mineral material can be high. Organic and mineral transformation also tend to be low, due to low temperatures and limited water availability. This also limits the transfer of material in solution, although transfer in suspension can be high due to rapid inputs of water during the melt season. Transfer by organisms is limited due to low temperatures, which restrict biotic activity, but transfers by frost action and slope processes can be very active. Losses can also be high due to active sediment transport processes, encouraged by frozen ground conditions and low vegetation covers.

Additions to mid-latitude soils are determined to a large extent by moisture; organic additions are greatest in humid areas where vegetation cover is complete, while mineral additions are often highest in semi-arid areas, where aeolian activity and flashy runoff regimes predominate. Organic and mineral transformations are generally more rapid than at high latitudes, due to higher temperatures. The transfer of material in solution and suspension is dependent to a large extent on moisture availability. Bioturbation is generally greater than in high latitudes due to higher levels of biotic

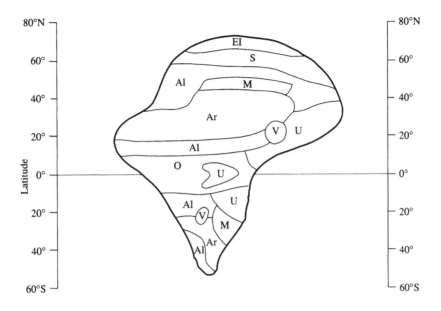

Figure 4.18 Schematic diagram of the distribution of Soil Survey Staff (1975) soil orders on an imaginary supercontinent (based on Strahler and Strahler 1989). Al = Alfisols, Ar = Aridisols, EI = Entisols and Inceptisols, M = Mollisols, O = Oxisols, S = Spodosols, U = Ultisols, V = Vertisols

activity, while the extent of other forms of mechanical transfer depends on the extent of temperature and moisture fluctuations, being favoured by wet–dry and freeze–thaw cycles. Subsurface losses are usually greater than at high latitude, although losses from the surface are generally lower because of denser vegetation covers and limited frozen ground conditions. Soils of mid-latitudes are characterised by a climatically related spectrum of types, ranging from Spodosols in the coolest, wettest areas, through Alfisols and Mollisols, to Aridisols in the warmest, driest regions.

At low latitudes, organic additions are usually high in humid areas, although much lower in arid areas due to lower vegetation covers. Conversely, mineral additions are lower in humid areas than in arid areas, where limited vegetation and flashy runoff regimes encourage sediment transport. Organic and mineral transformations also tend to be rapid due to high temperatures; this produces surface organic horizons of limited thickness, but often deeply weathered zones. In humid regions transfers by water can be extensive, resulting in laterites and indurated materials such as ferricrete, silcrete and plinthite, while in drier areas, the transfer of alkaline components can produce calcrete or saline soils. Transfers by bioturbation are also often high, as is mixing by cyclical shrinking and swelling in Vertisols, in which

Table 4.3 Relationship of Soil Survey Staff (1975) soil orders to bioclimatic zones

Bioclimatic zone	Soil orders	Annual precipitation range (mm)	Temperature patterns
Equatorial and tropical rainforest	Oxisols, Ultisols	1,800–4,000	Always warm (21–30 °C, mean 25°C)
Tropical seasonal forest and scrub	Oxisols, Ultisols, Vertisols, some Alfisols	1,300–2,000	Variable, always warm (> 18°C)
Tropical savanna	Alfisols, Ultisols Oxisols	900–1,500	No cold weather limitations
Mid-latitude broadleaf and mixed forest	Ultisols, some Alfisols	750–1,500	Temperate, with cold season
Needleleaf and montane forest	Spodosols, Histosols, Inceptisols, Alfisols	350–1,000	Short summer, cold winter
Temperate rainforest	Spodosols, Inceptisols	1,500–5,000	Mild summer and mild winter for latitude
Mediterranean shrubland	Alfisols, Mollisols	250–650	Hot, dry summers, cool winters
Mid-latitude grasslands	Mollisols, Aridisols	250–750	Temperate continental regimes
Warm desert and semi-desert	Aridisols, Entisols	< 20	Mean around 18°C, highest temperatures on Earth
Cold desert and semi-desert	Aridisols, Entisols	20–250	Mean around 18°C
Arctic and alpine tundra	Inceptisols, Histosols, Entisols	150–800	Warmest month < 10°C

Source: Christopherson 1992

expansive clay minerals are formed. Losses of solutes can be high due to extensive leaching, particularly in humid areas, while losses of mineral particles depend to a large extent on the protection afforded by vegetation cover. Low latitude soils are dominated by Oxisols and Ultisols in humid regions, Alfisols and Vertisols in seasonally dry areas, and Aridisols and Entisols in the driest areas.

Mountain soils can be considered separately from those of broad latitudinal zones because of the wide range of altitudinally related climatic and biotic conditions that can occur within a small area. At high altitude, organic additions are often limited, but mineral additions can be high due to active slope processes along with aeolian inputs. Organic and mineral transformations are generally restricted because of low temperatures. These can also limit transfers by bioturbation due to low organism populations, and by water due to frozen ground conditions, although mixing by freeze–thaw and unstable slope movement can be extensive. Steep and unstable slopes also produce large losses of material, often resulting in shallow, poorly developed profiles. Due to their many environmental restrictions, mountain soils generally show a decrease in development with increasing altitude.

5

SOILS AND THE PAST

5.1 INTRODUCTION

The previous chapter has been concerned with soil formation under present environmental conditions, but in all parts of the world, environmental conditions have changed through time, and soils very often reflect these changes. The magnitude of environmental change can vary greatly. For example, a broadleaf forest may experience a change in the relative proportions of its component species, which would represent a minor environmental change in terms of pedogenic processes and the resulting soil characteristics. In contrast, an area whose climate changed from warm and dry to cold and humid would be likely to experience a major pedogenic transformation. There can also be great differences in the length of time over which environmental change occurs; some changes may be rapid, for example those resulting from vegetation clearance or artificial drainage may be complete in less than a year, whereas others, particularly major climatic change, may occur over thousands or even millions of years.

Not only do soils respond to environmental change, but they can also provide information about past environmental conditions, if they are able to preserve their associated characteristics through to the present time. The study of soil formation in relation to the past is known as *paleopedology*, and is closely related to the discipline of geology, in which a rock type and the fossils preserved within it are used to provide evidence of past environmental conditions and to determine the time at which these conditions prevailed. With increasing time, the preservation of soil characteristics relating to the past decreases, due to the greater likelihood of disturbance by erosion or diagenesis (alteration following burial). As a result, most paleopedological studies have concentrated on the most recent geological period – the Quaternary – although there is evidence of soils dating back to the Paleozoic era, some 245–570 million years ago (Retallack 1992) (Table 5.1). The Quaternary, although very short in comparison with preceding periods, was a time of major global climatic fluctuation, perhaps demonstrated most forcibly by the presence of glaciers and ice sheets on a number of occasions in areas which today have temperate climates, and also by the presence of

132

Table 5.1 The geological time-scale

Eon	Era	Period	Epoch	Date (Ma BP)
Phanerozoic	Cenozoic	Quaternary	Holocene	0.01–0
			Pleistocene	1.6–0.01
		Tertiary	Pliocene	5.3–1.6
			Miocene	23.7–5.3
			Oligocene	36.6–23.7
			Eocene	57.8–36.6
			Paleocene	66–57.8
	Mesozoic	Cretaceous		144–66
		Jurassic		208–144
		Triassic		245–208
	Paleozoic	Permian		286–245
		Carboniferous		360–286
		Devonian		408–360
		Silurian		438–408
		Ordovician		505–438
		Cambrian		570–505
Precambrian				4,600–570

Source: Monroe and Wicander 1992

large lakes in low latitude regions which are now arid (Dawson 1992, Williams *et al.* 1993). These major climatic changes, accompanied by changes in biota, hydrology and geomorphology, have therefore exerted a marked influence on many soils. The Quaternary period also includes the time of human occupation of the planet, which has had a major effect on many soils since the end of the last major cold period of the Quaternary, about 10,000 years ago; this is known as the *Recent* or *Holocene* epoch (Table 5.1).

A change in environmental conditions can have a variety of effects on soils. In some instances it may superimpose a new set of pedological characteristics on the previous set. For example, a soil showing clay illuviation, developed under moist temperate conditions, may experience increased leaching with a change to cooler, wetter conditions. This may eventually change the soil to one of a podzolic type, but the zone of illuvial clay may remain partially intact. The pedological features preserved from former environments are known as *pedorelicts* (Brewer 1976), and in the case of the example given, these would probably take the form of illuvial clay deposits or argillans (section 3.2.3). Alternatively, environmental change may result in partial erosion of a profile, usually its upper portion, leaving the remainder of the profile intact. In this case, pedorelicts could occur on a larger scale, as fragments of the eroded soil incorporated in the parent material of a younger soil. Erosion could occur, for example, if a change from a temperate to a cold climate caused glacier ice to develop and cause erosion of an area, leaving only the lower portions of soil profiles remaining when

133

the ice melted as a result of subsequent climatic warming. Conversely, soils can become buried by a variety of methods, for example fluvial, glacial, aeolian, colluvial or marine deposition, volcanic or extrusive igneous activity, or human activity (Figure 3.14), and some buried profiles may re-emerge at the surface if the overlying deposits are removed by subsequent erosion.

A soil which preserves pedogenic features formed under environmental conditions different from those of the present time is called a *paleosol*. Traditionally this term has been applied to profiles which are complete or remain only partially intact, and which are preserved either at the surface or buried beneath various types of sediment. Various alternative or qualifying terms have been used to refer to these particular conditions. For example, a profile at the surface in which environmental change has caused a new set of pedogenic characteristics to be superimposed on the previous set, can be referred to as a *relict* soil, a soil which has become buried is simply termed a *buried* soil, while such a soil which has subsequently become exposed by erosion is termed an *exhumed* soil (Wright 1992) (Figure 5.1). However, Fenwick (1985) has argued that the term *paleosol* can be ambiguous, and has suggested that it should be confined to soils which are buried to a sufficient extent as to be isolated from present pedogenic processes and which are therefore fossilised, with soils occurring at the surface and displaying characteristics of past environmental conditions being described simply as containing *paleosolic elements*.

The first part of this chapter will consider the history of soils in relation to past environmental conditions, with reference to paleosols, or soils with

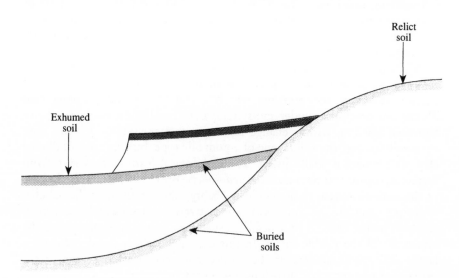

Relict
soil

Exhumed
soil

Buried
soils

Figure 5.1 Types of paleosol

134

paleosolic elements. This will be examined first with respect to natural environmental change, particularly that of climate, and then in terms of human activity. Once the relationship between soil characteristics and past environmental conditions is understood, this information can be used in environmental reconstruction, which will form the basis of the second part of the chapter. Finally, the use of soils in dating past pedogenic and other environmental events will be discussed.

5.2 SOIL HISTORY

5.2.1 Natural environmental change

Climatic change has occurred throughout geological time, and a variety of causes have been proposed, including variations in solar activity, changes in the Earth's orbit, continental drift and changes in atmospheric and oceanic circulation patterns (Goudie 1992). Former climatic conditions can leave their imprint on soils, not only directly, in terms of temperature, humidity and other climatic characteristics, but also indirectly because of other closely associated factors such as biota, hydrology, geomorphic activity and sea level. We shall examine the effects of these past environmental influences on soils in terms of changes in temperature and moisture, and their effect on the four groups of soil-forming processes recognised previously – additions, trans-formations, transfers and losses.

Organic additions may be initiated, or may increase, due to an increase in vegetation resulting from a change to a more humid or warmer climate. For example, an arid soil with little or no surface organic matter accumulation, could develop an organic-rich surface horizon following an increase in vegetation density due to an increase in humidity. Aeolian material often forms the main mineral input at a regional scale, with additions by surface wash and mass movement usually occurring at a more local level. The size of aeolian material varies according to wind speed and the size of particles available for transport, but material of predominantly silt size is common in many Quaternary soils; this material is known as *loess*. In cases where large quantities of loess are deposited rapidly, soil profiles become buried by sheets of sediment. Loess may be forming at the present time in arid environments, which are susceptible to aeolian transport processes (section 3.2.4), but where it occurs in more humid regions it is considered to relate to drier climatic conditions in the past. Such deposits are widespread in China, central Europe and central North America (Catt 1986a). Where loessic additions have been more minor, they have been incorporated into the soils by transfer processes, rather than burying them. This can be seen, for example, in many English soils developed over base-rich rocks. The soils have a characteristic silty texture, and show a westward decrease in modal particle size; this relates to the winnowing effect of easterly winds which transported the material during the dry conditions towards the end of the last major cold period of the Quaternary, around 15,000 years ago (Catt 1985a). Coarser-grained material, known as *coversand*, can be transported by stronger winds; for example, in the Netherlands and Belgium at least six separate episodes of coversand deposition have been recognised during this cold, dry period (Catt 1986a).

At lower latitudes, more arid phases during the Quaternary, perhaps accompanied by higher wind velocities, associated with changes in position of the major climatic belts, caused aeolian transport in areas where sediment movement is now restricted (Williams *et al.* 1993). This has resulted in burial of, and additions to, soils, as in the case of aeolian salt added to certain saline soils in Australia (Isbell *et al.* 1983). On a shorter time-scale, aeolian erosion and deposition has been reported in northwest Europe during the second half of the Holocene period, due to changes to drier conditions (Bell and Walker 1992).

In areas which have experienced former glacial conditions, some soil

Figure 5.2 A podzolic soil buried by glacial till

profiles may have become buried by till (Figure 5.2), although many have often become obliterated by glacial erosion. Climatic warming following glaciation causes sea-level rise due to melting of ice sheets and glaciers, which can also cause burial of coastal soils, as in the case of peaty organic soils buried by post-glacial marine clays in estuarine areas of England (Catt 1986a). In contrast, additions or burial by surface wash and mass movement can occur without any major climatic change. For example, soils on an alluvial flood plain can show layers of sediment deposited during times of flood, while soils on unstable slopes can become buried by sudden, random movement of material downslope (Gerrard 1981).

Organic transformations may have been different under past climatic conditions and this can be seen in the case of soils which show large accumulations of poorly decomposed material formed under a change to cooler or wetter conditions. For example, peat expansion in upland regions of northwest Europe occurred during the mid- to late Holocene in response to a deteriorating climate, leading to waterlogging of the soils due to reduced evapotranspiration (Macphail 1986, Bell and Walker 1992); this was probably also enhanced by human activity (section 5.2.2). A similar situation has also been reported in parts of Canada, by the growth of blanket peat during the last 6,000 years in response to a change to wetter conditions (Zoltai and Vitt 1990).

Transformations in terms of mineral weathering characteristics can also reflect past climatic conditions. Soils formed under warm climates in the past will often show enhanced weathering characteristics such as deep profiles, exotic weathering products and rubification (Catt 1989). Examples of deep weathering have been reported in cool-temperate soils in northeast Scotland, where it has been attributed to earlier warmer conditions (Hall *et al.* 1989). The weathering is associated with products such as kaolinite and hematite, which are not considered to form under the present climatic conditions. Indeed, hematite occurs in many mid-latitude buried and surface soils, in which it is attributed to weathering under warmer conditions during Quaternary interglacial periods or in pre-glacial times, and if present in sufficient quantity this causes the soil to be rubified (Catt 1986a) (Plate 14). However, rubification may not always be restricted to warmer conditions; for example, it can be inherited from hematite-rich parent materials, and has been reported to have formed under cool conditions in well drained, calcareous, coarse-textured soils in France and Germany (Kemp 1985).

Soils can preserve evidence of transfer both by water and mechanical processes. For example, soils in temperate regions which have experienced cold conditions during the past can show periglacial features such as vertically orientated stones, involutions and fragipans; smaller scale (micromorphological) features may also be preserved such as vesicles, structures formed by segregation ice lens growth, and the capping of particles by translocated silt (Van Vliet-Lanoë 1985, Catt 1989). A change to colder conditions may also be marked by the fracturing of cutans (Chartres 1980).

In lower latitudes, relict transfer features usually relate to different moisture conditions during the past, rather than to different temperature conditions. For example, in a study of Aridisols in California, argillic horizons were considered to result from wetter periods and carbonate accumulation from drier periods since the mid-Pleistocene (Eghbal and Southard 1993), while smectite formation in western India has been related to more humid conditions in the past (Bhattacharyya *et al.* 1993). On Lanzarote, Canary Islands, two phases of carbonate accumulation have been interpreted as reflecting more humid conditions, separated by a more arid phase at the height of the last major cold period of the Quaternary (Magaritz and Jahn 1992). In Australia, fossil laterites are widespread in arid and semi-arid areas, where they have often become subsequently eroded and dissected (Hubble *et al.* 1983). Here hardened crusts, known as *duricrusts*, formed of materials such as calcrete or silcrete, developed under humid, warm conditions of the Oligocene period (Table 5.1), and laterisation in desert regions ceased by the mid-Miocene period, when the climate became more arid (Van de Graaff 1983). Lateritic weathering relating to previously moister conditions has also been reported in southeast China (Singer 1993). However, because of their great age, and therefore the fact that they often show destruction as well as formation mechanisms, the relationship of lateritic soils to climatic change is difficult to determine in detail (Tardy and Roquin 1992).

The loss of soil material can occur due to changes in temperature or humidity. For example, a change from humid to drier conditions may lead to a decreased vegetation cover, thus exposing the soil surface to wind and water erosion. This can leave a residual accumulation of larger material at the surface, once the fines have been removed, as in the case of *lag gravels* in arid and semi-arid regions of Australia, which result from the erosion of laterite and silcrete (Hubble *et al.* 1983). In some cases, the eroded soil surface may subsequently become buried by sediment, which preserves the surface as a *stone line* within the soil profile (Gerrard 1981, Stoops 1989); these features may appear similar to the stone lines produced by bioturbation (section 3.2.3), although the latter are usually limited to within a metre or so of the surface and the soil material above and below them is essentially of the same type. More severe soil erosion will lead to incision, which produces landscape features, as in the case of eroded deep-weathering profiles in Australia (Churchward and Gunn 1983). Soil profile erosion can also occur as a result of the development of glacial conditions, which can cause truncation of profiles by ice sheets and glaciers. This may have occurred in parts of Scotland, where remnants of deep-weathered Tertiary profiles have survived, following removal of their upper parts by Quaternary glaciation (Wilson 1985).

Although the four groups of soil-forming processes have been considered separately, it is important to note, as in the case of the previous two chapters, that they usually operate in combination, and that a soil may therefore possess more than one feature relict of past climatic conditions. Because climate fluctuates through time, soils can also contain features which represent more than one set of former climatic conditions. This can be seen, for example, in the case of *Clay-with-flints*, which occurs on the chalk of southeast England (also known as *Argile à silex* on the chalk of northern France) and is one of the oldest and most complex soils in temperate regions. It is thought to derive from dissolution of chalk and mixing of the residue with a thin veneer of Tertiary sediment left on a sub-Tertiary erosion surface; the mixing probably occurred by cryoturbation during cold phases of the Quaternary and alternated with interglacial episodes of decalcification, clay illuviation and rubification (Catt 1986b).

Although climatic change is the most important natural environmental change in the case of most soils, other changes can also influence soil development, in particular tectonic and volcanic activity, and diagenesis. For example, dissection of land surfaces due to tectonic activity can result in soil erosion, as in the case of certain Australian laterites which have become eroded in response to Tertiary uplift (Beckmann 1983). Volcanic activity can bring about changes to soil profiles not only by their burial (Gerrard 1985, Limbird 1985), but also by the effects of heating. For example, the deposition of hot lava or ash at a soil surface can cause combustion of the organic horizon and baking of the mineral soil. The combustion of organic matter may, however, be incomplete in cases where the oxygen supply is restricted by rapid burial.

Buried soils can also be altered by diagenesis; this is particularly important in pre-Quaternary soils which have experienced high pressures or temperatures during or after burial. The changes are usually associated with texture and mineralogy, and the older the paleosol, the more likely it is to have become altered. For example, a comparison of Carboniferous and Precambrian paleosols by Nesbitt (1992) indicated that they were affected by similar diagenetic conditions, with the introduction of potassium resulting in the conversion of Al-silicates to illite, and the production of anomalously high K_2O values. Iron and magnesium trends were also thought to have been affected by diagenesis in both these soils.

5.2.2 Human activity

Human activity during the Holocene period has influenced soils both directly, and indirectly via its effects on vegetation and hydrology. These influences have in some cases been deliberate while in others they have been unintentional. For example, in the case of soil additions, the most obvious deliberate action is to add material to the soil to improve its quality for agricultural production. This will be examined in more detail in Chapters 7 and 8, with respect to modern agricultural systems, but historical aspects can be briefly illustrated here. For example, in low-lying estuarine areas, soil drainage has sometimes been improved by adding sediment to raise the level of the surface by a process known as *warping*, in which estuarine water is repeatedly channelled into embanked areas then allowed to drain away, causing layers of sediment to accumulate (Catt 1986a). These soils therefore comprise thin layers of sediment whose texture relates to the speed of water flow in the channels. Elsewhere seaweed or manure has been added to the soil in order to improve its productivity; this produces what are often known as *plaggen* soils (Mücher *et al.* 1990). Additions can also relate to the construction of earthworks or buildings, which in some cases has resulted in burial of the soil.

Transformation of organic matter in terms of its decomposition has been influenced in many parts of the world by cultivation techniques, principally tillage, which have increased the rate of oxidation, and in some cases have led to the degradation of soil structure and subsequent soil erosion (Bell and Walker 1992). Increased organic matter oxidation can also result from artificial drainage of peaty soils, as in the case of the Fenlands of eastern England, where drainage has been in operation since the seventeenth century (Curtis *et al.* 1976).

Historically, the greatest indirect human influence on soils has been via vegetation clearance, often associated with cultivation, which has influenced transformation, transfer and output processes in terms of both acidification and erosion. Clearance began at different times in different parts of the world, but its effects only became significant during the second half of the Holocene, as populations expanded and technology developed (Roberts 1989). The removal of vegetation, by felling, burning or the grazing of animals, reduces the interception of rainfall and losses of water to the atmosphere via evapo-transpiration, thus increasing the potential for infiltration and leaching of the soil. This can cause acidification as basic cations are lost, and as acid-tolerant vegetation covers become established, a moder or mor humus will form. Such processes are thought to account for the development of many podzolic soils in Europe south of the boreal (needleleaf) forest zone (Macphail 1986, Bell and Walker 1992). Evidence for this situation comes from studies of peat and lake sediments which have preserved past vegetation components such as pollen grains and seeds. Identification of these components allows vegetation assemblages to be reconstructed, and by dating the material in which they occur, a pattern of vegetation change through time can be seen. In Europe, for example, there is a marked decline in woodland during the second half of the Holocene, which, when considered in association with archaeological investigations, strongly suggests clearance (Bell and Walker 1992). It is also considered that in some wetter, upland areas of Europe the development of podzols eventually led to poorer drainage conditions due to the formation of

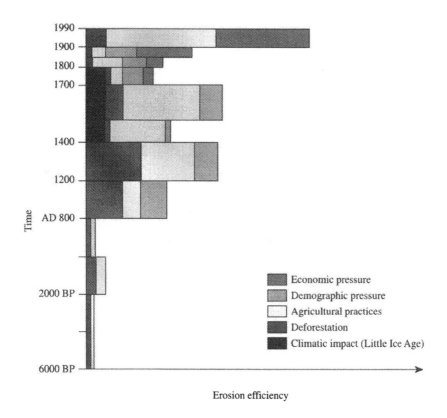

Figure 5.3 Soil erosion efficiency of various factors in western Europe (from Van Vliet-Lanoë *et al.* 1992)

impermeable iron-pans, which in turn encouraged peat accumulation (Roberts 1989).

Vegetation removal has also been historically important in causing soil erosion, and this has been further enhanced by cultivation of the soil (Van Vliet-Lanoë *et al.* 1992) (Figure 5.3). Evidence for erosion can be seen from the soils themselves, and also from the areas of deposition of the eroded material. In the former case, soil thinning can be seen by comparing the thickness of soil profiles buried *in situ*, for example beneath earthworks (Figure 5.4), prior to cultivation with those of adjacent areas which have suffered erosion during cultivation (Ellis and Newsome 1991). An estimate of soil thinning can also be obtained by dividing the volume of colluvial and alluvial deposits by the area from which they have been derived, although the possibility of post-depositional removal of some of the deposited material

Figure 5.4 Soil buried beneath the chalk rubble of a Bronze Age structure in eastern England

limits the accuracy of this method. In this way, Evans (1990a) derived average estimates of up to 2.5 m of topsoil removal in England and Wales resulting from woodland clearance and cultivation; it was suggested that rates of erosion were very varied, both temporally and spatially, with most of the erosion occurring from the Bronze Age onwards during times of rapid expansion of arable land, agricultural innovation and population increase. Because the onset of colluviation in Britain varies from about 5,000–1,000 years BP, with no obvious temporal grouping, it is therefore thought to relate largely to anthropogenic rather than climatic factors (Bell and Walker 1992).

Soil erosion has also been related to aggradation in river valleys. For example, the Mediterranean region has traditionally been considered to show two major sediment accumulation episodes relating to accelerated erosion – the *Older Fill*, dating to the last major cold phase of the Quaternary, and the *Younger Fill*, dating to around the end of the Roman civilisation – both episodes being related to climatic change (e.g. Vita-Finzi 1969). However, it has since been suggested that the Younger Fill event relates largely to landuse changes – for example, the abandonment of terraces following the collapse of the Roman civilisation could have led to accelerated colluviation, leading to aggradation of river sediments – although extreme rainfall events may also have been important (Bell and Walker 1992, Bintliff 1992). Soil erosion as a result of agricultural exploitation has also occurred in Central America, where forest clearance and cultivation of Mollisols by the Maya caused not only soil deterioration but also silting of drainage systems and reservoirs, which are associated with the decline of the civilisation (Olson 1981a).

Past soil erosion can also be recognised from studies of lake sediments, by the use of mineral magnetic or chemical analyses to identify episodes of soil inwash into a lake (Dearing *et al.* 1985, Heathwaite and Burt 1992). For example, Heathwaite and Burt (1992) observed increases in allogenic potassium and aluminium in the upper part of cores taken from lake sediments at Slapton Ley in southwest England; this was related to increased erosion of these elements from the lake catchment area, due to the ploughing up of grassland (Figure 5.5).

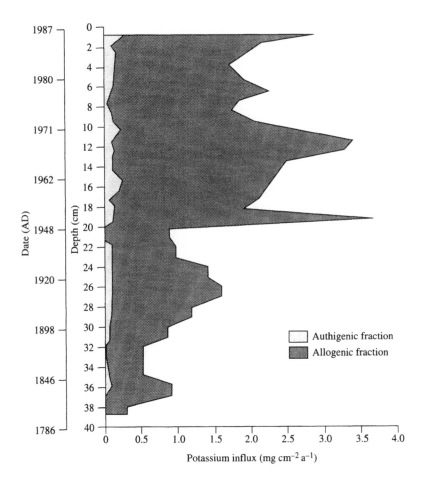

Figure 5.5 Potassium influx to the sediments of Slapton Ley, southwest England
(from Heathwaite and Burt 1992)

Human activity has also resulted in the erosion of soils by wind. For example, in Britain, the Netherlands and Germany aeolian deposits have been reported above occupation levels dating from the Neolithic onwards, relating to both vegetation removal and a change to drier conditions (Bell and Walker 1992). In addition to soil removal by erosion processes, more localised removal has occurred in the form of extraction, particularly of peat, which has for many centuries been cut for fuel or as an additive to improve agricultural soils (e.g. Curtis *et al.* 1976). Other smaller-scale soil features resulting from past human activity include abrupt boundaries, lateral discontinuities, mixed layers, pore infillings and compaction (Collins and Shapiro 1987, Macphail 1992).

5.3 ENVIRONMENTAL RECONSTRUCTION

Both surface and buried soils can preserve features which allow us to obtain information about past environmental conditions. This may be in the form of either fossil pedological characteristics, which relate to past conditions under which the soils developed, or biological material preserved in the soil from the time when these conditions prevailed.

5.3.1 Pedological characteristics

With an understanding of the relationships between present-day soil-forming processes and the factors that control these processes (Chapter 4), it is possible to obtain a great deal of information about the environmental conditions under which a soil formed in the past, particularly with regard to climate, on the basis of its fossil pedological characteristics. This information can be obtained from soils developed at the surface which contain paleosolic elements, but is often best preserved in soils buried to a sufficient depth so as not to be affected by post-burial processes operating downwards from the new surface, although diagenesis can cause complications in very old or deeply buried soils (Martini and Chesworth 1992). A further problem with the use of pedological characteristics concerns the assumption that the soils were in equilibrium with their environment and that the pedological characteristics used in environmental reconstruction are a direct response to the conditions prevailing at the time of their formation. For example, if environmental change was rapid, there may have been insufficient time for the soils to reach equilibrium, so features relict of past environmental conditions may be less developed than these conditions would otherwise allow. Indeed, rapid environmental fluctuations may completely prevent the development of certain pedological characteristics. It is therefore important to bear these problems in mind when using soils to reconstruct past environmental conditions.

An additional set of problems relates to the identification of paleosols; it may not always be clear from field observation what the pedogenic features of a paleosol are, or indeed that a paleosol is present at all; the soil may be visually little different from the parent material or the overlying material, particularly if the upper part of the profile has been removed by erosion or organic matter oxidation prior to burial, or if the horizonation has no contrasting colours. In order to confirm the presence of a soil in such cases it may therefore be necessary to subject the material to laboratory analysis, for example, to provide evidence for transformation such as changes in clay mineralogy or evidence for translocation in the form of illuviation cutans (section 3.3.1) (Fenwick 1985).

Buried soils can be used to obtain environmental information from many geological time periods. For example, in England red coloration of alluvial deposits in Upper Carboniferous mudstones has been interpreted as representing iron oxidation and dehydration under moist, tropical conditions shortly after deposition (Besly and Turner 1983), while albic (bleached eluvial) horizons in deposits of this age have been taken to indicate the occurrence of well-drained sites within a generally waterlogged deltaic plain (Percival 1986). In northwest France kaolinitic weathering profiles have been interpreted as indicating hot wet climates during the Cretaceous and Eocene periods (Esteoule-Choux 1983). Drier conditions in the past can be inferred from fossil calcretes, as occur in various Tertiary sediments in western and central Europe; calcretes generally form in areas with an annual rainfall in the order of 200–500 mm, so their presence in the geological record normally indicates a semi-arid climate (Goudie 1985).

Soils have become of great importance in the study of Quaternary environments. For example, paleosols in the loess deposits of central Europe demonstrate a number of glacial–interglacial cycles over a period of about 900,000 years. Interglacial paleosols are of the brown earth type, with argillic B horizons, formed under mixed deciduous forest, and are buried by loess of the succeeding cold phase; many interstadial soils (representing warm phases within major cold periods) are also present in the sequence, indicating grassland and open forest conditions (Kukla 1975, Morrison 1978). A similar situation occurs in the USA, where loess–paleosol sequences may extend back over the last 1.5 to 2 million years (Busacca 1989), and also in China, where such sequences have recorded climatic cycles over the last 2.5 million years (Bronger and Heinkele 1989, Zhongli et al. 1993). In some cases, over thirty paleosols have been reported in the Chinese loess sequences, and here loess thickness and extent of weathering have been used to interpret the intensity of climatic variations. In the upper and lower parts of the sequence the loess layers are relatively thick and unweathered while the soil layers are strongly weathered (Figure 5.6); this is considered to indicate phases of intense climatic variation, ranging from cold conditions of loess deposition to warm conditions of strong soil weathering. In contrast, the middle part of the sequence contains thinner and more weathered loess layers, indicating

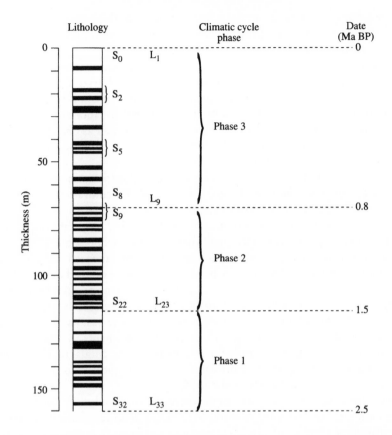

Figure 5.6 Sequence of loess and paleosols from Baoji in north-central China (based on Zhongli *et al.* 1993). Soils and loess layers are numbered sequentially from the top (S = soil, L = loess). Climatic cycle phases 1, 2 and 3 relate to the lower, middle and upper parts of the sequence respectively

less intense changes in climate between periods of loess deposition and those of soil formation (Zhongli *et al.* 1993). The relatively intense climatic cycles in the lower and upper parts of the sequence occurred mainly at intervals of around 100,000 years, while the less intense variations of the middle part are dominated by 40,000 year cycles.

In England, the buried Valley Farm Soil (Plate 14) indicates weathering and clay translocation over long time periods during temperate interglacial conditions, while the buried Barham Soil, which shows features characteristic of periglacial environments, indicates cold climate conditions prior to the Anglian glaciation (Rose *et al.* 1985). Interstadial soil formation has also been recognised in England from buried soils, as at Pitstone (Figure 5.7), where two thin humus layers have been interpreted as relating to short phases of temperate pedogenesis at the end of the last major Quaternary cold period (Rose *et al.* 1985).

Surface soils have preserved evidence of past climates in the form of paleosolic elements. For example, some brown earths in eastern and southern England show very high illuvial clay contents, along with rubified argillans. These features are considered to result from interglacial pedogenesis, and are known as *paleo-argillic B horizons* (Avery 1985, 1990). In some instances, a series of alternating warm and cold phases has been recognised where soils contain rubified argillans formed under interglacial conditions, which have become disrupted by cryoturbation in subsequent cold periods (Bullock 1985). In many temperate soils of western Europe and the USA, ice lens structures and fragipans are interpreted as fossil periglacial features associated with frozen ground conditions (Van Vliet-Lanoë 1985), while silty

Figure 5.7 The interstadial paleosol between layers of geliflucted chalk rubble at Pitstone, southern England. The top of the paleosol is marked by the dark layer below the trowel; the dark material above the trowel is Holocene colluvium

Figure 5.8 Fossil ice wedge cast preserved in a soil developed in glacial outwash, eastern England

textures are considered to represent periglacial aeolian additions (Catt 1986a). Soils may also preserve fossil patterned ground features such as ice wedge casts, formed by infilling of ice-filled thermal contraction cracks when the ice melts as periglacial climates change to more temperate conditions (Figure 5.8).

Although rubification in temperate soils is usually interpreted as a paleosolic feature, its meaning in climatic terms is uncertain. The presence of hematite is generally considered to indicate pedogenesis under warm and seasonally dry conditions (Catt 1986a, Kemp 1986, section 5.2.1), but it may also be able to form in temperate areas, and it may be that long periods of time are more important than critical climatic conditions in producing reddening (Schwertmann *et al.* 1982, Fenwick 1985, Kemp 1985).

Soil features relating to disturbance by human activity can also be preserved and used in environmental reconstruction. For example, Macphail

(1992) reports features such as pore infillings and the remnants of surface crusts, considered to result from prehistoric cultivation.

5.3.2 Biological material

There are various types of biological material which can be preserved in soils from the time of former environmental conditions, and these can provide useful evidence in environmental reconstruction. Of the macrofossils, mollusca have probably been the most widely used, while pollen and spores have received most attention from amongst the microfossils. There are, however, a number of problems associated with their use. First there is the extent to which fossils reflect conditions at the site where they are found. For example, pollen and spores can be transported several tens or even hundreds of kilometres by the wind, while molluscs can be washed from one site to another, an obvious problem in soils formed in fluvial deposits. Second, certain types of fossil may not preserve well and certain key environmental indicators may therefore be under-represented or even absent; while pollen can be preserved in acid, anaerobic soils in which biological activity is limited, acid conditions will clearly be detrimental to the preservation of calcareous molluscan shells. Abrasion and fracturing during fluvial transport may also make identification of fossil material difficult. Third, transfer processes operating in the soil will mean that the depth at which the fossil material occurs will not necessarily relate to the time at which it was originally incorporated into the soil, such that material relating to different climatic conditions over a period of time may become mixed at the same depth and will therefore be of little interpretative value. Finally, the use of pollen in the quantitative reconstruction of vegetation can be complicated by the differences between species in the quantity of material produced and its distance of transport, although this problem can be addressed by studies of modern pollen dispersal (Lowe and Walker 1984).

Molluscs have been used extensively to reconstruct Quaternary climate and vegetation. For example, Evans (1972) discussed the use of molluscan analysis of soils within an archaeological context, as in cases where soils have been buried beneath earthworks, from which the environmental conditions at the time of construction can be determined, along with changes in environmental conditions relating in particular to landuse changes. Molluscs occurring in colluvial soil material have also been used in a similar way (Preece 1992). Soil palynology has also been widely used in archaeology for similar purposes (Dimbleby 1985, Sakai and Kumada 1985), while Caseldine and Matthews (1985) were able to identify altitudinal changes in the alpine vegetation belts relating to climatic change in southern Norway during the last 5,000 years, by analysing pollen preserved in soils buried beneath a glacier end moraine. On a shorter time-scale, pollen analysis of a plaggen soil in the Netherlands allowed the agricultural history of the area to be examined

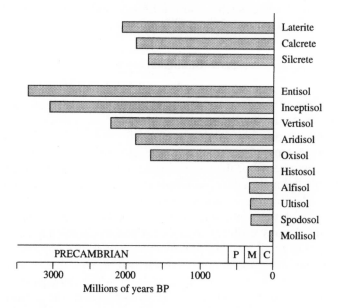

Figure 5.9 Geological time ranges of soil orders (Soil Survey Staff 1975) and related features (from Retallack 1986). P = Paleozoic, M = Mesozoic, C = Cenozoic (see Table 5.1)

over the last few hundred years (Mücher *et al.* 1990).

Many other types of macrofossil have been reported from soils, particularly pre-Quaternary buried soils, including stumps, leaves, roots, bones, teeth and opal phytoliths (silica absorbed by plants and precipitated in their cells). From these Retallack (1992) has proposed a sequence of vegetation development through geological time, starting with microbial earths and polsterlands, established by late Ordovician times (Table 5.1), and developing through breaklands to shrublands and woodlands from mid-Devonian time onwards. Linked to this is the development of soils themselves, with Entisols and Inceptisols being the first soil types to form, without the presence of vegetation, and other types developing much more recently with the establishment of higher forms of vegetation (Figure 5.9).

5.4 DATING

Both surface and buried soils can be used to date past pedological and environmental events, or the surfaces on which the soils have developed. This can be achieved by a number of methods, which allow three main types of dating – age estimation (sometimes known as absolute dating), relative dating and age correlation.

150

5.4.1 Age estimation

The principal age-estimation methods involving soils are those of radio-metric dating. These are based on the radioactive properties of certain elements in the soil. Probably the most commonly used method, radiocarbon dating, involves the isotope ^{14}C. This occurs naturally in all living matter, but once that material dies, decay commences. ^{14}C has a half-life of approximately 5,730 years, which means that after this time its activity has been reduced to half that of its original level, with the decay proceeding in a negative exponential manner (Figure 5.10). Because of the shape of the decay curve, the age of an organic sample becomes increasingly difficult to measure with increasing age, and the technique is not usually applicable to material older than about 40,000 years, although enrichment techniques can allow the dating of material almost twice this age (Bradley 1985). Radiocarbon dating

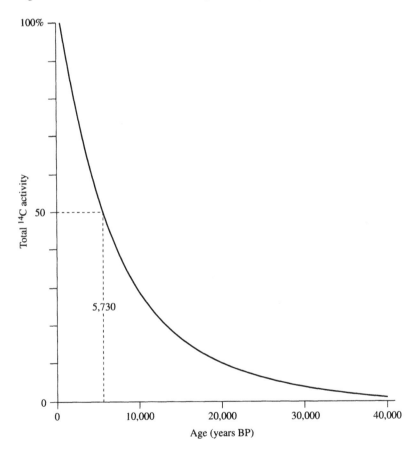

Figure 5.10 Decay curve for radiocarbon (from Roberts 1989)

151

of soils can be used in a variety of contexts. For example, the dating of former surface organic horizons of buried soils can provide an indication of the date at which burial occurred, which may have been triggered by a change in environmental conditions (Matthews 1980, Nesje et al. 1989). The dating of soil organic components allows the rate of organic matter decomposition to be examined (Stevenson 1986, Ayanaba and Jenkinson 1990, Page and Guillet 1991), while the dating of organic matter in illuvial horizons allows the timing of translocation to be considered (Ellis and Matthews 1984).

There are, however, a number of problems associated with the radiocarbon dating of soils (Matthews 1985). First there is the problem of contamination by carbon from other sources which can result in a date being either increased or decreased. Dates can be increased by the addition of minerogenic carbon derived from materials such as coal, which can be incorporated in the soil parent material, or carbonates dissolved in groundwater. Decreased dates are produced by the addition of carbon with higher levels of activity than those occurring naturally within the soil. In the case of surface soils, the principal source of contamination is atmospheric 'bomb' carbon produced as a result of nuclear testing since the 1950s, which severely limits their usefulness in dating. The main risk of contamination in buried soils comes from organic material derived from the present-day soil above, either in the form of roots penetrating down into the buried soil or organic matter moved downwards by various transfer processes.

A second problem is that within any horizon, the organic material may not all be of the same age, and indeed may be of widely differing ages. This is because organic horizons usually contain material in different stages of decomposition which has been added to the soil over a period of time. The dating of a horizon will therefore give a value which represents a combination of the different ages of the organic components; this is known as the *apparent mean residence time* or AMRT (Matthews 1985). In the case of a buried soil this will therefore increase the apparent age of burial. For example, a soil with an AMRT of 1,000 years which was buried 3,000 years ago will give a ^{14}C date of 4,000 years, so the AMRT must be subtracted from the ^{14}C age to give the time elapsed since burial. Because most organic matter is usually added to a soil via the surface, a sample taken from near the top of a horizon will normally possess a shorter AMRT than one taken from the base, while a sample comprising the entire horizon will give an intermediate value. The degree of difference between these values will depend on the rates of organic matter addition and decomposition, and the extent of mixing; rapid decomposition or extensive mixing will cause smaller differences.

Different ages can also be obtained from different types of organic matter. In the context of soil dating, three types of organic matter are normally recognised – the fulvic acid, humic acid and fine residual fractions (Matthews 1985). Although not always the case, the fulvic acid fraction often yields dates somewhat younger than those of the other two fractions, at any one depth.

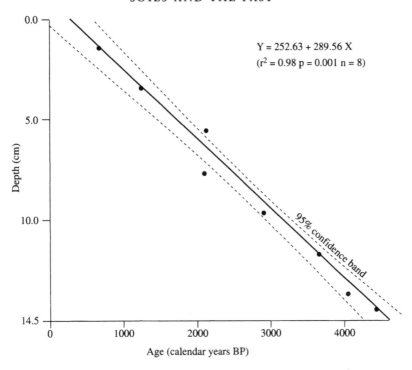

Figure 5.11 Age–depth relationships in a ^{14}C-dated former surface organic horizon of a buried podzol in southern Norway, based on the humic acid fractions (from Matthews and Dresser 1983)

Dates derived from the humic acid or fine residual fractions will therefore generally approximate more closely to the true age of the organic matter at any particular depth within a soil. Given that the AMRT generally increases with depth in soils which have not experienced excessive mixing, dating of the humic acid or fine residual fractions of a sample taken from the base of a surface organic-rich horizon will therefore give the closest approximation to the date at which the soil began to form. With respect to buried soils, however, dates from these fractions which are obtained from near the soil surface will give the closest estimate of the time elapsed since burial (Figure 5.11). These procedures obviously work best in acidic or poorly drained soils, where mixing will not have seriously affected the age–depth relationship. Erosion of the soil surface prior to burial can also affect the accuracy of determining the time elapsed since burial because if erosion has occurred, dates obtained from the uppermost part of the soil will be older than they would otherwise be.

A further problem with ^{14}C dating is that the radiocarbon time-scale is different from that of our own calendar. This is because the concentration of

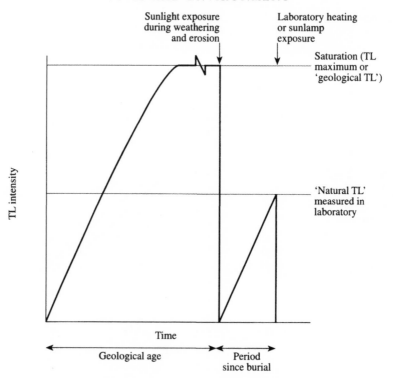

Figure 5.12 Schematic illustration of thermoluminescence (TL) intensity changes and their use in dating (from Bradley 1985)

^{14}C in the environment has fluctuated through time, producing a non-linear relationship between radiocarbon years and calendar years. A correction for this effect can, however, be made by comparing the age of materials measured in ^{14}C years with their calendar age derived by an independent method such as dendrochronology or reference to historical records, from which a calibration curve or chart can be constructed (Bradley 1985). In view of the above problems, it is clear that ^{14}C dating of soils must be undertaken with great caution (Matthews 1980, 1985).

Another, but less commonly used, radiometric dating method applied to soils is uranium-series dating, based on the decay products of ^{238}U and ^{235}U. Because these have half-lives of 4.51×10^9 and 7.13×10^8 years respectively, they can be used to date soils much older than those used in ^{14}C dating. The technique is usually applied to soil carbonates, and is based on the fact that uranium is precipitated with calcite or aragonite from drainage waters (Bradley 1985). However, while the method has been applied successfully to closed systems, its application to open systems such as soils, in which transfers of material occur, is much less secure (Fenwick 1985); this has caused problems, for example, in the dating of calcretes (Milnes 1992).

Other age-estimation methods include thermoluminescence and electron spin resonance. Thermoluminescence (TL) is the light emitted by the release of electrons trapped in defects within mineral crystal lattices, when they are heated to greater than 500°C or exposed to sunlight for more than 8 hours (Williams *et al.* 1993). Exposure of minerals to sunlight during weathering and erosion therefore releases the trapped electrons and sets the TL 'clock' to zero. When exposure ceases, following burial, electrons will again become trapped in a time-dependent manner, and subsequent measurement of the TL will therefore allow an estimate of the time elapsed since this event (Figure 5.12). This technique has been used to date loessic soils and sediment sequences (e.g. Wintle and Catt 1985, Stremme 1989). The electron spin resonance method is concerned with the energy states of electrons produced during time-dependent radioactive decay (Williams *et al.* 1993). Although its application to soils has so far been limited, it can be used to date soils with ages of several million years, as in the case of Australian silcretes (Radtke and Bruckner 1991).

5.4.2 Relative dating and age correlation

In certain situations it may not be necessary or possible to determine the absolute age of an event for a variety of reasons, for example the lack of suitable material, time or money. In this case relative dating and age correlation methods can provide a useful means of establishing a temporal framework. The principal methods involved with respect to soils are comparing degrees of soil development, using soils as stratigraphic markers, and the measurement of palaeomagnetism and amino acid racemisation.

On the assumption that soil development generally increases through time (section 4.2.5), it is possible to establish the relative age of soils or the surfaces on which they have developed by comparing their pedological characteristics. In making such comparisons, however, it is obviously important that the soils are developed under as near identical environmental conditions as possible, otherwise pedogenic differences may be due to factors other than time (section 4.2). It is also important that the time period under consideration is less than that required for soil development to reach equilibrium conditions, because in cases where equilibrium has been attained, the extent of soil development is no longer time-dependent (section 4.2.5). When dating surfaces, it is also necessary for soil development to have been continuous since the surfaces formed, otherwise soil conditions may not be related solely to surface age. For example, an old surface may have experienced erosion of the soil originally developed on it, and may now support a much less well developed soil.

Various pedogenic indices have been constructed for relative dating purposes, ranging from simple measures capable of use in the field to more complex expressions derived by laboratory analysis. Field measures are based on morphological properties usually related to weathering and translocation, while laboratory measures generally involve the use of mineralogical or chemical properties (section 3.3.1). Robertson-Rintoul (1986) used this approach to correlate river terrace fragments in Scotland, based on the extent to which their associated soils exhibited podzolisation. Five soil groupings were identified, allowing the fragments to be assigned to one of five phases of terrace development during the past 13,000 years. In the Ventura basin, western USA, river deposits with ages ranging over the last 200,000 years have been dated on the basis of colour and clay content characteristics (Rockwell *et al.* 1985). In this instance, deposits of unknown age were dated by fitting their associated soils into a chronosequence. Weathering rind development and mineral grain etching can also be used to date surfaces (e.g. Mahaney 1978, Burke and Birkeland 1979, Locke 1979, Colman 1986). In studies of this type it is important to ensure that the profiles are complete, and have not been disturbed or truncated during pedogenesis (Soller and Owens 1991).

The use of soils as stratigraphic markers is based on the principle that if a buried soil occurs in a number of sedimentary sequences, it can be used to distinguish the younger, overlying materials from the older, underlying materials irrespective of their absolute ages (Figure 5.13). For this method to be used successfully, the soil must be developed under closely similar

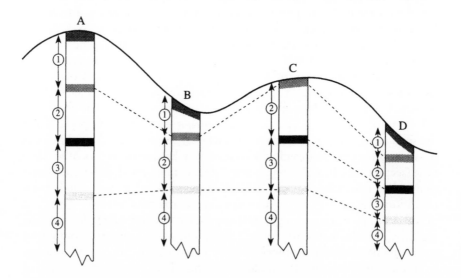

Figure 5.13 Principle of relative dating using paleosols. The sedimentary units are correlated on the basis of the soils (shaded areas) developed on them. Absence of units, as at locations B and C, can result from either erosion or lack of deposition

environmental conditions at each locality, and it must also possess features which allow it to be distinguished from any other buried soils which may be present. The stratigraphic approach to dating works best where the buried soil is laterally continuous or where its occurrences are separated by only small distances, otherwise the age correlations become more tenuous.

Some of the best-known examples of soils used as stratigraphic markers come from the mid-western USA, where they were first used a century ago (Follmer 1978). Here the Sangamon Soil, formed during the Sangamon interglacial, has been used as a marker between the Illinoian glacial deposits into which it is developed, and overlying loessic deposits of the Wisconsinan cold period. A number of other paleosols have also been recognised, such as the Aftonian and Yarmouth Soils relating to earlier interglacial periods (Follmer 1978, Catt 1986a). A more recent discovery is the widespread occurrence of the Valley Farm and Barham Soils of eastern England (section 5.3.1), which it has been suggested can also be used as stratigraphic markers in the reconstruction of synchronous buried land surfaces (Rose *et al.* 1985).

Soil stratigraphy can also be used to examine the chronology of geomorphological events (Gerrard 1981). For example, in Australia Butler (1959) developed the concept of the *K-cycle*, in which the youngest ground surface was designated K_1 and progressively older surfaces were assigned progressively higher K numbers. Each cycle had an unstable phase of erosion and deposition, followed by a stable phase during which pedogenesis occurred. Walker (1962) demonstrated the K-cycle approach in southeast Australia, where three ground surfaces were recognised (Figure 5.14). K_1 soils were formed under present conditions, while K_2 and K_3 soils occurred as relict soils at the surface and buried soils beneath younger materials.

Soils have also played a part in chronological investigations of erosion surfaces in western and central Africa, where a series of surfaces have been recognised, the altitude of which decreases with decreasing age; the highest three surfaces are bauxitic and are ascribed Jurassic, Cretaceous and Eocene ages from highest to lowest, while the lower four surfaces show ferricrete development and are thought to be of Pliocene and Quaternary age (Tardy and Roquin 1992). However, the relationship between soil development and erosion surface age must be regarded with caution for a number of reasons. For example, surfaces which occur through a range of altitudes are likely to have experienced different types of pedogenesis due to climatic and vegetational differences. Additional problems are brought about by the possibility of different lithological conditions, and the possibility that certain duricrusts formed below, rather than at, the surface and were exhumed at a later stage (Goudie 1985). An erosion surface may also be much older than the soils developed on it if it has experienced more than one cycle of erosion and pedogenesis, as recognised earlier. It is also possible that certain pedogenic processes may not have been operative throughout the entire period over which a surface has been exposed (Gerrard 1981).

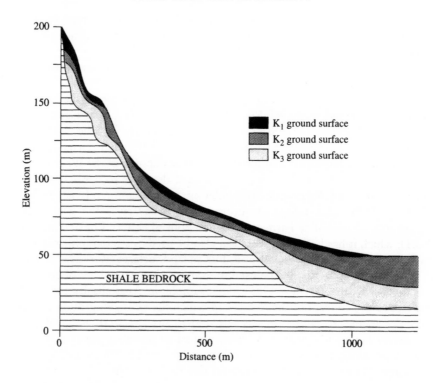

Figure 5.14 K-cycle soil layering in southeast Australia (from Walker 1962)

The use of palaeomagnetism as a dating method allows age correlation by comparing the record of variations in the Earth's magnetic field which are preserved by magnetic minerals in sediments, with standard palaeomagnetic reference curves which have been dated by another technique (Williams *et al.* 1993). This method can be used over long time-scales, as in the case of Chinese loess and paleosol sequences developed over the past 2.5 million years (Zhongli *et al.* 1993) (Figure 5.6). A further method of establishing relative age which has been applied to soils is amino acid racemisation dating. This involves the change in optical properties of amino acid molecules which occurs following the death of an organism, in terms of the direction in which they rotate plane-polarised light. Most amino acids in the proteins of living organisms have an *L (levo)* configuration, which changes to a *D (dextro)* configuration after death (Figure 5.15). Measurement of the D:L ratio therefore allows an estimate of the time elapsed since death (Williams *et al.* 1993). This dating method has been applied to Australian calcretes, where an increase in D:L ratios with depth indicated increasing age (Milnes 1992).

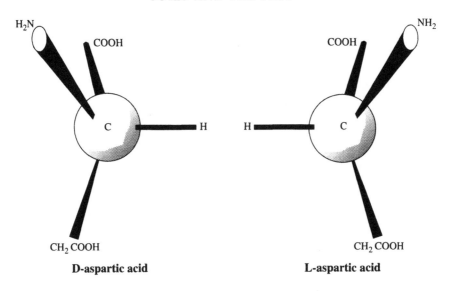

D-aspartic acid   L-aspartic acid

Figure 5.15 D and L configurations of aspartic acid (from Bradley 1985)

5.5 SUMMARY

In this chapter we have seen that soils can preserve characteristics formed under past environmental conditions, and that these characteristics can also be used to provide information about the past. Soils also allow the dating of past events, either in relative or absolute terms. Changing climatic conditions can result in organic additions to soils, as in the case of increased surface organic horizon thickening due to increased vegetation growth. Past mineral additions can take the form of loess or coversand inputs during previously drier climates, or in more extreme cases may result in profile burial, for example as a result of climatically induced ice sheet advance or sea level rise. Organic transformation responses to past climatic conditions may be seen in terms of peat accumulation due to increased wetness, while former mineral transformations often take the form of exotic weathering products or rubification, developed under warmer conditions in the past. Relict features of transfer include illuvial deposits resulting from former wetter conditions, and disturbance and fragipan characteristics produced under past periglacial conditions. Losses from soil due to past climatic conditions are best seen in profiles incised or truncated by erosion, or in those with residual accumulations of larger material at their surface as a result of greater surface runoff or aeolian transport.

Features of past human activity can be seen, in terms of additions, in warp and plaggen soils which have been improved for agriculture, or in more extreme cases by soils buried by material used in the construction of earthworks or buildings. The greatest influences of past human activity on soils is, however, expressed in acidification and erosion. Instrumental in these processes is vegetation clearance, which increases leaching and surface runoff, thus acidifying and thinning the soil. This has operated for several millennia in some parts of the world, although elsewhere it is a much more recent phenomenon. In addition to climatic change and human activity, soils may express characteristics of former environmental conditions resulting from tectonic activity, which can cause soil erosion, volcanic activity, which can produce baking, or deep burial which can result in diagenetic transformation of minerals.

Once the formation of relict features in soils is understood, the features can themselves be used to provide information about past environmental conditions. Buried soils are particularly useful in this respect, and these are preserved in some cases from the earliest periods of geological time. Their weathering and transfer characteristics, such as rubification and periglacial features, can be used to indicate past climatic conditions, although there is some uncertainty concerning the interpretation of rubified features. Soils can also preserve biological material, that can provide information about biotic and climatic conditions at the time the material became incorporated in the soil, although in some cases this may be complicated by problems such as transport, differential preservation and mixing.

Soils can be used in age estimation and for relative dating and age correlation purposes in order to establish the timing of past pedological or environmental conditions. In terms of age estimation, the radiocarbon method has been the most commonly used. This can provide estimates of soil age, rates of organic matter turnover and times of soil burial. Problems exist, however, in the form of contamination, residence time and calendar year conversion. Uranium-series dating can be used to date material much older than that capable of being dated by the radiocarbon technique, but the method lacks security in open systems such as soils. Thermoluminescence and electron spin resonance analyses provide additional dating methods, but have not been widely used within the context of soils. Soils can be used in the relative dating and age correlation of surfaces or deposits by comparing relative degrees of soil development, although this can be particularly problematic in low latitude regions because of altitudinal and lithological variations, and also the long timespans involved over which complex cycles of erosion and pedogenesis may have occurred. Soils can also be used as stratigraphic markers, particularly in Quaternary deposits, as in North America, and as markers of geomorphological events in the context of slope instability phases. Measurements of palaeomagnetism and amino acid racemisation have also been used in the age correlation and relative dating of soils.

6

SOILS IN NATURAL
ENVIRONMENTAL SYSTEMS

6.1 INTRODUCTION

Soils not only respond to environmental conditions, but can also influence these conditions and therefore play an important role in the operation of environmental systems. These systems can be divided into four main types – the hydrosphere, atmosphere, geosphere and biosphere (section 1.1). Although these environmental components are in many ways interrelated, this chapter aims to examine the role of soil in each system separately in order to provide a clearer understanding of the processes involved. We will start by examining the hydrosphere because soil moisture is one of the main keys to soil-environmental influences, and many of the aspects relating to the hydrosphere are therefore relevant to those of the other systems. Having first introduced some basic concepts, various components of the hydrological cycle will then be considered. Closely related to the hydrosphere is the atmosphere, with soils exerting an important influence on climate at or near the ground surface, and this is considered next. The geosphere is then examined with respect to the role of soil in geomorphic processes and landform development. Finally, we consider the biosphere, in which soils are an important factor in controlling the habitation and distribution of biota. This chapter will concentrate predominantly on natural environmental conditions; the important human aspect of soil-environmental relationships will be discussed separately in the following two chapters.

6.2 THE HYDROSPHERE

The activity of water at or below the ground surface is part of the hydrological cycle, in which water arrives at the surface via precipitation and is eventually returned to the atmosphere via evapotranspiration, having passed along a number of possible pathways in between. Some precipitation may be prevented from reaching the surface directly by obstructions such as vegetation or buildings, and some may not reach the ground at all, being intercepted and returned directly to the atmosphere by evaporation. That water which does reach the surface will either remain there as *surface water*,

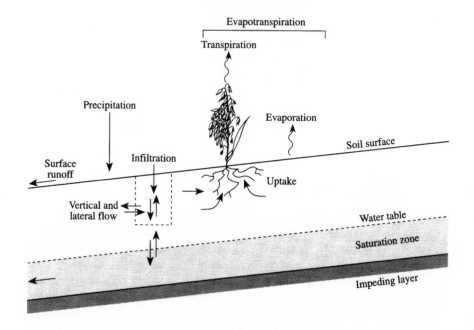

Figure 6.1 Components of the hydrological cycle in relation to the soil layer (from Singer and Munns 1991)

flowing over the surface or being stored in depressions, or it will infiltrate the ground to become *subsurface water*; in this form it can be stored or moved through the ground, and may eventually become surface water (Figure 6.1). According to Ward and Robinson (1990), subsurface water can be classified into four major zones (Figure 6.2). The *soil zone* lies nearest the ground surface and therefore controls the infiltration of water into the ground. The underlying *intermediate zone* is one in which percolation of water is the dominant process and this can vary enormously in thickness depending on the relief and rock type. This overlies the *capillary fringe* in which most of the pores are filled with water, and beneath this lies the *saturation zone*; the interface of these two zones is marked by the *water table*. While these zones may be separated on interfluves and valley sides, they usually converge downslope and may overlap on valley floors.

In this section we will examine the influence of soils on the hydrosphere under three main headings – the forces controlling water movement, soil water additions and losses in terms of infiltration, percolation and evaporation, and water storage and flow. It is important to recognise that soils can also influence the chemical characteristics of water, but this will be considered in the context of the geosphere (section 6.4).

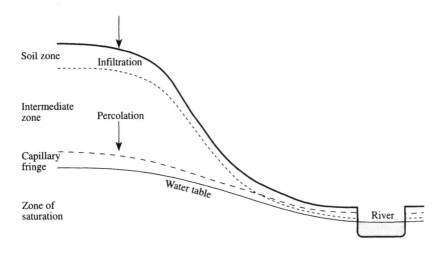

Figure 6.2 Classification of subsurface water (from Ward and Robinson 1990)

6.2.1 Forces controlling water movement

The movement of water into, through and out of soils is controlled to a large extent by gravity, and also by three types of force determined by soil properties which can either encourage or restrict water movement – adsorption, capillarity and osmosis, the combined effects of adsorptive and capillary forces being known as *matric suction* (section 2.3.4). The total suction in a soil will depend to a large extent on its texture and pore size. Higher clay or organic matter contents will have stronger adsorptive forces, and finer textures are also usually associated with smaller pore sizes, in which capillary forces will be greater. These factors therefore influence the retention and drainage of water. Coarse-textured soils generally drain more quickly and retain less water than finer-textured soils which, for any given water content, will have a greater suction. Water content will, however, also influence suction. Dry soils will have high values, but these will rapidly decrease as the pores become filled with water and the suctional effects are therefore lost (Figure 6.3).

The speed and direction of water movement in a soil will depend on the magnitude of these suctional forces compared with the gravitational force. The rate at which water can move through a soil is measured by *hydraulic conductivity*, which is determined to a large extent by soil water content and also by texture and its associated pore size. Water transfer is more effective in wetter soils than in drier ones because drier soils have a greater volume of air in their pores which inhibits the conduction of water from one location to another. Therefore at high moisture contents conductivity increases as

163

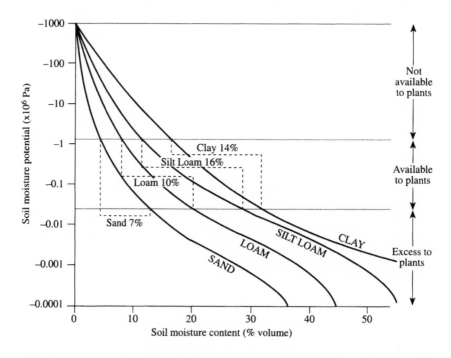

Figure 6.3 Relationship between matric suction and moisture content for soils of different texture (from Oke 1987)

texture coarsens; values can be less than 1 mm per day in the case of clays, 10 cm to 10 m per day in silty sand and over 100 m per day in gravel (Daniels and Hammer 1992). However, at low moisture contents conductivity increases as texture becomes finer, because of higher suction forces. The direction of movement under gravity is downwards, but soil water can move in other directions depending on the other forces present. For example, it can move laterally due to osmosis if there is a lateral variation in the concentration of salts dissolved in the soil water due to variations in parent material, or it can move upwards due to capillary forces operating as a soil dries out.

6.2.2 Infiltration, percolation and evaporation

When rainfall reaches the ground surface, some or all of the water will infiltrate the soil. Initially the matric suction gradient will be relatively high and infiltration will therefore be rapid, but as wetting increases the suction gradient lowers and the infiltration rate decreases as gravitational forces become more important and eventually a steady flow will be attained, approximating to the *saturated hydraulic conductivity* (Ward and Robinson

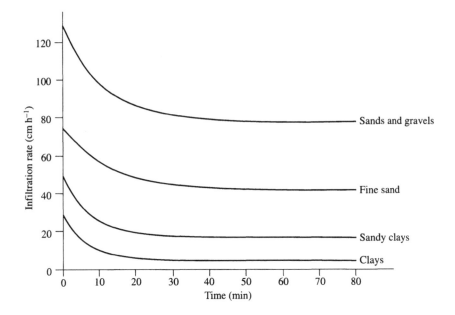

Figure 6.4 Typical infiltration curves for ponded infiltration on different materials
(from Gerrard 1981)

1990). The *infiltration capacity* is a measure of the ease with which water can penetrate the surface. For example, because of their larger pores, coarse-textured soils will allow more rapid infiltration than finer materials (Figure 6.4). Infiltration will also be reduced if a crust has formed, for example by aggregates breaking down under raindrop impact, or if fines are washed into the surface pores (Slattery and Bryan 1992, Messing and Jarvis 1993). The type and thickness of the litter layer can also affect infiltration; a thick cover of broadleaf litter with the leaves lying horizontally can reduce infiltration appreciably. Soil freezing will also impede infiltration as the pores become filled with ice, infiltration becoming zero in wet soils where the pores become completely ice-filled. Conversely, soils which develop cracks at the surface will have high infiltration rates, as in the case of Vertisols (section 4.3.3).

The percolation of water through the soil zone and intermediate zone towards the water table will continue while rainfall is infiltrating the surface, but once rainfall ceases the soil will start to dry out as water drains downwards and is also lost back to the atmosphere via evapotranspiration. Drainage under gravity is usually more or less complete after two or three days and the soil is then said to have reached *field capacity*. The movement of water through a soil is enhanced by the presence of macropores (section

2.3.3); these are relatively large pores which can result from shrinkage of a soil on drying or from the development of burrowing or root channels. They are particularly important in infiltration and percolation during intense rainfall, although the pores must be interconnected otherwise the continuity of flow is interrupted.

The loss of water from a soil by evaporation will be controlled by a number of factors, principally climate, moisture content and texture. High temperatures and windspeeds will enhance evaporation, as will high vapour pressure gradients between the soil and above-ground atmosphere (Barry and Chorley 1992). The moisture content of the top few centimetres of soil is important in controlling evaporation, which will decrease as the soil dries out, becoming zero once the soil is completely dry. In contrast the moisture content of the subsoil is considered to have little effect on evaporation because of the slow rate of soil moisture movement. Eventually a point is reached when little water is available for movement in the liquid form, and any remaining movement occurs by the diffusion of water vapour. At this stage the soil is said to be at *wilting point*. Vapour diffusion is controlled by matric and osmotic pressure to a minor extent, but the most important control under most soil conditions is temperature, with vapour diffusion occurring from relatively warm to cooler areas of a soil in response to the vapour pressure gradient (section 6.3.2). However, the movement of water vapour from the subsoil to the surface constitutes only a minor proportion of total evaporative losses (Ward and Robinson 1990). Hence, for any given volume of rainfall, soils which are regularly wetted at their surface will have greater evaporation values than those which are wetted more thoroughly but less frequently. Texture affects evaporation in that upward capillary movement of water is generally greater in fine-textured soils because of the greater suction forces. In some cases water can move upwards through a vertical distance of several metres, although in coarse-textured soils the distance is unlikely to exceed several centimetres. However, because the speed of capillary water movement is slow, this source of water does not usually contribute greatly to total evaporation except where the water table lies within a metre of the surface (Ward and Robinson 1990).

Other factors which will affect evaporation include soil colour and vegetation cover. Dark soils will generally have higher surface temperatures due to their lower reflectivities (section 6.3.1) and therefore experience higher rates of evaporation than lighter coloured soils. A vegetation cover can shade the soil surface and thus decrease surface temperatures and evaporation. It can also increase the relative humidity of the air near the surface (section 6.3.2), which will lower evaporation. However, these reduced moisture losses may well be offset by the loss of water via transpiration from the vegetation itself.

6.2.3 Water storage and flow

Water can be stored in or on a soil, or can flow over its surface or within it. Storage of surface water requires a soil of low permeability and a surface relief which will prevent lateral drainage. The storage may be either temporary, as in the case of puddles formed in microtopographic depressions, or it may be permanent if the supply of water is sufficiently great to exceed losses via evaporation and infiltration; in such cases ponds or lakes may form. The supply of water for surface storage can be provided directly by precipitation, water running downslope over the surface, or by ground water reaching the surface. Surface storage is therefore associated with fine-textured, compacted soils, and is favoured in cooler climates, where evaporative losses are limited. Water storage within the soil can occur where the regional water table approaches the surface, for example in low-lying areas or enclosed depressions. Storage can also occur where local conditions cause waterlogging near the surface, as in the case where a clay-rich Bt horizon or an iron pan impede the downward movement of water; this is known as a *perched water table*. However, where soils are located on a slope, water flow will usually occur. Three main types of flow can be recognised – *overland flow, throughflow* and *ground-water flow* (Ward and Robinson 1990).

Overland flow occurs when the rate of infiltration is exceeded by the rate at which water is arriving at the surface. This can occur, for example, in the case of frozen soil conditions, or where aggregate breakdown and surface sealing are caused by raindrop impact where soils lack a protective vegetation cover (section 3.2.4). Throughflow occurs when water flows laterally through the soil and is confined near the surface, as opposed to ground-water flow which involves water movement through the saturated zone beneath the water table (Figure 6.2), often at depths below the soil layer; this usually occurs more slowly than throughflow.

Throughflow can constitute a large proportion of total runoff from a catchment in cases where soils encourage infiltration and lateral water movement but restrict deeper percolation. Such conditions will occur, for example, if a soil has vertical cracks, extensive burrowing networks or other forms of macropore which allow rapid infiltration (McCaig 1985, Dabney and Selim 1987, Edwards *et al.* 1990, Booltink and Bouma 1991) and also a humified organic layer, Bt horizon or some form of pan which impedes vertical drainage (e.g. Blume *et al.* 1987, Parlange *et al.* 1989, Gafni and Brooks 1990). In extreme cases these conditions can lead to the development of piping (section 6.4.2), whose networks can provide major pathways for water movement through soils (Jones 1987, Walsh and Howells 1988). Soil thickness and slope angle are also important in that thin soils over impermeable parent materials will encourage rapid throughflow, as will steep slopes. Even in the absence of such favourable factors for throughflow, soil

(a)   (b)   (c)

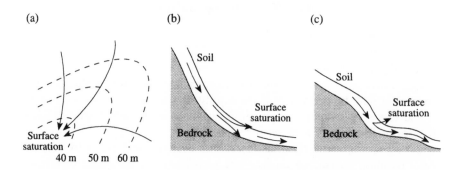

Figure 6.5 Principal locations of flow convergence to produce surface saturation: (a) hillslope hollow (dotted lines represent contours), (b) footslope, (c) soil thinning (from Ward and Robinson 1990)

hydraulic conductivity will usually be greater in the lateral than the vertical direction due to the development of structure and also to lower compaction than at depth, therefore on slopes, lateral flow will generally predominate over vertical drainage. Throughflow will be converted to surface runoff if the near-surface saturated zone reaches the surface. This can occur if the flow is concentrated in a hillslope hollow or towards the foot of a slope. Lateral thinning of the throughflow zone can also cause saturation to occur at the surface and therefore produce surface runoff (Figure 6.5).

The relative proportions of overland flow, throughflow and ground-water flow will determine the shape of the hydrograph of a catchment. A hydrograph is a plot of stream discharge against time, comprising one or more peaks, each of which has a rising limb and a recession limb. The rising limb represents an increase in discharge in response to an input of water to the catchment, usually by rainfall or snowmelt. The discharge will reach a maximum value beyond which it gradually decreases through time, and this represents the recession limb of the hydrograph (Figure 6.6). The steepness and height of the rising limb of the hydrograph will therefore depend on the rate at which water can reach the stream channels. Where overland flow occurs, water will reach the channels quickly, whereas throughflow and ground-water flow will delay its passage. The extent of the delay will be determined by slope angle and the way in which soil conditions influence vertical and lateral drainage, as discussed above. Therefore, soils on steep slopes and which impede infiltration will produce the most peaked hydrographs; soils which allow some infiltration but which have subsurface compaction, pans, Bt horizons or frozen subsurface layers, will tend to produce less peaked hydrographs, and soils which do not possess these features and therefore allow greater vertical infiltration will have the least steeply sloping and lowest hydrograph peaks (Figure 6.6). It is important to

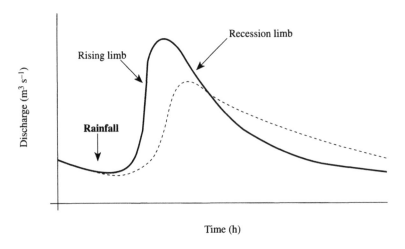

Figure 6.6 Examples of hydrograph peaks. The solid line represents a rapid discharge response to rainfall, while the dotted line represents a more gradual response

recognise, however, that these trends can be complicated by antecedent moisture conditions. For example, dry soils will often produce a delayed discharge response compared to wet soils, regardless of their effects on the type of flow, although under very dry conditions a rapid response may be encouraged by hydrophobic coatings of soil particles, such as clay or organic coatings of mineral grains or peds (Wilson *et al.* 1990).

6.3 THE ATMOSPHERE

Soils have their own climate and can also influence conditions in the lowest part of the above-ground atmosphere. The effect is, however, only on a small scale, and the study of soil–atmosphere interactions therefore forms part of the discipline known as *microclimatology*. The atmospheric components involved in this study will be considered under three headings – solar radiation and temperature, atmospheric moisture and air movement.

6.3.1 Solar radiation and temperature

Radiation reaches the ground surface from the sun, and leaves the ground surface, as electromagnetic energy of differing wavelengths. The wavelengths of the incoming radiation are predominantly shorter than those of the outgoing radiation, and these two types are therefore known respectively as *short-wave radiation* and *long-wave radiation*. Some of the incoming short-wave radiation will be reflected off the ground surface back into the atmosphere; the amount reflected will depend on the reflectivity, or *albedo*, of the surface. The net all-wavelength radiation (R_n) can be shown as follows (Barry and Chorley 1992):

$$R_n = [S(1-a)] + L_n$$

where S = incoming short-wave radiation, a = fractional albedo of the surface and L_n = net outgoing long-wave radiation. On unvegetated ground the albedo is determined by the nature of the soil. For example, dark soils, such as those rich in organic matter or derived from a dark-coloured parent material, can have an albedo of as little as 0.05 (or 5 percent) and will therefore reflect little incoming radiation, whereas soils which are low in organic matter or derived from a light-coloured parent material can have values as high as 0.60 (60 percent) and will therefore reflect much greater quantities of incoming radiation. Albedo also decreases with increasing water content due to internal reflection at the surfaces of water menisci in soil pores; a 20 percent water content can decrease soil reflectivity to a quarter of its dry value (Monteith and Unsworth 1990). The angle of incidence of incoming radiation can also have a marked effect on reflectivity, particularly in the case of a soil with standing water at its surface, as will microrelief. If the angle of incoming radiation is low, a water surface will be much more reflective than for higher incidence angles; the albedo of water is only around 0.05 for incidence angles greater than 45°, but rises rapidly for decreasing angles (Monteith and Unsworth 1990). Where soils have a marked micro-relief, some of the incoming radiation will be absorbed by other parts of the surface, following the initial reflection, and hence the overall albedo of the surface will be reduced (Oke 1987, Potter *et al.* 1987).

The exchange of energy between the atmosphere and a surface can be shown by the surface energy budget equation (Barry and Chorley 1992):

$$R_n = G + H + LE$$

where R_n = net all-wavelength radiation, G = ground heat flux, H = turbulent sensible heat flux to the atmosphere and LE = turbulent latent heat flux to the atmosphere. The *ground heat flux* represents the transfer of heat into the ground, while *turbulent sensible heat flux* to the atmosphere represents the transfer of heat by fluid (air) flow, and *latent heat* is heat absorbed during evaporation or released during condensation. During the day the available net radiation has a positive value and is balanced by turbulent fluxes of sensible and latent heat into the atmosphere and by conductive heat flux into the soil (Figure 6.7). Soils with a low albedo will attain higher temperatures at and close to their surfaces than those which reflect a higher proportion of incoming radiation, as seen in extreme cases for artificially whitened and blackened soils (Figure 6.8). Higher surface temperatures will usually give higher above-ground temperatures in the first metre or so of air due to sensible heat transfer. Some of the heat will also be transferred by evaporation if the soil contains moisture, and some will

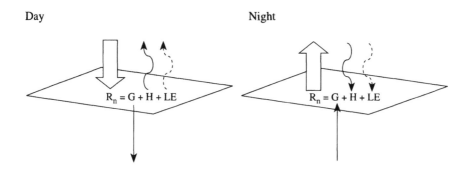

Day   Night

$$R_n = G + H + LE \qquad R_n = G + H + LE$$

Figure 6.7 Energy flows involved in the energy balance of a simple surface
(from Barry and Chorley 1992)

be transferred deeper into the soil by conduction.

The rate at which heat transfer occurs through a material is expressed by its *thermal diffusivity* (**k**). This is a function of its ability to conduct heat, known as its *thermal conductivity* (*k*), and the amount of heat necessary to cause a temperature change, expressed by its *heat capacity* (C). The relationship of these factors can be shown as: **k** = *k*/C, indicating that the rate of heat transfer is directly proportional to the ability to conduct heat, but is inversely proportional to the heat capacity (Oke 1987). Typical values of these properties are given for different soils in Table 6.1. When saturated, a

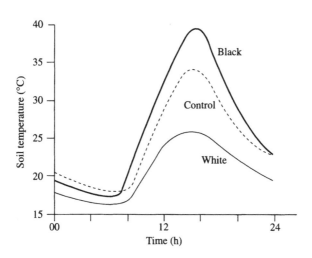

Figure 6.8 Effect of albedo change on near-surface (10 mm) soil temperatures (from Oke 1987). The control is short grass over fine sandy loam

Table 6.1 Thermal properties of soils

Material	Moisture	Thermal conductivity (k) (Wm⁻¹K⁻¹)	Heat capacity (C) (Jm⁻³K⁻¹ × 10⁶)	Thermal diffusivity (k) (m²s⁻¹ × 10⁻⁶)
Sandy soil	Dry	0.30	1.28	0.24
(40% pore space)	Saturated	2.20	2.96	0.74
Clay soil	Dry	0.25	1.42	0.18
(40% pore space)	Saturated	1.58	3.10	0.51
Peat soil	Dry	0.06	0.58	0.10
(80% pore space)	Saturated	0.50	4.02	0.12

Source: Oke 1987

sandy soil therefore has a higher thermal diffusivity than a clay soil, which in turn has a higher value than a peat soil, but these values decrease when the materials are dry (Ghuman and Lal 1985). However, the greatest diffusivity will occur at intermediate moisture contents, because at high contents it is reduced due to the high heat capacity of water. Although a dry peat soil will have a low albedo because of its dark surface, little heat will therefore be transmitted deeper into it and the surface will become very hot. Sensible heat transfer will also cause the overlying air to heat up. Conversely, a moist sandy soil is likely to have a higher albedo, reflecting more incoming radiation, and will also transmit more heat into the ground, therefore its surface temperature will be lower than that of the peat, as will that of the overlying air, but its subsurface temperature will be higher. At night, the net radiation becomes negative due to the loss of outgoing long-wave radiation, and this is balanced by conductive heat supplied from the soil plus turbulent heat from the air (Figure 6.7). Under these conditions the surface of the peat will become colder than that of the sand because the peat is less able to replenish surface heat loss from below due to its lower diffusivity; this can cause a strong temperature inversion in the overlying air. Most soils show greater diurnal and annual fluctuations in surface and overlying air temperatures than at depth (Nullet *et al.* 1990) (Figure 6.9), but soils with low diffusivity will have more extreme fluctuations in temperature at and near the surface than those with a higher diffusivity, and less extreme fluctuations at depth (Williams and Smith 1989). For wet soils the depth over which diurnal and annual temperature fluctuations occur is around 0.5 m and 9.0 m respectively, whereas in dry soils the values are around 0.2 m and 3.0 m (Barry and Chorley 1992).

Soil and near-ground temperatures will also be determined by vegetation cover (Hayhoe *et al.* 1990). This may have a different albedo from the soil and therefore reflect different amounts of incoming short-wave radiation. For example, grassland, heathland and scrub have albedos of around

Plate 9 A rubified latosol.
The photograph shows the
top 4 m of the profile

Plate 10 A regosol, showing a
thin A horizon above an
undifferentiated till parent
material

Plate 11 A rendzina. The mull surface horizon overlies a parent material of chalk rubble (scale in dm)

Plate 12 A gleyed soil. The mottled zone results from periodic saturation, while the underlying blue-grey zone indicates permanent saturation. The photograph shows the top 0.6 m of the profile

Plate 13 An arctic brown soil (scale in dm)

Plate 14 The Valley Farm Soil in southeast England. The rubified paleosol, approximately 1 m thick, lies above fluvial sands and is buried by a thin grey till

Plate 15 Landsat Thematic Mapper false colour composite satellite image (bands 4, 5, 3) of an area approximately 32 × 26 km in the Alpujarra, southern Spain. The distinct boundary running from upper right (northeast) to lower left (southwest) separates the ridge and valley terrain on resistant metamorphic rocks to the north and west from the 'badland' terrain on softer sedimentary rocks to the south and east. Soils in the former area are shallow but relatively stable, and land use is characterised by irrigated agriculture on the lower valley sides and floors (orange-yellow) and needleleaf forestry on the higher slopes (deep red). The 'badland' terrain is dominated by semi-natural shrubland with highly erodible soils (green). The transition within this area from green through blue to purple corresponds with increasingly sparse and lower-growing vegetation, resulting from decreasing water availability; white denotes areas virtually devoid of vegetation. Courtesy of G.J. Park. Original data © ESA 1995, distributed by Eurimage/NRSC, Farnborough

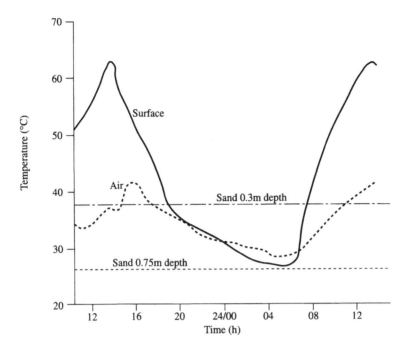

Figure 6.9 Diurnal soil and air temperatures in the central Sahara Desert in mid-August (from Peel 1974)

0.15–0.25, while forests have values of around 0.05–0.20 (Oke 1987). Vegetation also allows heat loss via transpiration in addition to evaporation. The effect of vegetation cover is to lower the diurnal range of soil surface temperatures; these are suppressed during the day, due to radiation reflection and heat absorption by the overlying vegetation, but enhanced at night due to long-wave emission and limited transpiration from the vegetation (Figure 6.10). Similarly, the effect of soil on air temperatures is reduced by the presence of a vegetation cover. Snow cover will also affect soil radiation and temperature conditions. Because of its high albedo (0.95, decreasing to 0.40 with age), large amounts of incoming radiation will be reflected, and because of its low thermal diffusivity, similar to that of dry peat soil, heat exchange occurs at its surface with little heat being transferred to or from the soil. Consequently, the soil is insulated by a snow cover from extreme temperature changes in the air (Ping 1987), and the thicker the snow cover the greater this effect will be (Figure 6.11). Similarly, the effect of the soil on air temperatures will be suppressed under snow covers. Soil structure also influences thermal characteristics of the soil, with structured soils showing greater heat conduction than unstructured soil (Kaune *et al.* 1993).

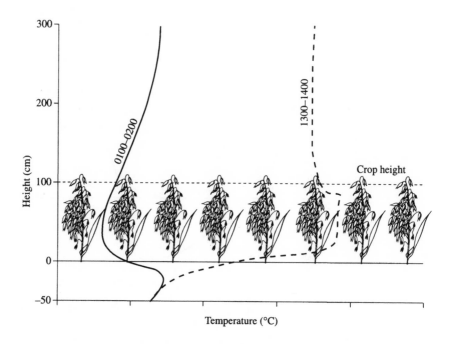

Figure 6.10 Temperature profile in a cereal stand in southern England (from Long *et al.* 1964)

6.3.2 Atmospheric moisture

Water occurring in a liquid form in the soil can influence the moisture of both the soil and above-ground atmosphere in terms of its water vapour characteristics. The amount of moisture which can be held as vapour in the air increases with temperature, and when the air is saturated with water vapour, condensation will therefore occur if cooling takes place. The temperature at which this occurs is known as the *dew point*. Because air in pores of moist soils is in close contact with the water, it is usually close to saturation. However, vapour gradients will develop due to variations in soil temperature with depth. During the day soils are usually warmer towards the surface (Figure 6.9), therefore more water vapour can be held in the pores in the upper part of the soil than in those lower down. The resulting vapour concentration gradient produces a net flow of vapour down into the soil. Conversely, soils cool towards the surface at night which produces a net vapour flow upwards. If the vapour is cooled to its dew point on reaching the soil surface, condensation will occur; this process is known as *distillation* (Oke 1987).

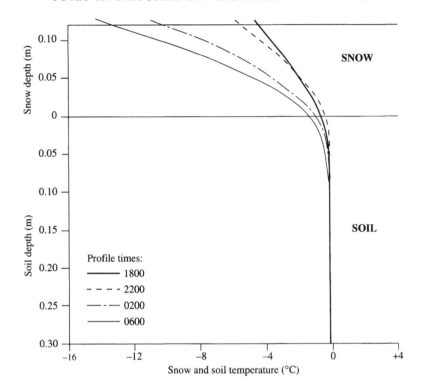

Figure 6.11 Temperatures in soil with a snow cover (from Oke 1987)

Moisture can also be added from the soil to the above-ground atmosphere via evaporation (section 6.2.2). This is greatest where temperatures are high, and is also enhanced by the desiccating effect of air moving across the soil surface, because fresh air is usually associated with lower humidities. Most evaporation occurs from the top few centimetres of soil, and once this water has been lost, further evaporation occurs by water being moved upwards by capillary suction, although this constitutes a relatively minor proportion of soil evaporative losses. The extent of upward capillary movement depends on pore size, small-pored fine-textured soils being more conducive to this process than larger-pored coarser-textured soils (section 6.2.2). Moisture lost from the soil via evaporation will generally increase the vapour pressure of the above-ground atmosphere, the increase being greatest close to the surface and decreasing with increasing altitude. The increase will be greatest during the day when temperatures, and therefore evaporation, are at a maximum. If at night the temperature of the soil surface falls to the dew point, condensation occurs on the surface; this process is known as *dewfall* (Oke

1987). The exchange of moisture from the soil to the above-ground atmosphere will also be influenced by vegetation cover. In addition to direct evaporative losses, moisture will also be lost to the atmosphere via transpiration, and within the vegetation layer itself the reduction in turbulent transfer allows vapour to accumulate close to the soil surface (Oke 1987).

6.3.3 Air movement

Soils can influence the movement of air above the ground in two main ways. Air movement occurs when it is heated which causes it to rise, a process known as *free convection*, and variations in soil surface types will cause different amounts of heating and therefore different extents of thermal uplift. For example, low albedo soils usually have higher surface temperatures during the day than adjacent higher albedo types, and areas of dry soil attain higher surface temperatures than adjacent wetter areas because of differences in thermal diffusivity (section 6.3.1). Because of such lateral variations in heating, the most favourable areas will experience greatest thermal uplift, which causes convectional circulation cells to develop, thus mixing the air and sometimes leading to the formation of cloud by condensation of the upward moving air (Figure 6.12). This mixing reduces vertical differences in temperature, humidity and wind speed, and if surface heating and convectional uplift are strong, the mixing effect allows faster-moving upper air layers to be brought down nearer the surface, thus increasing wind velocities close to the surface (Oke 1987). During the night a decrease in temperatures is accompanied by a decline in convectional activity, and therefore wind velocities near the surface also decrease.

The second way in which soil surfaces can affect air movement is by them causing horizontally moving air to be set in turbulent motion, a process known as *forced convection*. This is a smaller-scale process than the previous one, and the extent to which it occurs will depend on the aerodynamic roughness of the surface. This can be expressed as *roughness length* and will be determined by factors such as the size and shape of aggregates if the soil surface is unvegetated, or by the type of vegetation cover. For example, a sandy desert soil surface has a roughness length of 0.0003 m, while soils more typically have values of 0.001–0.01 m; in contrast, short grassland has values of 0.003–0.01 m, while taller grassland has values of 0.04–0.10 m and forests have values of 1.0–6.0 m (Oke 1987). Surface roughness will increase turbulence, although vegetation will reduce air movement within it; for example, forests can reduce wind speeds by over 90 percent beneath the canopy (Barry and Chorley 1992).

Air movement within a soil occurs by *diffusion* whereby oxygen diffuses into the soil and carbon dioxide diffuses out. It can also occur by *mass flow*, in which gas molecules move from an area of higher to lower pressure (Singer and Munns 1991). This can occur for a number of reasons. For example,

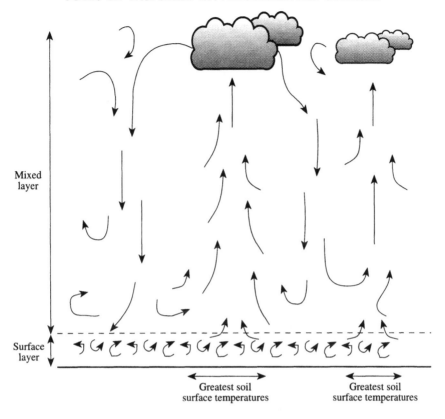

Figure 6.12 Airflows in the daytime surface layer and mixed layer (from Oke 1987)

turbulence can produce pressure differences which are equalised by air movements, some from above the ground and some from within the soil. Pressure differences can also occur within the soil due to diurnal temperature variations with depth (section 6.3.1). During daytime surface heating, warm air will rise from the upper part of the soil and this will be replaced by cooler air from below. Conversely, cooling of the surface during the night will cause relatively dense air to sink into the soil. In contrast, diurnal variations in vapour concentration will cause downward movement of air during the day and upward movement at night (section 6.3.2). Plant roots can also extract water from pores, causing it to be replaced by air. Air movement can be restricted where the topsoil is wet or frozen, causing increased levels of carbon dioxide (Magnusson 1992). Slower gaseous exchange with the above-ground atmosphere at depth can also lead to higher carbon dioxide concentrations than near the surface (Fernandez *et al.* 1993). Carbon dioxide concentrations also vary seasonally, with greater quantities being evolved during higher biotic activity at higher temperatures (Fernandez and Kosian 1987, Kiefer and Amey 1992).

6.4 THE GEOSPHERE

A variety of processes operate in the formation of a landscape, as shown in the hypothetical 9-unit landsurface model (Dalrymple *et al.* 1968, Conacher and Dalrymple 1977), in which a slope profile is divided into 9 units, each characterised by a set of pedological and geomorphological processes (Figure 6.13). The interfluve (unit 1) is dominated by vertical movement within the soil, while the upper slope (units 2 and 3) shows both vertical and lateral soil

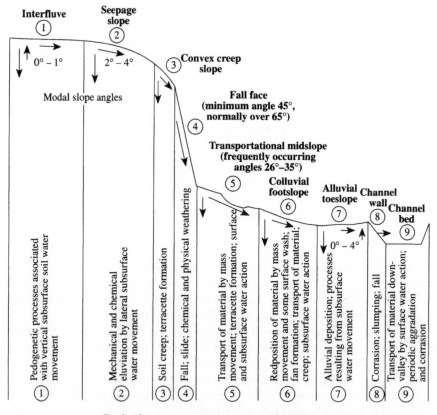

Figure 6.13 Hypothetical 9-unit landsurface model (from Dalrymple *et al.* 1968). Arrows indicate direction and relative intensity of movement of weathered rock and soil materials by dominant geomorphic processes

movement. The free face (unit 4) is characterised by weathering and rapid mass movement, and the midslope (unit 5) represents an area of downslope transportation, involving both mass movement and water flow. Unit 6 marks the zone of net deposition on the footslope while unit 7 is an area of deposition of material from both upslope and upvalley, via the river channel. Units 8 and 9 are dominated by river channel processes operating down-valley. In this model, soils are therefore important in three principal ways – they influence the nature and intensity of weathering, surface and subsurface sediment and solute transport, and mass movement, and each of these aspects will now be examined in turn.

6.4.1 Weathering

Soils are an important factor in the supply of moisture which is required for many weathering processes. For example, thick soils, or those of a fine texture or on gentle slopes, generally hold more moisture and are often associated with greater extents of bedrock weathering (Daniels and Hammer 1992). However, during its development a soil may reach a critical thickness beyond which little bedrock weathering occurs because the soil acts as a protective mantle rather than as a catalyst to weathering (Gerrard 1981). The variation in water availability can also be important in the case of weathering processes such as wetting and drying and salt crystal growth. Soils with a fluctuating moisture regime are therefore likely to allow more effective weathering of the underlying bedrock in these cases than continuously dry or wet soils. Soil thickness can also be important in terms of protecting the bedrock from fluctuating temperatures which can cause freeze–thaw weathering. Although climate will obviously play an important role in determining soil moisture and temperature characteristics, properties of the soil itself can also have a marked effect (sections 6.2 and 6.3).

The chemical composition of water can influence its weathering effectiveness, and this can be determined by the soil through which the water has passed. Of particular importance in this respect is the pH of the soil and the speed at which water moves through it. For example, water percolating through a soil of very high or low pH will generally have a greater weathering capability than that passing through a more neutral soil (Figure 3.4). Also, a fine-textured soil will hold the water for a longer period than a coarse-textured soil, and therefore allow a longer reaction time over which weathering can occur.

6.4.2 Sediment and solute transport

The transport of sediment as part of the denudation process can occur via the soil surface or subsurface, while solute transport tends to be confined within the soil. Surface sediment transport operates by aeolian processes, or by surface water movement; these processes have already been considered in the context of losses during soil formation (section 3.2.4), but soils can also influence the way in which these processes operate. For example, soils with a high silt or fine sand content are most susceptible to wind erosion, which is also enhanced by low moisture contents and sparse vegetation covers as occur in arid and semi-arid regions (Warren 1979, Selby 1993). Soils themselves can also provide a source of material for aeolian abrasion of exposed bedrock. The relative hardness of these materials will determine the effectiveness of abrasion, and since soils often contain many minerals with hardness values in the range 5–7 on the Mohs scale, these can be effective abrasives of relatively soft rock types such as shales and limestones (Gerrard 1988).

Soils influence the erosion of sediment by water via their control on runoff. Low permeability soils will encourage rapid runoff, which can take the form of overland flow or throughflow (section 6.2.3). Overland flow is encouraged by soils with low permeability surfaces, particularly where vegetation cover is low so that interception of precipitation is limited and infiltration rates are therefore more likely to be exceeded by precipitation rates. A limited vegetation cover will also expose aggregates at the soil surface to the direct impact of raindrops, which may cause disaggregation, resulting in blocking of the surface pores and therefore a decrease in permeability (Ekwue 1990, Daniels and Hammer 1992). Shallow soils will also encourage overland flow because of their limited water storage capacity. Unconcentrated flow or *sheetflow* will only occur on smooth surfaces, but on most surfaces their roughness causes the flow to be concentrated in the form of small ephemeral channels or *rills*, particularly on silty or clayey soils (Gerrard 1981). If these channels become sufficiently large to stabilise, gullies may develop which will then act as locations for concentrated erosion. Silt and fine sand will generally be most easily transported by surface wash, because clays and colloidal organic matter will resist movement due to their cohesiveness, while sand and gravel particles will be more difficult to move because of their greater weight (Thornes 1979, Martz 1992). However, in soils with a high sodium salt content, clays often become easily dispersed on wetting and drying, and this can encourage erosion and gullying, for example in southern Africa (Watson 1992).

In the case of subsurface sediment and solute movement, the hydraulic conductivity of a soil will exert an important control on the ease with which movement can occur (section 6.2.1). The potential for sediment movement will also depend on the size of material available for transport, the extent of its aggregation and the size of the soil pores, while solute movement will also be determined by the ease of dissolution of the soil components and the acidity of the drainage waters; in some cases, soils may inhibit solute loss by buffering the acidity, or by providing exchange sites which can remove dissolved ions from the drainage waters (section 2.4.2).

As in the case of surface runoff, material may be removed from a slope by subsurface processes in either an unconfined or a concentrated manner. Unconfined movement can occur throughout the soil as a whole, whereas concentrated movement operates in zones which develop in soils often in response to topographic conditions, forming patterns similar to those associated with surface runoff, although usually with more diffuse boundaries. The smaller features are sometimes known as *percolines*, as distinct from the larger and more easily recognisable *seepage lines* (Gerrard 1981). These zones can be important in the movement of solutes and fine particulate material. The removal of larger material can lead to the development of subsurface pipes (Figure 6.14), particularly in soils which are susceptible to vertical cracking, which allows water to penetrate rapidly into the subsoil; organic soils and those rich in expanding lattice clays are particularly prone to this condition. Piping will also be encouraged in soils which have an

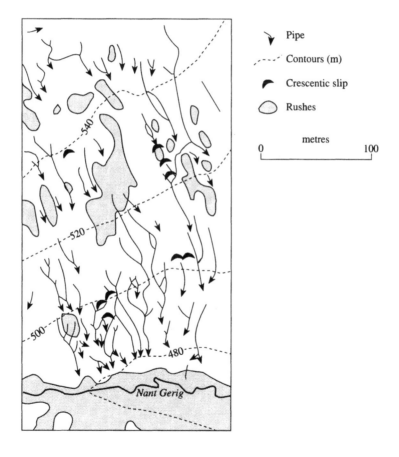

Figure 6.14 A pipe network in central Wales (from Gilman and Newson 1980)

abrupt decrease in porosity at some point below the surface, as this will inhibit downward water percolation and promote lateral subsurface flow (Jones 1981, McCaig 1985). Pipes can range from a few centimetres to several metres in diameter, and can therefore be responsible for transporting large quantities of water down hillslopes, along with sediments and solutes (Jones 1987, Walsh and Howells 1988). If a pipe develops to a size which exceeds that capable of being supported by the overlying soil, the roof will collapse, exposing the channel to the surface. A similar situation may result from the extension of vertical cracking down to the level at which the pipes occur. These can be important mechanisms in the formation of gullies, especially in arid and semi-arid environments (Gerrard 1981) and also in some peat soils.

Soils can influence landforms by their resistance to sediment and solute transport. This is particularly the case in low latitude regions where soils have become indurated by the formation of pans (sections 3.3.1 and 4.3.3). In these instances they can behave like resistant rock types and a number of geomorphological features can result from erosion (Goudie 1985). For example, indurated layers, or *duricrusts*, can form cappings over softer materials, producing flat-topped interfluves, benches along valley sides or pavements on valley floors; they can also form escarpment features if inclined at an angle to the horizontal, or irregular relief due to differential erosion (Churchward and Gunn 1983).

Studies of sediment and solute transport can be made at a variety of scales. For example, within a river catchment in mid-Wales, Reynolds *et al.* (1987) reported that physical denudation of the catchment was occurring at a rate of 3.5 mm per 1,000 years, based on measurements of river suspended sediment and bedload, while input–output budgets for solutes indicated that chemical denudation was operating at a rate of 2.9 mm per 1,000 years. At a smaller scale, rates of sediment and solute transport can be measured by hillslope plot studies; for example, material eroded from a known area can be collected as it moves downslope (e.g. Burt and Trudgill 1985, Morgan 1986), or erosion rates can be calculated from the redistribution patterns of radioisotopes within the soil (e.g. Pennock and de Jong 1990, Walling and Quine 1991).

6.4.3 Mass movement

The influence of soils on mass movement depends largely on its mechanical and hydrological properties. Of particular importance are cohesion and friction, which can be described by the Mohr-Coulomb equation:

$$\tau = c + \sigma_n \tan \phi$$

where τ = shear stress at failure, c = cohesion, σ_n = normal stress on the shear plane and ϕ = angle of internal friction (Whalley 1976). The extent of cohesiveness and friction depends mainly on soil texture and moisture

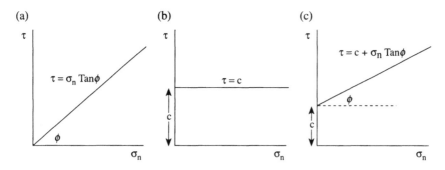

Figure 6.15 Shear stress–normal stress diagrams for three types of soil condition:
(a) c = 0, (b) ϕ = 0, (c) ϕ and c both present (from Whalley 1976)

content. In sandy soils there is little cohesion, so the shear stress at failure will depend on the normal stress and angle of internal friction (Figure 6.15a). Conversely, in clayey soils cohesion will be high but friction low, therefore the shear stress at failure will depend on cohesion (Figure 6.15b), while most soils have a combination of both friction and cohesion (Figure 6.15c). However, moisture content will also affect stability because of the effect of the normal stress on pore water pressure. By taking this into consideration, the previous equation can be rewritten as:

$$\tau = c + (\sigma_n - u) \tan \phi$$

or:

$$\tau = c' + \sigma'_n \tan \phi'$$

where u = pore water pressure, and c' and ϕ' are the cohesion and friction angle with respect to the effective stress. If pore water is therefore compressed by a soil on a slope, it will come under pressure and the value of σ'_n will decrease, thus reducing the shear strength (Whalley 1976). Conversely, negative pore water pressure, or suction (section 2.3.4), can increase soil strength; for example, damp sandy soil in which capillary forces are acting will be more stable than dry sandy soil in which there will be no such effect.

Mass movement comprises three main sets of processes – heave, flow and slide – and various types of movement can be identified according to the relative combination of these processes (Figure 6.16). Transfer by rivers, rockslide and talus creep are not strictly relevant in the context of soil influences, but soils can have important effects on the remaining processes. Soil creep can occur in one of two forms – *continuous creep* and *seasonal creep*. Continuous creep is mainly confined to materials with low shear stress values such as clay-rich soils. Movement is usually most rapid near the surface and decreases progressively with depth. Gerrard (1981) reports movement to depths of 10 m and rates of movement up to 22 cm a^{-1} in the upper layers. Seasonal creep is more common and results from forces which cause variations in shear stress such as expansion and contraction caused by wetting and drying or freeze–thaw, plus disturbances due to bioturbation (section 3.2.3) (Finlayson 1985). Movement is often confined to the top metre of soil in which these processes are most marked, and can occur in any direction. Rates of downslope movement are therefore difficult to establish, although values in the order of 0.5–2.0 mm a^{-1} have been suggested for humid temperate regions and up to 15 mm a^{-1} for temperate continental climates (Selby 1993). Soil creep can produce small step-like features known as *terracettes*, which run roughly parallel with the contour. The origin of these features has been the subject of some debate, and suggested alternative mechanisms to soil creep have included micro-scale slumping and trampling by animals (Gerrard 1981).

Gelifluction (solifluction) occurs by movement of material over a frozen subsoil. When ice lenses in the upper part of a soil melt, pore water pressures are increased if the water is unable to drain away because of the frozen subsoil, and this can reduce the shear stress at failure of the material, as described above, causing it to become unstable and move downslope (French 1988). Expansion and contraction by freeze–thaw may form an additional component of the

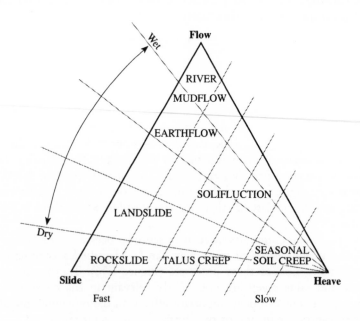

Figure 6.16 Classification of mass movement processes (from Carson and Kirkby 1972)

Figure 6.17 Gelifluction lobes downslope of a late-lying snowpatch

movement, as in the case of soil creep, as may the process of *frost creep* (Lewkowicz 1988). The latter has been suggested to occur when particles on a sloping soil surface are displaced upwards at right angles to the slope when needle ice forms, and then on melting are returned to the surface, under gravity, via a more vertical path. Repeated freezing and thawing therefore produces a zig-zag movement of particles down the slope in the vertical plane. However, the use of the term *creep* to describe this process has been criticised on the grounds that it has nothing to do with deformation of material in relation to shear stress, and its operation within a soil as a whole, as opposed to simply the surface particles, has also been questioned (Williams and Smith 1989). Silty soils are often most susceptible to gelifluction as these are prone to large quantities of segregation ice development so that large volumes of water will be released into the soil on melting (Williams and Smith 1989). Geliflucted material can occur as a featureless sheet if moving at a similar rate over an entire slope, or as terraces or lobes if moving at different rates or confined to particular parts of a slope. In the latter case gelifluction may be confined to areas downslope of late-lying snow patches which can increase the water content of the soil and therefore contribute to its instability (Figure 6.17). Rates of gelifluction vary widely, but are generally greater than those of seasonal soil creep, being typically in the order of 0.5–10 cm a^{-1} (French 1988), while depths of operation will depend on the extent of thawing.

185

Figure 6.18 Landslides resulting from heavy rainstorms on a slope with low-permeability podzolic soils, central Norway

Landslides, earthflows and mudflows are generally more rapid movements than gelifluction and soil creep, and usually result from instability caused by the reduction of shear stress at failure by high pore water pressures. They are therefore often associated with clay-rich soils with low angles of internal friction, or with soils which receive sudden influxes of large volumes of water via heavy rainfall; if removal of the water is impeded by an impermeable bedrock layer beneath the soil this will enhance instability (Carling 1986). Soils with pans or other forms of low-permeability illuvial horizon can also act in a similar way, as for example in the case of podzol Bs horizons (Figure 6.18) (Ellis and Richards 1985). Rates of rapid mass movement vary enormously from 1 cm day^{-1} to over 1 m s^{-1} (Daniels and Hammer 1992). Well-developed root networks in soils can increase the stability of slopes which are prone to slippage (Terwilliger and Waldron 1990).

In addition to the forms described above, mechanical transfer processes operating within the soil itself can produce small-scale microrelief and patterned features. For example, frost-susceptible soils can produce earth hummocks and sorted patterns in cold environments (sections 3.2.3 and 4.3.1), while Vertisols can also produce mounds and hollows (gilgai) (section 4.3.3).

6.5 THE BIOSPHERE

Soils are influenced to a great extent by biota in terms of pedogenic processes (section 4.2.2). However, this is by no means a one-way relationship – soils can have an important effect on biota in terms of the extent to which they encourage or constrain habitation. The principal influences of soil on biota are via nutrients, moisture and aeration, although a number of other controls, both chemical and physical, are also important. For ease of discussion these factors will be examined individually, although it should of course be recognised that in reality many of the factors operate in combination.

6.5.1 Nutrients

The nutrients required for plant growth fall into two types – the *macro-nutrients*, which are used in relatively large quantities, and *micronutrients* or *trace elements*, used in very small amounts (Table 6.2). Plant growth can be retarded if these occur in insufficient quantities, are not balanced by other nutrients or do not become available sufficiently quickly (Brady 1990). Micronutrients are required by plants in only small amounts, but they are no less important than macronutrients. While many nutrients are derived from the weathered mineral soil, organic matter can also be an important source, particularly in the case of nitrogen, phosphorus and sulphur, nutrients which

Table 6.2 Typical concentrations of essential nutrients in mineral soils, annual plant uptake and ratio of content to uptake

Nutrient	Content (% by weight)	Annual plant uptake (kg ha^{-1})	Ratio of content in 10 cm layer to annual uptake
Macronutrients			
Calcium	1	50	260
Potassium	1	30	430
Nitrogen	0.1	30	50
Phosphorus	0.08	7	150
Magnesium	0.6	4	2,000
Sulphur	0.05	2	320
Micronutrients			
Iron	4.0	0.5	100,000
Manganese	0.08	0.4	3,000
Zinc	0.005	0.3	2,000
Copper	0.002	0.1	1,000
Chlorine	0.01	0.06	200
Boron	0.001	0.03	400
Molybdenum	0.0003	0.003	1,000

Source: Foth and Ellis 1988

187

Table 6.3 Release of nutrient reserves in soils

Soil forms (reserve)	Release process	Solute forms
Nitrogen		
Organic matter	Microbial decay	NH_4^+
Interlayer NH_4^+ in mica and vermiculite	Weathering	NH_4^+
Exchangeable ammonium	Cation exchange	NH_4^+
Sulphur		
Organic matter	Microbial decay	S^{2-}, SO_4^{2-}
Sulphides (e.g. FeS) and sulphur in anaerobic soils	Microbial oxidation	SO_4^{2-}
Gypsum	Dissolution	SO_4^{2-}
Phosphorus		
Organic matter	Microbial decay	$H_2PO_4^-$
Attached to Fe and Al oxides and $CaCO_3$	Ligand exchange and dissolution	and HPO_4^{2-}
Potassium, Calcium, Magnesium		
Silicate minerals	Weathering (dissolution)	K^+
Exchangeable cations	Exchange	Ca^{2+}
Carbonates and sulphates	Dissolution	Mg^{2+}
Iron, Manganese, Zinc, Copper		
Precipitated as hydroxides	Dissolution	
Adsorbed on Fe, Al and Mn oxides	Desorption	Cations and
Bound on humus (chelated)	Dissociation	soluble chelates
Exchangeable	Cation exchange	
Boron, Molybdenum		
Adsorbed on Fe and Al oxides and other clay minerals	Desorption	H_3BO_3 MoO_4^{2-}

Source: Singer and Munns 1991

are often in short supply (Table 6.3). Nutrient content will therefore depend on the type of mineral and organic components which make up a soil. For example, potassium, calcium and magnesium will be plentiful in soils rich in feldspars, mica and hornblende, while nitrogen will occur in higher quantities in organic material rich in proteins and amino acids. Sulphur can occur in minerals such as pyrite and gypsum, in addition to organic matter, and phosphorus can occur in the mineral apatite as well as in nucleic acid and other organic forms. Consequently, nutrient-rich soils are generally associated with base-rich parent materials such as limestone, marl, basalt and some sandstones, and also with grassland or broadleaf trees; soils developed from parent materials such as granite, gneiss and quartz-rich sandstones, or those supporting heathland or needleleaf trees are usually nutrient-deficient.

However, nutrient availability can be equally important as nutrient content. For example, nutrients may be easily lost by leaching in a coarse-textured soil, even if developed from a nutrient-rich parent material. Nutrient availability is also determined by the quantity held in an exchangeable form on clay and organic colloids, from which they can be removed by roots. For example, calcium is normally held in this form in much larger quantities than potassium and magnesium, but consequently it is also more prone to loss by leaching, especially in freely draining soils. In the case of nitrogen, phosphorus and sulphur, their availability will also depend not only on the type of organic matter but on its rate of decomposition; if this is slow, due for example to low temperatures or poor drainage, these elements may not be released in sufficient quantities for certain plants.

Nutrient deficiency can be manifested in plants in four main ways. These are *chlorosis* (yellowing of leaves due to a lack of chlorophyll), *purpling* (darkening of leaves due to the accumulation of anthocyanin pigments), *local necrosis* (death of tissue) and *stunting* (reduced growth) (Singer and Munns 1991). Chlorosis can result, for example, from a deficiency in nitrogen, phosphorus, magnesium, iron, zinc or copper, while purpling is often associated with phosphorus deficiency. Local necrosis can result from potassium or molybdenum deficiency, and stunting of leaf growth can occur due to zinc deficiency.

Nutrients also exert an important influence on soil mesofauna and microorganisms. In terms of mesofauna, the principal source of nutrients is organic matter occurring as either living or dead plant or animal material. In the case of herbivorous fauna, the nutrient status of the plant-derived organic matter will determine population levels, with higher levels usually present in soils with high nitrogen content organic matter, such as those associated with many grassland and broadleaf tree types. Carnivorous mesofauna will occur in greatest quantities in soils which have large populations of animals on which they prey, and these are therefore usually the same types of soil. For soil micro-organisms, carbon and nitrogen are the nutrients which are most commonly deficient, and various adaptations have been developed in order to overcome this problem; these include the fixation of atmospheric nitrogen, the ability to survive long periods without nutrition, and the release of antibiotics which inhibit the growth of competitors (Singer and Munns 1991). Another adaptation is autotrophy, or by the oxidation of ammonia, sulphur or sulphide (section 2.2.2).

6.5.2 Moisture and aeration

Soil moisture is required for the growth of most plants, and the moisture content of a soil is determined to a large extent by drainage conditions (Nisbet *et al.* 1989, Wright *et al.* 1990). For example, a coarse-textured soil, or one with a well-developed open structure, will usually drain more easily than a finer-textured or compacted soil because of its greater hydraulic conductivity (sections 6.2.1 and 6.2.2), and will therefore be more prone to drought. However, clay-rich soils such as Vertisols, which contain expanding lattice minerals, may also have low moisture contents because of large vertical shrinkage cracks which form during periods of dry weather, causing rapid infiltration and percolation through the upper part of the profile. Similarly, shallow soils or those on steep slopes often have low moisture contents. In arid environments, slope position is particularly important, with footslope soils usually providing the best water supply (Smettan *et al.* 1993).

The availability of moisture to plants is, however, determined not only by the moisture content of a soil, but also by the suction at which the water is held (Figure 6.3). For a given moisture content, the forces holding water in the soil, which must be overcome if a plant is to obtain water, are lowest for coarse-textured soils which have relatively large pores, and greatest for fine-textured soils in which the matric suction effects of the small pores are much higher (section 6.2.1). For any given texture, the forces also decrease as moisture content increases; Figure 6.3 shows that sand has the least water available to plants (only 7 percent) while silt loam has the most (16 percent). In order to survive in moisture-deficient soils, plants have developed a variety of adaptations, for example deep and closely spaced roots to maximise water acquisition, and narrow, waxy or hairy leaves to minimise water loss from transpiration. In addition to plant uptake, moisture is required to allow penetration of roots and burrowing organisms; when soils are dry they often become hard and compact, thus restricting these activities. In contrast to plants and mesofauna, micro-organisms do not generally experience problems in dry soils, and indeed some types, especially spore-forming bacteria, can withstand drought conditions for many years (Davis *et al.* 1992), although bacterial growth rates usually increase when dry soils become moist (Hartel and Alexander 1987).

Excessive moisture can cause different, but equally severe problems for biota. Soils can be prone to excessive wetness for a variety of reasons; for example, heavy rainfall, low permeability or poor drainage due to topographic location. Under such conditions, most pores are filled with water and therefore the root zone becomes anaerobic, which can have serious effects on the metabolism and growth of plants not adapted to such conditions, even if the conditions last for only a few days. Root systems of these plants require oxygen partial pressure of at least 0.02 bar, with nutrient uptake increasing as partial pressure increases, up to a value of around 0.2 bar (Singer and Munns 1991). When root cells become affected by a lack of oxygen, they start to convert glucose into ethanol (C_2H_5OH), causing cell membranes to become prone to leakage, and ion and water uptake to be impaired (White 1987). Consequently wilting can occur, accompanied by chlorosis (section 6.5.1). Also under waterlogged conditions, anaerobic bacteria produce ethylene (C_2H_4), which can inhibit root growth at concentrations of 1 ppm or more, and volatile fatty acids can also be produced during carbohydrate decomposition, which can accumulate to harmful concentrations (White 1987).

Anaerobic micro-organisms can cause denitrification, reducing nitrate to nitrites, or in more extreme cases to nitrous oxide and nitrogen, therefore causing the loss of this important nutrient. An additional problem with this reaction is that an accumulation of nitrite nitrogen can be toxic to plants. In poorly drained gley soils, manganese and iron become reduced, in which form they are readily soluble and can be taken up by plants in toxic concentrations. Soils which experience periodic or prolonged flooding can pose an additional problem to plants, but those adapted to such conditions are able to transfer oxygen rapidly to their roots via spongy stem tissue, thus maintaining an aerated rhizosphere.

Soil micro-organisms are also inhibited by a restricted oxygen flow, as can occur in clayey soils (Higashida and Nishimune 1988), although certain groups are adapted to anaerobic conditions, but their growth is usually slower than those found in aerobic soils. Soil meso- and macrofauna, however, experience problems in extremely wet soils due to a lack of oxygen and instability of burrowing passages; in such soils the fauna are therefore concentrated near the surface, with few, if any, burrowing animals.

6.5.3 Additional factors

Although nutrients, moisture and aeration are important influences on most biota, a number of other factors, both chemical and physical, can exert an influence on these properties and also act as additional influences in their own right. The chemical factors relate principally to soil pH and toxins, while the physical factors include temperature, soil depth and ground stability, and soil pore size, compaction and induration. Soil pH can affect biota both directly and indirectly via its influence on nutrient availability and toxicity. The direct influence is often not great because most plants can tolerate a wide range of pH values as long as nutrients occur in sufficient quantity, although at very low pH, hydrogen ions become toxic to plants. However, the effect of pH on nutrient availability and toxicity can be major. For example, iron, manganese and zinc are less readily available at alkaline than acid pH values, while calcium and molybdenum availability is greater at higher pH values (Figure 6.19). Phosphorus is often in short supply because it occurs in forms which are not readily soluble, but it is most easily extracted by plants at around pH 6.5. At pH values below about 5.0, aluminium, iron and manganese can become soluble to an extent which can make them toxic to certain plants. For example, aluminium toxicity inhibits cell division in root tips, causing them to become stunted, while manganese toxicity produces leaf distortion and yellowing. Low pH is also associated with calcium deficiency, which can cause leaves to wilt and collapse. Conversely, at very high pH values, bicarbonate ions can occur in concentrations which reduce nutrient uptake. Carbonates can also accumulate as precipitates around roots, which inhibits water and nutrient uptake. The type

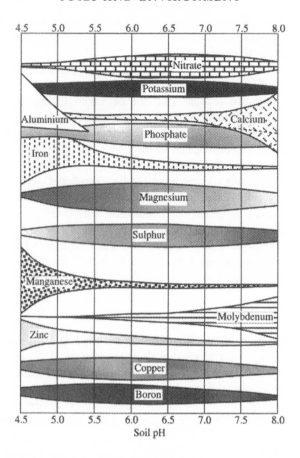

Figure 6.19 Relationship between soil pH and plant nutrient availability in mineral soils (from Foth and Ellis 1988)

of organisms inhabiting a soil can also be influenced by pH. For example, earthworms, bacteria and actinomycetes prefer neutral to slightly alkaline soil conditions, whereas fungi generally require more acid conditions (e.g. Boettcher and Kalisz 1991). There are exceptions, however; iron- and sulphur-oxidising bacteria can occur in extremely acid conditions, with pH values as low as 1 or 2 (Davis *et al.* 1992).

Some substances can also occur at toxic levels independently of pH conditions. Perhaps the most important of these are sodium salts, which increase the osmotic gradient that plant roots have to overcome in order to absorb water. If this gradient cannot be overcome, *plasmolysis* will occur in which movement of water out of plant cells towards the salt solution will cause the cells to collapse. Soils with high sodium concentrations, such as those occurring in arid or coastal areas, therefore support plant communities which are specially adapted to high osmotic pressures (Franco-Vizcaino *et al.* 1993). Other naturally occurring toxins include copper, chromium, arsenic and mercury, although these are normally found at toxic concentrations only in soils developed from metalliferous parent materials, or close to sources of industrial contamination (section 8.3.4). Toxins derived from vegetation can lead to lower populations of soil fauna, as in the case of earthworms (Boettcher and Kalisz 1991). Micro-organisms are also susceptible to natural toxins, although they are probably more tolerant of salinity than most plants (Singer and Munns 1991).

Soil temperature will have an important influence on vegetation in terms of seed germination and growth, and the growth of roots (Cutforth *et al.* 1986, Kaspar and Bland 1992), with both water and nutrient uptake increasing with increasing temperature up to an optimum value. For temperate species, optimum growth occurs at soil temperatures of around 20°C, while for tropical species it occurs at 30°C or more. Soils which warm up rapidly will therefore encourage growth to commence early. Rapid warming will be favoured by high thermal diffusivity; low diffusivities will encourage surface heating but inhibit heat penetration of the subsurface layers (section 6.3.1). If soil temperatures fall below freezing point this may be injurious to certain types of plant, because the expansion which accompanies freezing can damage organic tissue. This problem is reduced if a soil is protected from sub-zero air temperatures by a cover of snow (Figure 6.11), although very thick snow covers can both delay soil warming and cause physical damage to vegetation by compression. Soil temperature can also affect mesofauna and micro-organisms. Many mesofauna are susceptible to frost and also to drought conditions often associated with high soil temperatures and evapotranspiration rates. Many micro-organisms have an optimum temperature range of around 20–35°C, but some specialised groups have much lower (5–10°C) or higher (45–50°C) optimum temperatures (Davis *et al.* 1992). Soil temperature will therefore affect not only the populations of micro-organism groups, but also the rate at which processes such as organic matter decomposition and nitrogen fixation occur (Ross 1989). At extreme temperatures caused by fire, combustion of vegetation and organic accumulations in the soil can produce a marked increase in nutrient content, and therefore in soil fertility, although this can quickly decline due to subsequent rapid leaching and erosion (Kutiel and Inbar 1993).

Shallow soils or those on steep slopes can limit moisture availability (section 6.5.2), but can also limit vegetation growth in extreme cases due to restricted root penetration or instability of the ground. Plants adapted to such conditions are therefore usually not only drought-resistant but also shallow-rooted and fast-growing. However, shallow soils do not always limit root depth; in some cases roots can penetrate great distances into the underlying bedrock (Kalisz *et al.* 1987). Shallow soils or those on steep slopes may also discourage habitation by burrowing macrofauna because of the potential instability of burrowing passages. Instability on more gentle surfaces can also limit plant growth, as in the case of the removal of topsoil by wind or water erosion, or mechanical disturbances such as cryoturbation.

Soil pore size and interconnectivity, and soil compaction and induration, in addition to their effects on soil moisture and aeration, will influence plant growth in terms of the ease of root penetration through the soil (Hasegawa and Sato 1987, Hatano *et al.* 1988). Many roots exceed 60 μm in diameter, but in fine-textured soils many of the pores are much smaller than this (Figure 6.20), therefore these will be impenetrable; root penetration is therefore confined to the larger spaces between individual particles and aggregates (Russell 1977, Ross 1989). Soils with compacted or indurated layers will similarly restrict root extension. This can be seen for example in the case of iron pans, which can cause shallow tree root systems to develop above them, leading to instability in strong winds.

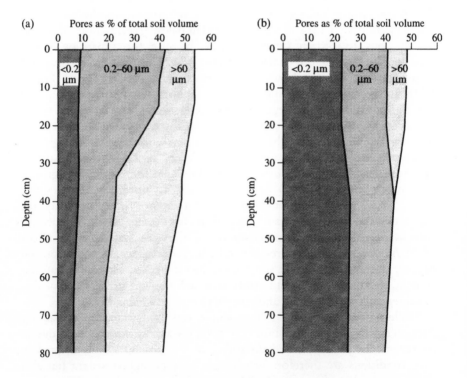

Figure 6.20 Percentage of total soil volume occupied by different sized pores in (a) sandy loam and (b) clay loam soils (from Russell 1977)

6.6 SUMMARY

This chapter has shown that soils can exert a wide variety of important influences on natural environmental systems. In terms of the hydrosphere, soils influence the movement of water within them via three types of force – capillarity and adsorption, whose combined effect is known as matric suction, and osmosis. These forces, which are controlled by pore size, colloidal surface attraction and differential solution concentrations respectively, influence both the rate and direction of water movement. The infiltration of water into a soil will depend on factors such as matric suction, pore size, nature of the litter layer, extent of freezing and vertical crack development. The loss of water from a soil by evaporation will be controlled by climatic factors, but also by soil moisture content, texture, colour and vegetation cover. The storage of water can occur in soils of low permeability where relief prevents lateral drainage, but on sloping surfaces water flow will occur. Overland flow is associated with restricted surface permeability, while throughflow occurs as a result of impeded infiltration close to the surface, which can lead to the development of piping. Ground-water flow occurs within the saturated zone, below the water table, often at depths below the soil layer, and usually operates more slowly than overland flow and throughflow. The relative proportions of these flow types determine hydrograph shape, with the most peaked hydrographs being associated with low infiltration soils.

Atmospheric systems are influenced by soils at a micro scale in terms of solar radiation partitioning, temperature, moisture and air movement. The albedo of a soil surface will influence the amount of incoming radiation reflected from the ground, and low albedos are associated with higher temperatures at or close to the surface. The thermal conductivity and heat capacity of a soil will influence the rate of heat transfer through it, and therefore the extent to which heat is distributed throughout the soil. Temperatures in and above the soil are also controlled by vegetation and snow covers. The concentration of water vapour held in soil air will vary on a diurnal basis due to temperature variations, and this leads to vapour flow from zones of high to lower concentration. Soil moisture can also be added to the atmosphere via evaporation. Soils influence the movement of air both above the ground and within the soil itself. Above-ground effects relate to free convection, associated with soil surface temperature, and forced convection, relating to surface roughness. Air movement within the soil occurs by diffusion and mass flow, the latter relating to pressure, temperature or vapour concentration variations with depth, or to the replacement by air of water taken up by plant roots.

Soils influence the geosphere via their role in geomorphic processes. In terms of weathering processes, soil thickness, moisture content, acidity and hydraulic conductivity are particularly important. Sediment and solute transport can be influenced by soils in a variety of ways. For example, soil texture will control the ease of wind erosion, while the effectiveness of associated abrasion will be determined by mineral hardness. The surface transport of sediment by water will depend on the infiltration characteristics of a soil surface, soils with low infiltration rates being more susceptible to erosion by sheetflow, or by rill or gully development. The subsurface transport of sediment and solutes is influenced by soil texture, pore size, solubility and hydraulic conductivity. Movement can occur either throughout the soil in an unconfined manner, or in discrete zones which take the form of percolines, seepage lines or pipes. Where soils are resistant to sediment and solute transport, they can form erosion surfaces in the landscape. Mass movement occurs by three main processes – heave, flow and slide – which combine to produce a variety of types of movement, including soil creep, gelifluction, landslides, earthflows and mudslides, all of which are influenced by soils via their mechanical and hydrological characteristics, particularly with respect to cohesion and friction.

The principal influences of soils on the biosphere relate to nutrient supply and availability, moisture and aeration, and additional chemical and physical factors such as pH, toxins, temperature, soil depth and ground stability, and soil pore size, compaction and induration. Nutrients fall into two groups – macro- and micronutrients – and deficiency in one or more of these can lead to chlorosis, purpling, local necrosis and stunting in plants, or a lowering of mesofaunal and micro-organism populations. Moisture is important for plant growth, but excessive moisture can lead to defects in ion and water uptake by roots and restriction of root growth. Anaerobic conditions can lead to a loss of nitrogen, and toxicity problems caused by the solution of manganese and iron. Soil pH can also have an important effect on nutrient availability and toxicity, while some toxins occur at high levels independently of pH. Soil temperature can influence seed germination and growth, the type and population levels of organisms and the rate of operation of the functions they perform in the soil. Plant growth and habitation by organisms are also inhibited by limited soil depth and unstable ground conditions, while small soil pores and compaction or induration provide additional limitations to plant growth by restricting root penetration.

7

SOILS IN LANDUSE SYSTEMS

7.1 INTRODUCTION

Landuse systems are open systems characterised by inputs, outputs, transfers and transformations of material and energy, and operating in parallel with hydrological, nutrient and energy cycles (Figure 7.1). In a general context, landuse systems may be classified as agricultural or non-agricultural. Agricultural systems may be further subdivided into arable, pastoral and forestry practices, while non-agricultural systems focus on urban, industrial, recreational and leisure activities.

The diversity of landuse practices reflects the influence of a complex array of physical and human factors, and soil occupies a central position within this array. The choice of land use in a given location is therefore based largely on the extent to which soil characteristics match landuse requirements. Other important physical controls include climate and relief, while important human controls include socio-economic status of the land user, the nature of

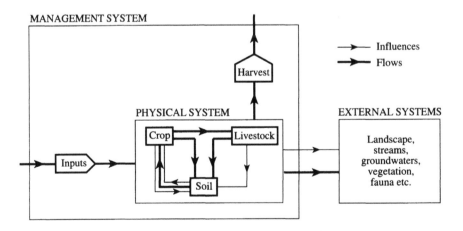

Figure 7.1 Generalised structure of an agricultural system and its relationship with external environmental systems (from Briggs and Courtney 1989)

land tenure, and the political and infrastructural characteristics of the country and region concerned.

Landuse practices vary dramatically not only in their operational details, but also in the scale of their operation. Individual activities may vary from a few hectares in area, as in the case of small forest clearances for shifting cultivation, to several square kilometres for mechanised cereal farms, and several hundred square kilometres for grazed rangelands. In addition, landuse activities are often integrated in various ways in an attempt to increase their flexibility and long-term sustainability. Examples include strip grazing in fields with arable crops, combinations of trees with arable crops (agroforestry) and integrated arable, pastoral and forestry activities managed at the watershed scale.

Land not developed for productive agricultural purposes, such as mountain, moorland, wetland and other wilderness areas, may be used for recreational activities, or it may be protected for conservation. Non-agricultural land may also be used for urban and industrial development. In some circumstances, pressure on land is so great that reclamation of previously unused land is necessary. This has been particularly widespread in low-lying coastal, estuarine and floodplain areas in heavily populated regions, including the Netherlands, and valleys of the Nile, Indus and Mississippi rivers. Similarly, there is increasing pressure in many areas to restore to its former quality, disused and derelict land, which has been disrupted by activities such as opencast coal mining, mineral extraction, waste disposal and construction.

In addition to their inherent spatial diversity and complexity, landuse systems are the product of numerous changes and developments which have taken place over a variety of temporal scales, more particularly during the last half century. In relation to arable systems, these changes include the evolution of advanced crop rotation systems, increased mechanisation, development of new crop varieties, improved agrochemical treatments, and the development of more effective drainage and irrigation technology. More recently, energy- and cost-saving developments include reduced and zero-tillage practices, and low external input systems. In pastoral systems, new breeds of livestock have been developed, together with improved varieties of grass, and more effective pasture improvement strategies. In relation to forestry, highly productive plantations of rapid growing species have been established in many areas, and in recent decades agroforestry and social forestry programmes have also been developed. Inevitably, all of these changes and developments have important implications for the maintenance of soil quality.

Clearly, the topic of soil–landuse relationships is enormous in its scope. It can be examined over a range of spatial scales, and is of interest to a variety of different disciplines. For example, an agronomist may wish to study the performance of crop plants in different growth media using pots or trays in a glasshouse experiment. Similarly, a soil physicist may be interested in the impact of different tillage regimes on the physical characteristics of soils; this could be tested at the farm scale using clearly identified plots of land which are similar in all respects apart from their tillage treatments. At the larger scale, an environmental manager may be concerned with sediment yield or soil acidity variations between catchments under different land uses, while a geographer may be interested in the spatial relationship between soils and land use at the regional or national scale where broad soil types, rather than specific soil properties, may be important.

In this chapter, the factors which influence landuse systems will be examined, with particular emphasis on the role of soils. A variety of examples of practices within each of the major landuse systems – arable, pastoral, forestry, urban/industrial and recreation/leisure – will then be considered in terms of their effects on the soil. Emphasis will be placed on broad soil–landuse relationships rather than on details of the processes involved. For a more comprehensive discussion of this topic more specialist texts should be consulted (e.g. Davidson 1980, Briggs and Courtney 1989). The environmental problems associated with the above landuse activities will be examined in the following chapter.

7.2 INFLUENCE OF SOILS ON LAND USE

Soil plays a central role in the effective operation of landuse systems. In agricultural systems it has an influence on the yield of produce from arable, pastoral and forested land, particularly through its control over conditions within the root zone (rhizosphere). It also affects the nature and timing of mechanical operations and animal stocking levels. In addition, soil has an influence on the behaviour of fertilisers and pesticides, and other related agricultural chemicals; consequently the susceptibility of certain crops to pests and diseases is influenced by soil characteristics. In non-agricultural landuse systems, the role of soil is just as important as in agricultural systems. For example, in the engineering and construction industries, the success of mechanical operations, and the stability of resulting structures, depends on the mechanical characteristics of soils and their parent materials. Soil characteristics also have an important bearing on waste disposal practices, and on land reclamation and restoration procedures.

Many of the soil characteristics which influence landuse practices are interrelated to a greater or lesser extent (Chapter 2). Hence, statistical associations between individual soil properties and landuse parameters, such as crop yield, are often unclear (Pitty 1979). Nevertheless, it is possible to examine some of the broader influences of the more important soil characteristics on landuse potential.

7.2.1 Soil physical factors

Texture is one of the most stable soil properties and has a major influence on landuse capability. Clay soils display relatively good water and nutrient retention but are often characterised by poor drainage and aeration. Furthermore, because such soils are often wet, they respond only slowly to changes in air temperature (section 6.3.1). Consequently, in clay soils of temperate environments, crop germination and emergence may be delayed in spring, although the onset of autumn frost is also delayed. Clay soils are frequently difficult to cultivate and are thus said to have a 'heavy' texture. Workability and trafficability are particularly restricted during the autumn and winter periods (Figure 7.2). As clay soils are characterised by the development of large aggregates, however, they tend to be relatively resistant to erosion.

In contrast to clay soils, sandy soils display relatively poor retention of water and nutrients, and tend to be freely or excessively drained, and well aerated. Consequently, they respond rapidly to changes in air temperature (section 6.3.1). Sandy soils are generally easy to cultivate and they are said to have a 'light' texture. Aggregation is poor in such soils, however, and they may be susceptible to erosion, particularly if they are dominated by fine sand and contain little organic material. Similarly, silty soils are prone to crusting and capping, which results in reduced infiltration, increased frequency of overland flow and increased risk of erosion (sections 3.2.4 and 6.4.2). In an agricultural context, medium-textured soils, especially the finer silt and clay loams, are the most desirable. Such soils display optimum conditions in terms of water and nutrient retention, drainage and aeration, response to changes in air temperature, resistance to erosion and ease of cultivation.

In terms of landuse capability, soil structure is one of the most important soil characteristics, and is best developed in soils which contain appreciable quantities of clay and humus (section 2.3.2). Aggregates provide the basis for nutrient and water retention, while inter-aggregate pore networks facilitate aeration, rapid drainage of excess water, and extension of plant root networks (Figure 7.3). Similarly, well-developed soil structure plays a crucial role in the reduction of erosion risk. Plant growth tends to be most effective in soils where crumb and granular structures predominate; optimum growth conditions are often associated with crumb aggregates of 1–10 mm diameter.

Soil structure should not be confused with the *tilth* of the soil. Provided the appropriate soil constituents are present, structure will develop in response to natural processes, such as wetting and drying and freeze–thaw cycles. In contrast, tilth is created by tillage operations, such as harrowing and discing, which break down large aggregates to form a finer seedbed suitable for the germination and emergence of crops (Figure 7.4). The production of a good tilth requires great skill and practice. If the soil is too wet, then compaction may lead to impeded drainage and poor development

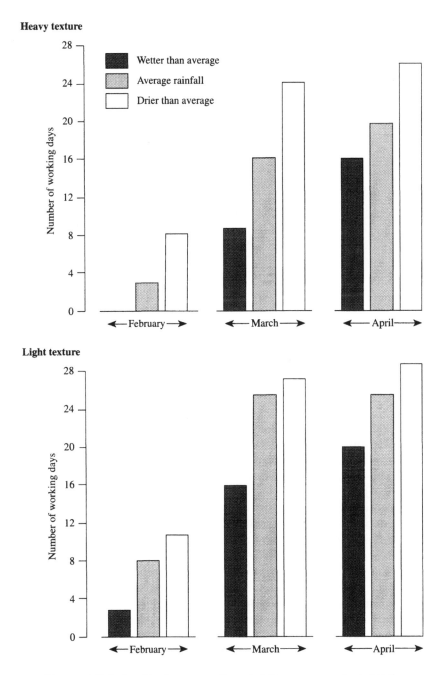

Figure 7.2 Relationships between soil workability, texture and rainfall (from Simpson 1983)

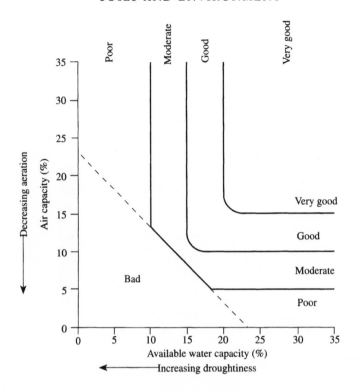

Figure 7.3 Classification of the structural quality of topsoils (from Hall *et al.* 1977)

of plant root networks. Furthermore, the resulting structural damage may increase the susceptibility of soils to erosion. These problems may be overcome to some extent through the adoption of minimum or zero-tillage options, although these are not suitable for all soil types and conditions (section 7.3.1).

The porosity and pore size distribution of soils are influenced directly by texture and degree of structural development. They are also closely associated with bulk density and have a major influence on water retention and aeration characteristics of soils (section 2.3.3). For example, if porosity is reduced due to compaction by heavy agricultural machinery, then increased waterlogging may lead to poor germination and emergence of crops. Restricted root network development may also lead to poor crop performance in such conditions. In fine-textured soils with naturally poor drainage, however, mechanised tillage operations may lead to improved drainage due to increases in macroporosity. The beneficial effects of ploughing on soil structure and macroporosity were illustrated by Mackie-Dawson *et al.* (1989) in a farm-scale study of poorly drained clay soils in southern England. Here, macroporosity was found to be consistently higher in ploughed soils,

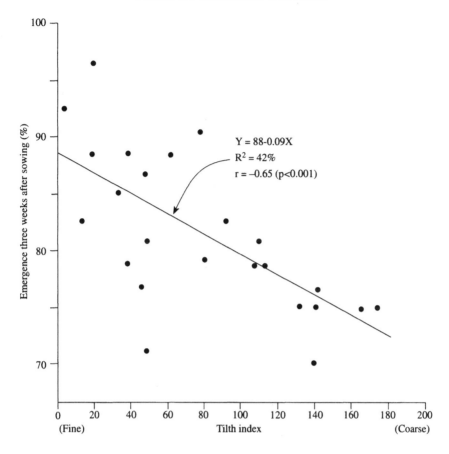

Figure 7.4 Relationship between soil tilth and seedling emergence for oats. Tilth index is a measure of the coarseness of soil aggregates (based on Thow 1963)

even though they were traversed by agricultural machinery, than in direct-drilled soils where the use of such machinery was limited (Figure 7.5).

The content of plant-available water is a vitally important soil characteristic in relation to cropping potential. The amount of plant-available water depends largely on texture, porosity and pore size distribution, degree of structural development, and organic content, and is at its maximum in soils that are near field capacity (section 6.2.2 and Figure 6.3). In coarse-textured, sandy soils, amounts of plant-available water are small as water drains rapidly, under the influence of gravity, from relatively large pore spaces. Such soils tend to be drought susceptible (droughty) and irrigation water is often needed to supplement soil water reserves. In fine-textured, clayey soils, amounts of plant-available water are also small. Here, much of the water is held in very small pore spaces at tensions greater than permanent wilting

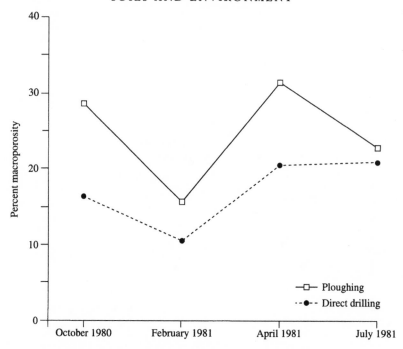

Figure 7.5 Macroporosity changes in topsoil (0–10 cm) under ploughing and direct drilling (from Mackie-Dawson *et al.* 1989)

point (> 15 bars) and is thus unavailable to plants. Such soils tend to suffer from restricted drainage and are susceptible to waterlogging. In these circumstances, removal of excess water is encouraged through drainage practices. Amounts of plant-available water are greatest in soils with a silt loam texture (Figure 6.3).

When the water content of a soil is less than field capacity, it is said to be in a state of soil moisture deficit (SMD). The potential SMD is calculated on the basis of a growing season and represents the accumulated sum of potential evapotranspiration minus precipitation. A cereal crop which yields 10 t ha^{-1} of dry matter, for example, could transpire 2,000–5,000 t ha^{-1} of water during a growing season, which is equivalent to 200–500 mm of rainfall (White 1987). This value makes no allowance for drainage losses, surface runoff or evaporation directly from the soil surface. Thus, in drier parts of eastern and southern Britain, for example, where mean annual rainfall is near 500 mm, crop yields may be adversely affected and irrigation is required to supplement soil water reserves on arable land. When the water content of a soil is above field capacity, cultivation using heavy machinery, and excessive animal stocking, may cause compaction, particularly in clay soils (section

8.2.2). In most instances, the field capacity period sets broad limits to good ground conditions for use of agricultural machinery and for animal stocking (Jarvis *et al.* 1984).

An understanding of the hydrological characteristics of soils is important not only on agricultural land, but also on land used for urban and industrial activities. For example, drainage and aeration problems are commonly experienced on land which has been restored following opencast mining programmes. Usually, these problems are overcome through the installation of appropriate drainage systems (e.g. Younger 1989). Similarly, the behaviour and migration of solutions are of critical importance in relation to contaminated land, and land which is used for waste disposal purposes (e.g. Bridges 1991a).

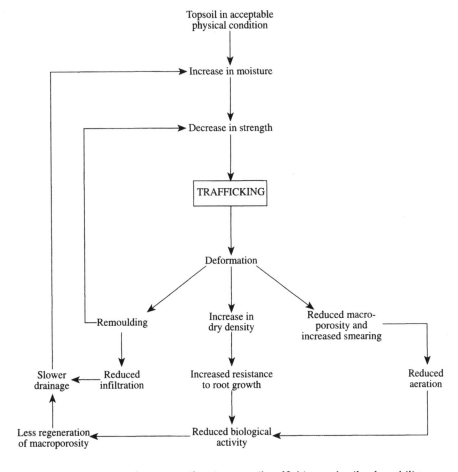

Figure 7.6 Interaction between soil moisture, soil trafficking and soil vulnerability to damage (from Hodgson and Whitfield 1990)

The mechanical characteristics of soils (section 2.3.6) have a major influence on their trafficability, and determine the nature and timing of mechanical operations. If agricultural machinery is used on soils which are wetter than their plastic limit, severe structural damage, compaction and the development of cultivation pans may occur (Figure 7.6). These problems will in turn result in impeded drainage and increased susceptibility to water-logging, and under these conditions crop performance may be severely restricted (Figure 7.7).

Problems may also arise in soils with unfavourable shrink–swell characteristics. Expansion of swelling clays on wetting and shrinkage on drying result in churning and cracking of the soil. These changes cause disruption of soil structure and tilth, and are likely to have an adverse effect on crop yields. Vertisols in particular undergo swelling and shrinkage, in response to

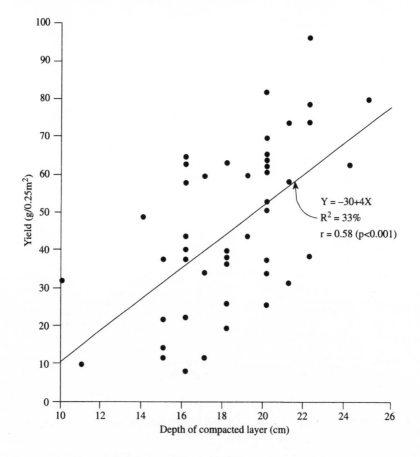

Figure 7.7 Effect of soil compaction by machinery on yield of spring barley in a single field in Devon (based on Briggs 1977b)

seasonal wetting and drying cycles (section 4.3.3), and are thus very difficult to cultivate. In a micromorphological study of Vertisols in southeast Australia, McGarry (1989) concluded that shearing, compaction and damage to soil structure were minimised using dry, as opposed to wet, cultivation.

In pastoral systems, the mechanical characteristics of soils have an important bearing on the timing and intensity of animal stocking. Once again, under wet conditions, particularly when the soil is above its plastic limit, animal trampling and associated compaction results in poaching and puddling of the soil. Similar mechanical problems also occur in forested systems where heavy machinery is used in land preparation prior to planting, and during felling operations. In many areas of the world, forestry is practised on land which is unsuitable for arable cultivation. Soils are often developed on steep slopes, or are highly organic or have high water contents, and therefore tend to be mechanically weak. Under these conditions, severe disturbance may result in compaction, increased rates of organic matter breakdown and enhanced soil erosion (sections 8.2.1 and 8.2.2).

On land used for urban and industrial activities, the mechanical characteristics of soils, particularly load-bearing capacity and consistency limits, have a major influence on the stability of buildings and their foundations. Similarly, if the substrate contains large amounts of swelling clays, stability may be adversely affected. This has been illustrated in many parts of southeast England where, in recent years, drought has resulted in severe shrinkage of clay soils and parent materials, and consequent subsidence of buildings (Greenwood 1993).

Soil temperature has a major influence on plant growth at all stages (section 6.5.3). At very high temperatures, however, the main limitation to growth is intense evaporation, and the resulting moisture stress, rather than temperature alone. Under these conditions, plants may suffer from physiological drought. Here, plants are able to cope with daytime moisture deficits, induced by intense evapotranspiration, by making best use of night-time moisture excesses.

7.2.2 Soil biochemical factors

Soil organic matter plays a fundamental role in agricultural land use, largely through its influence on water content, nutrient status and structural stability (e.g. Bell 1993). The humus fraction in particular has a very high water-holding capacity and can retain up to four times its own dry weight of water; about 50 percent of this is likely to be plant-available. In fact, the presence of 5 percent humus will increase the plant-available water content of a sandy loam by more than 50 percent, and that of a clay loam by about 30 percent, above levels in comparable organic-free soils (Simpson 1983). In addition to its water retention properties, organic matter is often added to soils as a mulch to reduce water losses by evaporation. Organic matter is also an

Table 7.1 Average organic content of topsoils under different land uses in southeast Scotland

Type of agriculture	% organic matter	Range
Arable	3.8	2.0–6.2
Arable with 25% grass	4.4	1.9–7.4
Arable with 50% grass	5.0	3.6–6.8
Permanent pasture	8.5	7.9–9.5

Source: Simpson 1983

important source of essential plant nutrients, particularly nitrogen. The humus fraction in particular has a very high CEC and is therefore able to retain nutrients such as base cations (section 2.4.2), which are available for plant uptake. In a mechanical context, organic matter plays a crucial role in the development of soil structure (section 2.3.2). For this reason, soils with appreciable quantities of organic material are less susceptible to erosion than comparable soils with low organic contents.

The organic content of soils varies dramatically in response to differences in land use (Table 7.1). Contents are often lowest in soils under intensive arable cultivation, as relatively little organic matter is returned to the soil after harvesting. Moreover, with the advent of inorganic fertilisers, the importance of organic material as a source of nutrients has declined. In southern and eastern areas of Britain, where arable cultivation is widespread, soil organic contents are often below 2 percent by weight, although values vary somewhat depending on the type of crop rotation system adopted, and position within the rotation cycle. In soils under permanent pasture, levels of organic material are considerably greater than in arable soils, and are often in excess of 10–15 percent. Furthermore, soils under continuous arable cultivation often suffer a rapid decline in organic content over time, unlike those under pasture where organic contents remain relatively high (Figure 7.8). In forested soils, particularly those under needleleaf trees, organic contents can be very high. In many parts of upland Britain, for example, needleleaf plantations are often established on soils which are peaty in character.

The size and diversity of soil organism populations are just as important as organic contents in terms of land use. Earthworms in particular are known to have a beneficial effect on nutrient levels, and on structural stability, in agricultural soils. The number and diversity of soil organisms are often greatest in soils under permanent pasture where relatively high organic and moisture contents are maintained. In soils under arable cultivation, however, populations are often restricted as the food supply is relatively low. Similarly, in forested soils, particularly where needleleaf plantations are widespread, soil organism populations are restricted due to high levels of soil acidity and

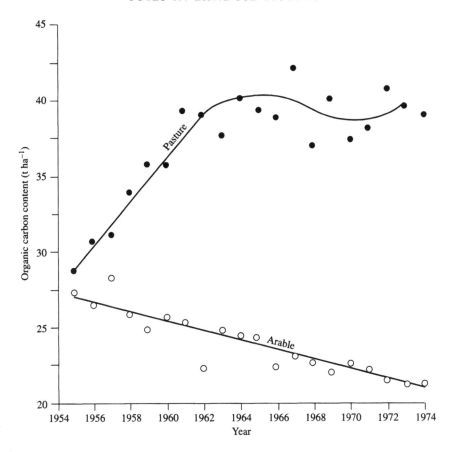

Figure 7.8 Changes in soil organic content over time on arable land and pasture (based on Garwood *et al.* 1977)

the prominence of unpalatable organic material.

Plant nutrient status is defined in terms of a number of soil properties, including exchangeable base content, nitrogen and phosphorus contents, CEC, percent base saturation and acidity (sections 2.4.1, 2.4.2, 2.4.3 and 6.5.1). In fertile, nutrient-rich soils, CEC is dominated by exchangeable bases and the percentage base saturation is high. In acid, nutrient-poor soils, however, CEC is dominated by H^+ and Al^{3+} and base saturation is low. Here, it may be increased by liming, where adsorbed H^+ and Al^{3+} ions are replaced by Ca^{2+} ions. Similarly, levels of K^+ ions in the soil may be increased by the addition of NPK (nitrogen-phosphorus-potassium) fertilisers. It should be noted, however, that prolonged dominance of the soil exchange complex by one or two specific cations is detrimental to the availability of other cations to plants. Such competition between different ions for plant uptake is known as 'antagonism'. High levels of Na^+, for example, are common in saline and sodic soils of semi-arid environments (section 4.3.3). Na^+ is toxic to many plants and tends to inhibit crop performance. Furthermore, high levels of Na^+ encourage aggregate dispersion and consequent breakdown of soil structure (section 8.3.2).

In addition to the plant macronutrients, amounts of soil micronutrients, or trace elements, such as molybdenum, boron and copper, may need to be controlled. Deficiencies or toxic excesses are often evident in crops and grazing animals, and can be regulated either through soil amendments or by direct treatment of crops and animals. Many anions are just as important as cations, in plant nutrition. Such anions include nitrate, phosphate and various forms of sulphur. As AEC is restricted in most soils, particularly in temperate environments, anion retention is limited. Consequently, losses through leaching tend to be rapid and levels which are sufficient for productive arable cultivation can only be maintained through fertiliser additions, or perhaps by mulching (e.g. Addiscott *et al.* 1991) (section 8.3.3). The close relationship between nutrient status and organic content of soils was illustrated by Jenkinson and Johnston (1977), in a study of changes in the nitrogen content of arable soils in response to different manurial treatments (Figure 7.9); nitrogen contents were found to be considerably greater in soils which received larger inputs of farmyard manure (FYM).

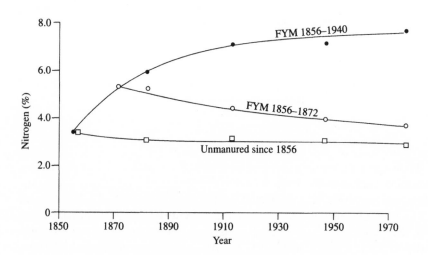

Figure 7.9 Changes in nitrogen content of soils under continuous arable cropping under different manuring strategies (from Jenkinson and Johnston 1977)

Acidity plays a crucial role in the availability of plant nutrients. If pH is too high (> 8.0), or too low (< 5.0), availability is likely to be restricted for certain nutrients (section 6.5.3). For most arable crops, growth is optimal at pH 6.0–7.0. Soil acidity is, however, rather unstable and susceptible to change. Excessive addition of nitrogen fertilisers, for example, may result in soil acidification (Chalmers 1985) (section 8.3.1). Grassland vegetation on pastoral land, particularly that in humid, cool temperate environments, grows most effectively at soil pH values of 5.0–6.0, while needleleaf trees often grow most effectively where soil pH values are even lower (3.5–4.5); few other productive crops can tolerate this level of acidity.

7.2.3 Soil type

Because soil type is defined by a combination of numerous soil characteristics, its influence on land use will be more general and less direct than that of an individual soil property, such as organic content, texture or acidity. Consequently, the relationship between soil type and land use is observed most clearly in geographical investigations of relatively large areas, particularly at the regional or national scale where a range of soil types and land uses may be identified. At the more local scale, individual soil properties are likely to show much greater spatial variation than soil type, and therefore to have a much greater influence on land use. The broad geographical relationship between soil type and land use will be examined with reference to soil orders of the Soil Survey Staff (1975, 1992) classification system; approximate major soil grouping equivalents according to the FAO-Unesco (1989) classification system are shown in Table 3.5.

Entisols

These are shallow, immature soils which are found in a variety of situations, including steeply sloping sites with colluvial parent materials, and floodplain sites on alluvial deposits. They may also exist in cold or exposed environments where soil formation is restricted. Entisols often have a low water-holding capacity and low nutrient status, and can rarely be used for productive arable cultivation. They may be used for grazing although the quality of pasture is likely to be limited unless a pasture improvement programme is adopted. Entisols are rarely deep enough to sustain productive forestry, although shallow-rooted species may grow well in such soils.

Inceptisols

These are deeper and rather better developed than Entisols, although the morphological expression of soil development may not be particularly strong. They often have a relatively low nutrient status but with careful management they can be highly productive; this might include maintenance of organic matter levels and protection against erosion. Inceptisols are widespread in their occurrence and are particularly common in floodplain areas of the River Ganges in India, and in many other parts of southern Asia where they are often used for high yielding rice production. They are also common in the Amazonian region of Brazil, and in parts of the USA they are used extensively for sugar-cane production. In the Sahelian zone of Africa, however, where Inceptisols are common, severe drought combined with widespread overgrazing has resulted in severe erosion problems (section 8.2.1).

Alfisols

These are particularly common in humid temperate environments and are widespread in western Europe, including eastern and southern areas of Britain, and in eastern parts of North America (section 4.3.2). Alfisols, particularly those with a fine loamy texture, are often highly fertile and contain significant quantities of mull humus. They also have a high water-holding capacity and stable soil structure, and therefore tend to be well-suited to productive arable cultivation, scoring highly in terms of landuse capability (section 9.2.2). In Britain, they are often used for intensive cropping of both winter and spring cereals and exhibit few limitations to agriculture. In low-lying areas, however, particularly where profiles have a pronounced clay enriched B (argillic) horizon (section 3.3.2), waterlogging may present problems during the autumn and winter months. In such circumstances, trafficking is restricted to prevent compaction and the formation of cultivation pans, and installation of artificial drainage is often necessary.

Mollisols

Mollisols are particularly common in relatively dry temperate environments, under prairie or steppe grasslands of continental interiors, such as central North America and Eurasia, and Argentina. These soils are rich in mull humus and have a well-developed crumb structure in their upper horizons (section 4.3.2). They also contain high levels of calcium and large earthworm populations, and are potentially very fertile. They are, however, regularly exposed to drought conditions and, because they are often developed in loess parent materials, they tend to be susceptible to wind erosion, particularly if the surface organic-rich layer is degraded or removed. If carefully managed, using appropriate irrigation and crop rotation practices, Mollisols are highly productive and are widely used for arable cultivation in areas where population densities are relatively low. In such areas, cereals, especially wheat, are widely grown; collectively, these areas constitute the world's 'bread basket'. In Argentina, however, extensive rearing of beef cattle is also practised in areas dominated by Mollisols.

Histosols

These soils may develop in a variety of situations from upland bog and moorland sites, to lowland fen and carr environments. Some Histosols are highly acidic and nutrient-poor, while others are nutrient-rich, and all tend to be waterlogged for long periods of time. Acid, nutrient-poor Histosols are of limited agricultural use, while nutrient-rich peats can be of great value for arable farming, once they have been drained. However, drainage leads to rapid decomposition (oxidation) of organic matter which has a detrimental effect on structural stability, water-holding capacity and nutrient status. Furthermore, conservationists are particularly concerned about the destruction of wetland habitats, caused by extensive drainage of these soils (Briggs and Courtney 1989).

Spodosols

These often have a peaty surface horizon consisting predominantly of mor humus and poorly decomposed plant remains. In addition, they are often coarse-textured, highly leached and acidic in character. For the most part, Spodosols have a low water-holding capacity, which renders them drought susceptible, and also have a low nutrient status. Consequently, they are rarely used for productive agricultural purposes except in the case of forestry. Spodosols are widespread in cool temperate, humid regions, such as the boreal forest zones of North America and Eurasia, particularly where soil parent materials are low in base-rich, weatherable minerals (section 4.3.2). In many parts of eastern USA, between the Great Lakes and the Atlantic seaboard, much of the virgin land colonised during the late eighteenth and early nineteenth centuries was underlain by Spodosols. This was of low fertility, and crop yields were so low that much of it had been abandoned by the mid-nineteenth century (Figure 7.10) (Foth 1990).

Ultisols

Ultisols are at the 'ultimate' or end stage of weathering and contain few weatherable minerals. These highly weathered and leached soils are acidic and of low nutrient status, and are dominated by kaolinitic clays. They are

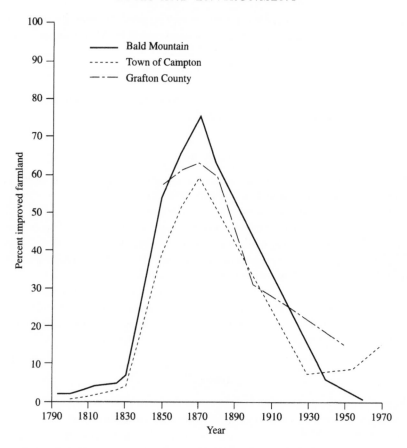

Figure 7.10 Trends in farmland improvement in New Hampshire, AD 1790–1970
(from Foth 1990)

mature soils which have developed over a very long time period, in humid warm temperate to tropical environments (section 4.3.3). Ultisols occur most commonly in southeastern parts of the USA, eastern and southern parts of Brazil, and in parts of southeast Asia. Although the inherent fertility of these soils is rather low, they can be utilised for productive arable cultivation if carefully managed. In southeast USA, Ultisols used to be subjected to a type of shifting cultivation regime, based largely on tobacco and cotton, the soils being cropped for a few years and then abandoned as fertility declined. However, this practice became unnecessary with the development of effective crop rotation systems which incorporated the use of lime, manures and legumes (Foth 1990).

Oxisols

Oxisols are very highly weathered and leached, and have developed over long time periods. In addition, they are often acidic and of low nutrient status. They are characterised by high contents of oxides and hydroxides of iron and aluminium, and occur most commonly in humid tropical environments, particularly in Africa and South America (section 4.3.3). In rainforest areas of central Africa and Amazonia, where rainfall is evenly distributed throughout the year, Oxisols are used largely for shifting cultivation. Here, small-scale plots of land are cleared for cultivation and then abandoned after a few years due to declining fertility and weed invasion. The land then reverts to forest-fallow and nutrient levels are gradually restored (section 7.3.1). In a regional context, however, shifting cultivation can only sustain a relatively low population density. In savanna and savanna-woodland areas, where rainfall is more seasonal in character, Oxisols are not well suited to shifting cultivation, and pastoral activities are more common. Plots of land used for arable cultivation can then benefit from applications of manure derived from grazing animals. In spite of their low fertility, Oxisols can be highly productive with intensive management, as illustrated by the high yielding production of pineapples in Hawaii (Foth 1990).

Vertisols

Vertisols are characterised by the presence of swelling clays and are subject to severe cracking on drying, and to swelling on wetting. Such physical disturbance often leads to tillage and trafficability problems, and to restricted water availability. In spite of these problems, however, these soils often have a high nutrient status and are highly fertile. Vertisols occur most frequently in warm temperate to tropical environments, and are particularly common in Australia, India and Sudan (section 4.3.3). They also occur in southern coastal regions of the USA where they are developed in marine clays.

Cultivation of these soils is difficult due to their adverse physical characteristics. They tend to be very hard when dry and highly plastic when wet. Consequently, the timing of various cultivation operations is critical. In India, for example, planting of summer crops, which should ideally take place just before the monsoon rains, was virtually impossible in soil which was severely cracked and hardened following the long dry season. Similarly, once the monsoon rains commenced, planting was also difficult as the soils soon developed plastic characteristics. Thus, planting had to take place just after the monsoon rains had finished. This practice was rather inefficient as the crops were unable to take full advantage of available water in the soil and depended primarily on the presence of stored water. This problem of timing has now been overcome using a different approach to tillage, developed from research at the International Crop Research Institute of the Semi-Arid

Tropics (ICRISAT). Soils were tilled immediately after the harvest of the winter crops, before the soil became too dry. Farmers could then establish the summer crop in loose dry soil just before the monsoon rains commenced (Foth 1990).

Aridisols

These develop in arid and semi-arid environments, where soil water is severely restricted and productive agriculture is virtually impossible without irrigation. Poor management of Aridisols often results in soil degradation by wind erosion, salinisation and sodification (sections 8.2.1 and 8.3.2). High levels of sodium in such soils are of particular concern. Sodium is toxic to plants, although the degree of tolerance varies from species to species. High levels of sodium can also lead to aggregate dispersion and consequent breakdown of soil structure. Effective management of soil salinity and sodicity depends on sensible irrigation and drainage practices, together with evaporation control, selective chemical treatments, and the use of salt tolerant crops. With careful management, Aridisols are able to support a range of arable and pastoral activities. In the irrigated 'western range region' of the USA, for example, grazing of cattle and sheep is widely practised, together with cultivation of arable crops.

Andisols

These develop in parent materials of volcanic origin, notably ash and related deposits. They tend to have a relatively fine texture, with large amounts of fresh, weatherable minerals and low bulk density. They also contain significant quantities of organic matter and have a high ion exchange capacity. Consequently, they are often very fertile and tend to be intensively cultivated, particularly in tropical environments where they are renowned for their flexibility. In some instances, however, aluminium saturation and low phosphate availability may present problems. Moreover, Andisols are commonly developed on rather steep slopes in mountainous regions, such as parts of southeast Asia, central America, east Africa and North Island New Zealand, and are fairly localised in their occurrence. Consequently, these soils may be particularly susceptible to erosion if poorly managed.

7.2.4 Additional factors

A number of additional factors have an influence on landuse potential. These include climate, relief, parent material, biota and human activity. As these factors are closely interrelated, however, and because they act, in part, via the soil, evaluation of the effect of individual factors is rather difficult. Perhaps the most important factor is climate, which has a direct bearing on landuse capability. Climatic parameters include mean annual precipitation and potential evapotranspiration, which determine the level of plant-available water and extent of the soil moisture deficit (SMD). In Britain, for example, the 800 mm isohyet forms a distinctive boundary between predominantly arable cultivation in the drier east and south, and greater pastoral activity in the wetter west and north.

In terms of agricultural land use, the seasonality and intensity of precipitation are just as important as the mean annual total. Many areas of the world have a pronounced seasonal distribution of rainfall, notably the monsoon areas of southern Asia, savanna regions of Africa, Mediterranean countries and western USA. These have a prolonged dry season when SMD is a problem and irrigation is often necessary. Rainfall is concentrated into a few months of the year and is often very intense (Figure 7.11). If agricultural land is not carefully managed, or the planting of crops is badly timed, this type of rainfall can be particularly destructive, resulting in severe soil erosion and consequent loss of productivity. In many parts of India, for example, planting of the summer crops is timed to coincide with the pre-monsoon rains so that a protective vegetation cover (ideally > 30 percent) is established before the main monsoon begins. In many cool temperate areas, soils are considerably wetter during the winter months than at any other time of year. Under these conditions, workability and trafficability are severely restricted. If soils remain wet during the spring and early summer periods, many crops become increasingly susceptible to pests and diseases.

In addition to precipitation, temperature plays a crucial role in agricultural landuse capability. Temperature has a direct influence on growing season length, and on rates of crop growth and maturation. Although mean annual temperature is important in this respect, seasonal variations in temperature are often more significant, particularly in temperate regions; in tropical environments temperatures are always high, except in mountainous areas, and plant growth continues throughout the year. The accumulated temperature above 0°C for the first six months of the year (January–June) is frequently used as a measure of heat energy available for plant growth. Average (median) values of this index appear to correlate well with the length of growing season (Jarvis et al. 1984). For some crops, however, the more usual index of accumulated temperature above 5.6°C is preferred (Bendelow and Hartnup 1980). Another important climatic parameter, closely allied to temperature, is incidence of frost. Although frost plays an important role in the development of soil structure, and in the control of pests and diseases, it can present problems if it occurs in the early stages of crop growth. Fruit trees in particular are very sensitive to late frost during the spring when they are in blossom. Similarly, grape crops are sensitive to early frost during late summer and early autumn. This may inhibit ripening of the fruit, which tends to swell and split.

In addition to climate, relief has a major impact on landuse activities.

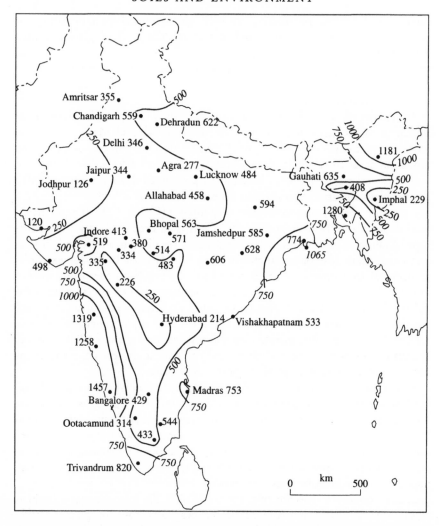

Figure 7.11 Rainfall erosivity in India: mean values of the rainfall erosion index (R) (from Central Soil and Water Conservation Research and Training Institute 1977). $R = EI_{30}/100$ where E represents kinetic energy of the rainfall and I_{30} represents the maximum 30-minute rainfall intensity (see section 8.2.1)

Important parameters include altitude and aspect, both of which have an influence on climate, and slope angle. Generally, limitations on land use increase with increasing altitude, particularly in temperate environments, where pastoral activities replace arable as conditions become cooler and wetter. The effect of aspect is felt most strongly in hilly and mountainous terrain of high latitude areas. Here, the angle of incidence of the sun is

218

relatively low for much of the year and large areas may be shaded. Thus, in the northern hemisphere, soils of north-facing slopes tend to be colder and wetter than those of south-facing slopes, and the duration of snow cover in winter is very much greater. In areas where the climate is already cool and wet, unfavourable aspect may severely restrict landuse options. At lower latitudes, however, where the climate is warmer, soils on a well-shaded slope often have higher moisture and organic contents than comparable soils on slopes which are exposed to direct sunlight. Here, evaporation is intense and organic matter is rapidly broken down, thus restricting landuse options at such sites.

One of the most important limiting factors in terms of land use is slope angle. This has a major influence on soil depth and stability, and on workability of the land. On steep slopes, soils are often shallow and unstable, and are particularly susceptible to erosion. This problem is compounded by the use of agricultural machinery. On low-angled slopes, which are less than about 10°, land is usually cultivated parallel to the contours, thus reducing the risk of soil erosion. On steeper slopes, however, agricultural machinery is usually operated perpendicular to the contours in order to minimise the risk of overturning. It is well known that erosion risk increases dramatically on land which is cultivated in this way, as overland flow is concentrated by compacted wheel tracks and plough furrows. Slope angle also has an important bearing on soil conservation practices. As slope angle increases above about 10°, contour cultivation practices such as strip-cropping and contour bunding are usually replaced by terracing. If slopes steeper than about 30° are terraced, however, the risk of land slippage increases dramatically, thus posing a threat to property, communication systems and life itself.

Parent material has an effect on land use, although this operates via the soil and is thus less direct in its influence than climate and relief. Soil characteristics which are most strongly associated with parent material include stone content, texture, mineral nutrient levels and acidity. Vegetation and soil organisms existing prior to cultivation may give an indication of landuse potential. Good quality grassland and deciduous woodland, for example, indicate soils of relatively high nutrient status. Such soils usually have high earthworm populations and significant quantities of mull humus, and if they are well drained, they could be used for productive arable cultivation. However, the rapid expansion and intensification of arable cultivation in recent decades has led to overproduction, particularly in the developed world, and to soil degradation, often at the expense of areas with natural and semi-natural vegetation covers (see Chapter 8). Thus, conservation of natural and semi-natural vegetation, and the re-establishment of such vegetation in formerly cultivated areas, are often encouraged in preference to clearance and cultivation.

Today, many of the physical limitations on land use can be overcome, to some extent, by human intervention, particularly in agricultural systems. Since settled agriculture began, several thousand years ago in some parts of the world, there have been tremendous developments in crop rotation practices, seed and crop varieties, levels of mechanisation, tillage practices and agricultural chemicals. These developments have been coupled with advances in our understanding of soil and water conservation, drainage and irrigation, and pollution control, and all of these factors have had a dramatic impact on the performance and efficiency of agricultural systems, particularly in the developed world. Many would argue, however, that intensification of agriculture and increased productivity are not sustainable in the long term, and are likely to have an adverse effect on the environment (see Chapter 8).

In addition to the above, a number of other factors have an important bearing on the operation of landuse systems. These include economic considerations, the nature of land tenure, farm policies, political and infrastructural support systems (e.g. transport, communications, education and incentive schemes), and social and cultural structures. Their impact can perhaps best be appreciated through consideration of the dramatically contrasting situations in the developed and developing world. In the developed world, populations are relatively stable in terms of their size and spatial distribution. There is also more productive land per head of population and a wider choice of landuse options than in most developing countries. These factors alone often ensure the most effective and appropriate use of land (Hudson 1992). Furthermore, the relatively high economic status of land users in developed countries facilitates the establishment of high input agricultural systems which utilise high yielding crop varieties, are highly mechanised, and which rely on a range of agricultural chemicals. Land users are also able to invest in long-term conservation strategies, a practice which is often supported at national level by government policies, such as the 'Set Aside' scheme in Britain. Here, land users are given a financial incentive to plant trees and hedgerows and to preserve various endangered habitats by setting aside land formerly used for productive arable cultivation. Not only does this contribute to environmental conservation, but it also helps to reduce excessive agricultural production, and the cost of maintaining excessive food reserves.

In the developing world, population is growing rapidly and is often unstable in terms of its spatial distribution. Rural–urban migration is widespread and the growing number of poor, peasant farmers are forced onto increasingly marginal land which is unable to sustain intensive use. In other areas, exhausted land is abandoned and becomes severely degraded (e.g. Millington *et al.* 1989). Extensive exploitation of woodfuel resources in many developing countries has also lead to widespread land degradation (section 8.2.1). In extreme circumstances, famine and warfare have resulted in large-scale migration with consequent increases in refugee populations, particularly in parts of Africa. The majority of land users suffer extreme poverty and are forced to eke out a living from tiny small-holdings on land which they do not own (Thapa and Weber 1991). Peasant farmers are driven by short-term needs rather than by long-term goals (Hudson 1992). Rarely can they afford to invest in high input agricultural practices which require purchase of expensive seed, fertilisers and pesticides, and the use of agricultural machinery. In fact, such practices and associated technology are inappropriate in many developing countries, where agriculture is often particularly labour intensive. Instead, farmers have to be resourceful and must rely on cheaper alternatives which are closer to hand; these include animal manure, fire ash, crop residues and household waste. In addition, poverty and the small-scale of many landuse operations do not allow establishment of large-scale, elaborate conservation schemes like those often found in the developed world. The environmental implications of these problems are examined in more detail in Chapter 8.

7.3 AGRICULTURAL LAND USE

7.3.1 Arable cultivation

Methods of arable cultivation have evolved in different ways in response to the vast array of contrasting global environments. In many humid tropical regions, where soils are of low fertility, shifting cultivation, sometimes known as *slash and burn*, is commonly practised, although this often sustains only a small fraction of the population in such regions. Essentially, shifting cultivation involves a three-phase rotation. This commences with the clearance of existing vegetation, often by burning which helps to improve soil nutrient status (e.g. Ulery *et al.* 1993), particularly in tropical ecosystems where much of the nutrient stock is stored in the biomass rather than in the soil (Stromgaard 1992). Following clearance, the land is cultivated, but for only a few cropping cycles (cultivation phase), as the decline in soil fertility is rapid and weed invasion is a major problem. The period of cultivation is influenced to some extent by rainfall, with wetter areas being cropped for shorter periods than drier areas. Generally, however, a cropping period of about one to five years is commonly adopted. After cultivation, the land is abandoned and former vegetation is allowed to re-colonise (bush-fallow phase). This phase is essential for the restoration of soil nutrient levels (Figure 7.12), and its length again depends to a large extent on rainfall. Usually, the recommended period is five to twenty years, although it may be slightly longer in wetter areas and a little shorter in drier areas. An example of shifting cultivation is the Chitemene system of northern Zambia. Here, felling of trees is limited and crops are grown in ash gardens created by burning lopped branches (Stromgaard 1992) (Figure 7.13). Similarly, the Jhum system practised in Bengal, India comprises five-, ten- and twenty-year

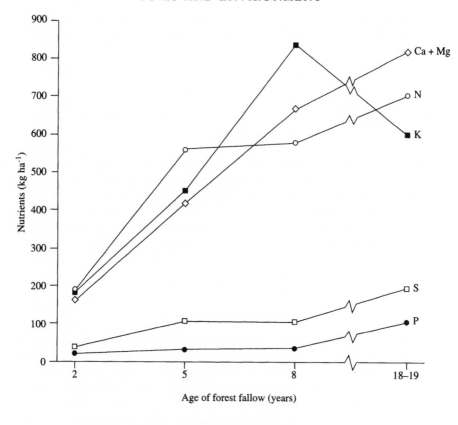

Figure 7.12 Nutrient build-up in forest fallow vegetation, Zaïre
(based on Foth 1990)

cycles with a long fallow and a variety of mixed cropping. A related practice is the Khoriya system of Nepal (Thapa and Weber 1991).

In temperate areas, land has often been more closely settled, and competition for land has been greater, than in tropical regions. Furthermore, land in temperate areas cannot be cropped for significant periods of the year, due to the restricted growing season, and consequently more intensive cropping systems have evolved (Briggs and Courtney 1989). One of the oldest recorded systems in Britain, practised as early as the Celtic period, was a three-year rotation of autumn cereal, spring cereal and fallow. This was largely replaced during the eighteenth century by other rotations such as the Norfolk four-course system. Here, grass-clover pasture, root crops and cereals were alternated, and arable practices were closely integrated with livestock rearing. The nitrogen for crop growth could therefore be fixed in the soil by clover, which is a legume, and released from grass residues and manure.

During the nineteenth century, other cropping practices evolved, largely

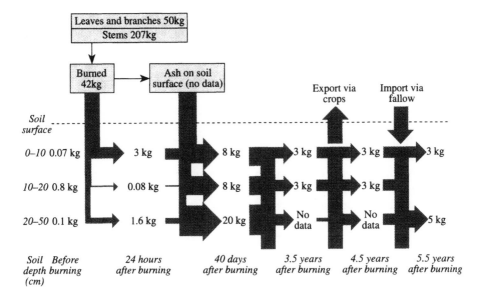

Figure 7.13 The Chitemene shifting cultivation regime in Zambia: changes in the soil
nutrient store of phosphorus (from Stromgaard 1992)

in response to depressed cereal prices and the increasingly accepted belief in
the beneficial effects of grass as a soil 'conditioner' (e.g. White 1987, Francis
and Kemp 1990). At this time arable–pasture (or arable–ley) rotations, with
a relatively long period of ley (up to three years), became increasingly
common. It was during this period that mixed farming became widespread,
with livestock being reared on farm-produced grass, cereals and root crops.
This type of farming was based on a nutrient supply from farm-derived
organic materials such as farmyard manure (FYM), which consists of animal
dung and urine mixed with straw, slurry (dung suspended in urine and
washing water) and compost. Sewage sludge from local sewage treatment
works has also been used on agricultural land. Unfortunately, however, this
material often contains high levels of heavy metals, such as lead, zinc and
cadmium, derived from industrial effluent (e.g. Thornton 1991) (section
8.3.4). Green manure crops, sometimes known as *catch crops*, have been an
additional source of nutrients. Here, a quick-growing leafy crop is ploughed
into the soil, releasing nitrate for the benefit of the following cash crop. This
approach reduces the risk of nitrate leaching (section 8.3.3), but its effects are
short-lived and it contributes little to soil organic matter levels (e.g. Saull
1990).

In recent decades, due largely to increasing population pressure, demand
for food and competition between different land uses, there has been a move
away from the more traditional agricultural practices outlined above, in both

tropical and temperate regions. During the Second World War in particular, the demand for food increased and agricultural activity intensified. Arable and livestock rearing became increasingly separated so that FYM was no longer available where it was needed. At this time, greater emphasis was placed on cereal monoculture at the expense of arable–ley rotations. This change became feasible as the traditional sources of nitrogen were replaced by inorganic nitrogen fertilisers. However, excessive use of such fertilisers led to increased soil acidification (section 8.3.1), and to nutrient enrichment of runoff from agricultural land with consequences for surface water quality and aquatic biota (section 8.3.3).

In addition to the chemical changes, the intensity and scale of mechanisation in arable cultivation has increased dramatically. This has contributed to increased soil compaction which in turn has led to impeded drainage and restricted development of plant root networks (Briggs and Courtney 1989) (section 8.2.2). In an attempt to overcome these problems, minimum and zero-tillage options have been developed, where the use of heavy agricultural machinery is minimised. An example of such a practice is direct drilling where crops are sown directly into the existing vegetation cover. This is now practised in many parts of the USA, particularly in the central corn/maize belt, and in many parts of tropical Africa, South America and Australia. The advantages of minimum and zero-tillage include improved topsoil organic content, aggregate stability, moisture retention, erosion control and workability (e.g. Edwards et al. 1993, Van Vliet et al. 1993). In addition, energy requirements are reduced and excessive heating of the soil in hot environments is prevented. In a comparison of zero-, minimum and conventional tillage practices in northern Italy, Mbagwu and Bazzoffi (1989) reported an improvement in a number of soil characteristics, particularly those relating to aggregate stability, with reduced tillage (Figure 7.14). However, the disadvantages of minimum and zero-tillage include increased bulk density, reduced total porosity and slow warming of the soil during the spring. Moreover, phosphorus and potassium become concentrated in the surface layers (0–5 cm) of the soil which may restrict their availability to plants if the soil dries out. This is in contrast to tilled soils where the nutrients are more evenly mixed throughout the topsoil (0–20 cm). In addition, accumulation of volatile fatty acids around fermenting crop residues may inhibit seed germination, and the increase in perennial weeds is difficult to control with herbicides alone.

In many tropical and subtropical areas, particularly in the developing world, low external input agriculture is becoming increasingly common as a component of sustainable development (e.g. Moyo et al. 1993). Here, farmers use crop varieties which are adapted to the main soil constraints, while minimising the use of purchased inputs, and maximising nutrient recycling. This is very different from the high input systems from which much of the world's food supply is derived, but which involve intensive use of agricultural chemicals to overcome soil constraints. An example of a low input agricultural system for acid soils in Peru was examined by Sanchez and Benites (1987). This system was developed as part of the transition process between shifting and continuous cultivation. The principal features include slash and burn clearance, rotation of acid-tolerant upland rice and cowpea cultivars, maximum residue retention, zero-tillage, and absence of lime and fertilisers. As yields declined, due to nutrient deficiencies and increased weed pressure, a kudzu fallow was grown for one year. Subsequent options included pasture, agroforestry and fertiliser based continuous cultivation. It

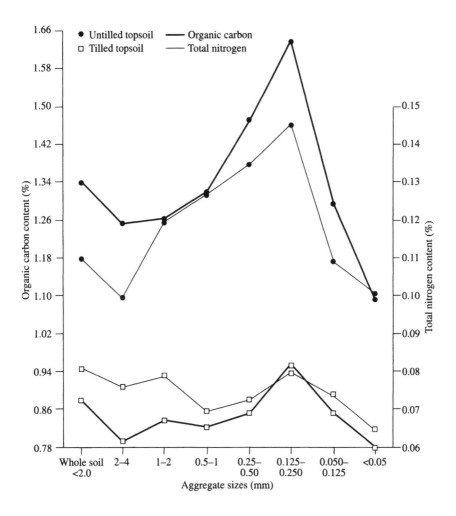

Figure 7.14 Comparison of selected soil characteristics under conventional and no/minimum tillage (from Mbagwu and Bazzoffi 1989)

225

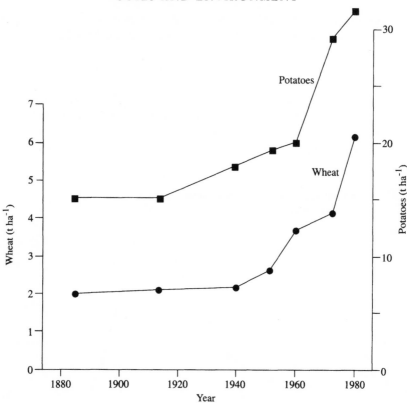

Figure 7.15 Yield increases over time (3-year averages) for wheat and potatoes in the UK (from Cooke 1984)

was suggested that this system helped to preserve some agro-ecosystem diversity, while contributing towards sustainable production and income for farmers.

All of the above developments are clearly reflected in increased crop yields, particularly in the developed world (Figure 7.15). In western Europe, for example, the amount of grain produced per head of population increased from about 250 kg in 1950 to nearly 500 kg in 1990, an increase of 2.5 percent a^{-1} (Hudson 1992). To a large extent, these increases have been achieved through the intensification of agriculture, rather than by increasing the area of land under cultivation. In the developing world, however, production has increased only slightly, often by increasing the area of marginal land under cultivation, and in some areas production has declined. In Africa, for example, the amount of grain produced per head of population decreased from approximately 150 kg in 1950 to about 120 kg in 1990, a decrease of 1.6 percent a^{-1} (Hudson 1992). This decrease, coupled with rapid population

226

growth and the extension of cultivation onto marginal land in many areas, is clearly not sustainable in the long term.

7.3.2 Pastoral activities

Grazing as a landuse practice involves the rearing of animals mainly for transport, meat, milk or wool. In many areas, natural and semi-natural grassland and scrub are used for grazing. These include parts of the prairies and steppes of the temperate continental interiors of North America and Eurasia, the pampas of South America, and the savanna and rangeland areas of Africa and Australia. Even some of the most fragile ecosystems are used to support grazing livestock, including semi-arid regions of Africa and Asia, and tundra areas of northern Scandinavia.

In other parts of the world, natural and semi-natural vegetation has been cleared or at least substantially modified to allow good quality grassland to become established. The quantity and quality of herbage can be improved either by replacement of existing vegetation, in conjunction with fertiliser and lime treatments, or by improvement of existing grasses (e.g. Newbould 1985). Although the former approach is often considered to be the most effective, it takes longer to achieve and is more costly than the latter. Once grazing has commenced, re-establishment of pre-existing vegetation is prevented by the consumption of new shoots.

Until the 1960s, almost all hill land improvement in Britain involved ploughing and reseeding, which proved to be appropriate for the brown earth and podzolic soils most commonly associated with grazing prior to this time. More recently, however, with the extension of pasture improvement onto peaty gleyed podzols and humic gleys, the more gradual improvement of existing grasses appeared to be a more appropriate option (Adams and Evans 1989). Consequently, in order to achieve the most effective solution, in economic and biological terms, an understanding of soils, vegetation and grazing strategy is essential (e.g. Newbould 1985). One of the most widely used pasture improvement strategies in the British uplands still involves the replacement option. Soil pH is raised to about 5.5 or more by liming (c. 7 t ha^{-1} ground dolomitic limestone with high levels of calcium and magnesium). Soil nutrient levels are also increased by fertiliser treatments and a seed mixture of grasses and white clover. Clover, being a legume, helps to fix nitrogen in the soil, a process which is enhanced by inoculation of clover seed with nitrogen-fixing bacteria such as *Rhizobium* spp (e.g. Newbould 1985). Sowing is usually carried out in April or May, with limited cultivation and light grazing initially. Grazing can also be controlled effectively using appropriate fencing strategies. Pasture quality must be maintained through the application of dressings of lime and fertiliser, particularly at wetter sites. In addition, livestock should be monitored for problems relating to trace element deficiency or toxicity.

A number of environmental problems have been associated with pastoral activities, including erosion, compaction, salinisation and acidification of soils, and leaching of organic wastes into potable water supplies (Chapter 8). Some of these problems were illustrated by Aweto and Adejumbobi (1991) in a study of the impact of grazing on soils in the southern Guinea savanna zone of Nigeria. Here, topsoils (0–10 cm depth) in grazed plots were compared with those in ungrazed savanna woodland. Properties examined included organic carbon, total nitrogen, exchangeable bases, CEC and phosphate content. Mean values of all properties were significantly lower in soils of the grazed plots than in those of the ungrazed sites (Table 7.2).

The decline in soil organic content and nutrient levels in the grazed plots was explained in terms of a number of interrelated causal factors. Removal of the protective vegetation cover by grazing livestock and by burning, together with soil compaction caused by animal trampling, resulted in increased soil exposure and enhanced surface runoff. These increases in turn resulted in accelerated soil erosion, organic matter decomposition and leaching losses. It appeared that the return of organic matter and nutrients to the soil by grazing livestock was insufficient to counteract the deterioration in soil herbage quality. Suggested responses to these problems included rotational grazing, incorporating the planting of a supplementary fodder of legumes in grazing areas. This practice may reduce the frequency of overgrazing, particularly during the dry season. Hay storage for use during the latter part of the dry season was also encouraged. In addition, it was suggested that the timing and amount of burning should be reviewed. Burning helps to reduce pest levels and to restrict bush invasion, but it also destroys soil organic matter. It was suggested that burning should be carried out early in the dry season, before the ground becomes tinder dry, rather than late in the season when it can be particularly intense and destructive.

Table 7.2 Mean values for a selection of soil properties on grazed and ungrazed plots

Soil property	Grazed (n = 10)	Ungrazed (n = 10)
Organic C (%)	1.3	3.4
Total N (%)	0.07	0.17
Ca (Me/100g)	3.13	10.60
Mg (Me/100g)	0.86	0.89
K (Me/100g)	0.38	0.86
Na (Me/100g)	0.12	0.15
CEC (Me/100g)	4.9	12.74
pH	6.82	7.49
Phosphate (mg kg^{-1})	8.2	21.2

Source: Aweto and Adejumbobi 1991

Another problem with grazing in dryland environments is that it has traditionally been transhumant, with pastoralists moving to floodplain areas in dry seasons and away from them in wet seasons. Commercial irrigated farming in floodplain areas has prevented this, leading to the over-use of other, more marginal areas for grazing.

7.3.3 Forestry

Many areas under natural and semi-natural forest and woodland have been, and are currently being, exploited for their timber, often using large-scale, mechanised logging operations. Such areas include the tropical forests of central Africa, Amazonia and southeast Asia, and the boreal forests of North America and Eurasia. It should be noted, however, that reliable estimates of deforestation rates are difficult to establish. In recent years, satellite imagery has been used to assess the extent of deforestation, particularly in tropical forest areas (e.g. Skole and Tucker 1993). Ideally, this approach should be used in conjunction with ground verification otherwise estimates may be misleading and possibly erroneous. For example, areas of deforestation have sometimes been determined from aerial estimates of recent fire damage. However, this can lead to overestimates of deforestation rates, as smoke plumes from recent fires are likely to cover a far larger area than that which has been deforested. Conversely, underestimates may result if forest areas are thinned rather than clear-felled; in this situation low or intermediate canopy trees may be removed, while the upper canopy remains intact.

The reasons for deforestation are largely economic. Softwoods from areas under boreal forest are used in the paper and construction industries, while tropical hardwoods are used in the manufacture of furniture and for construction purposes. In many developing countries, forest and woodland areas are also heavily exploited for woodfuel, which is the main source of energy (Kirkby and Sill 1991, Moyo et al. 1993), and for agricultural land. Clearly, the exploitative decline in forest and woodland resources, is not sustainable in the long term and a number of environmental problems, including soil erosion, salinisation and biodiversity losses, have developed as a result (Chapter 8). In addition, improved road access into deforested areas is likely to encourage population influx and increased shifting cultivation. A variety of approaches to these problems have been suggested (e.g. Poore 1989, Trudgill 1990). These focus on the sustainable use of forests, which involves selective logging of several areas each year, followed by forest closure and regrowth. This approach allows for sustainable timber production, conservation of soil and water resources, and maintenance of ecosystem diversity. It also ensures the survival of indigenous populations within forest areas.

An example of sustainable forest management is agroforestry, where trees and arable crops are combined. This has been particularly successful in

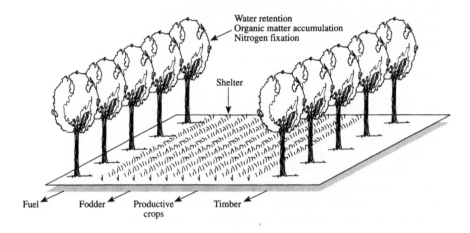

Figure 7.16 The benefits of agroforestry (based on Trudgill 1990)

tropical and subtropical regions where the degree of shading is less than in temperate areas (e.g. Szott *et al.* 1991). The trees are usually planted in rows, between which a variety of crops are established. Although the trees may compete with crops for nutrients, there are a number of benefits associated with the incorporation of trees into cropping systems (Figure 7.16). As well as providing shelter, trees help to control runoff and erosion. They also help to improve soil nutrient status through the addition of organic material and maintenance of soil structure. Moreover, the trees are an important source of timber, woodfuel, animal fodder and fruit crops. The most successful agroforestry projects use indigenous tree species which are best adapted to local soil and climatic conditions. In India, for example, species of *Leucaena* are considered to be particularly beneficial as they are able to fix nitrogen in the soil. In contrast, eucalyptus trees are less favoured because they have a high water demand and their leaf litter is toxic. Studies by the Soil Conservation Institute of the Indian Council of Agricultural Research have shown that eucalyptus trees planted close to crops may reduce crop yields by 10–15 percent (Trudgill 1990). In spite of these problems, eucalyptus trees grow rapidly, providing a useful source of woodfuel, and consequently they are often incorporated into agroforestry systems, but are usually planted in stands well away from crops.

Another approach to sustainable forestry and woodland management is the promotion of small-scale social forestry programmes, where local communities are involved in the establishment of tree nurseries. Farmers are then encouraged to transplant saplings from the nurseries onto their own land. Multipurpose tree species, which provide animal fodder, fruit and woodfuel, and which grow rapidly, are particularly favoured. Since the late 1970s, social forestry schemes, established at the local *panchayat* (village lands) scale, have been successful in parts of Nepal, where they have been supported by the FAO, UN Development Programme, World Bank and Community Forestry Development Programme (CFDP). Sensible forestry and woodland management has also been facilitated by switching energy sources from woodfuel to cheap and sustainable alternatives such as bio-gas, which is derived from the decomposition of animal and human excrement and related waste products (Kirkby and Sill 1991).

In many temperate regions, forestry programmes often utilise rapidly growing softwood species. Here, forestry practices are usually relegated to steeply sloping uplands which are of limited agricultural value. In the last few decades, many areas of the British uplands have been converted from moorland and rough pasture to needleleaf forestry. The main species planted include Sitka spruce (*Picea sitchensis*), Scots pine (*Pinus sylvestris*), Lodge-pole pine (*Pinus contorta*), Japanese larch (*Larix kaempferi*) and Norway spruce (*Picea abies*). These species are well adapted to the cool climatic conditions, poor soils and steep slopes often found in upland areas, and are ready for harvesting within a few decades.

The increase in needleleaf forestry in Britain has arisen primarily from the work of the Forestry Commission and government-aided private forestry which, before the tax year 1988–89, received subsidies of up to 75 percent; these subsidies were withdrawn in 1990. Productive forestry in the UK increased from 573,000 ha in 1919 to 2,207,000 ha in 1985 (9 percent of the UK land area, of which 60 percent consisted of needleleaf plantations) (Denne *et al.* 1986). Although expansion of the forestry industry may be appropriate in economic terms, the environmental consequences may be serious, especially if expansion is poorly managed. In particular, coniferous afforestation has been closely associated with soil and surface water acidification (Hornung 1985, Miller 1985) (section 8.3.1).

As the soils in many planted areas are often wet and acidic, drainage and nutrient amendments are frequently required before planting can commence. Drainage ditches are usually constructed parallel to the direction of slope, and frequently feed into deep check drains which run across the slope. These serve to reduce rates of water flow and soil transport into adjacent streams (Soutar 1989). Following these land pre-treatments, saplings are planted on the ridges between drainage ditches and fertilisers may be added to stimulate growth in the important early phase of tree development. About ten to twenty years later, thinning takes place. This practice encourages stronger growth, thus reducing the susceptibility of plantations to damage by windthrow which is a serious problem in exposed areas (Anderson *et al.* 1989).

The trees are usually harvested after about forty to sixty years, towards the end of the optimum growth phase. After this, the economic rate of return diminishes as growth rate slows. The harvesting process is highly mechanised and may be carried out in one of two main ways. First, only the trunk of the tree is removed, whilst the thin top and branches (brash) are left to protect the soil (lop and top felling). Second, all parts of the tree are removed (whole-tree harvesting) and the soil is left with minimum protection. In a study of whole-tree clear-cutting on soils in the eastern USA, Johnson et al. (1991) observed increases in bulk density caused by mechanical compaction; for high-value timber, however, this can be alleviated by helicopter logging. Decreases in pH, CEC and percent base saturation were also observed, particularly in the upper soil horizons. In a similar study, Goulding and Stevens (1988) compared the effects of conventional clear-felling (lop and top) and whole-tree harvesting of Sitka spruce on future productivity of forestry at a site in North Wales. Soil potassium reserves were used as an indicator of productivity, and although both methods of felling resulted in the return of some potassium to the soil, amounts were greater following conventional clear-felling.

7.4 NON-AGRICULTURAL LAND USE

7.4.1 Urban and industrial activities

Urban and industrial land use both influence, and are influenced by, the soil and its parent material. For example, construction and related engineering programmes are determined to a greater or lesser extent by the physical and mechanical characteristics of soils and their substrates. Such characteristics include load-bearing capacity, shear strength, consistency, shrink–swell potential and drainage capacity. These parameters will influence the strength and stability of foundations, as well as the cost of construction operations.

Another aspect of urbanisation and industrialisation is waste disposal, much of which takes place at landfill sites. Seepage and diffusion of toxic liquids and gases may lead to soil contamination at such sites. The behaviour of toxic constituents in the soil will depend to a large extent on its physical characteristics. Variations in texture, porosity and pore size distribution, in particular, will influence rates of liquid seepage and gaseous diffusion. These processes will also be influenced by the specific site characteristics and management techniques adopted at the landfill site. In addition to the burial of waste materials at designated landfill sites, surface disposal of low-toxicity, biodegradable waste is commonly practised. An example of this is the application of sewage sludge, from sewage treatment works, onto agricultural land on the outskirts of urban areas. In spite of its organic- and nutrient-rich properties, sewage sludge is often found to contain high levels of heavy metals, such as lead, zinc and cadmium, derived from industrial effluent (e.g. Logan 1990, Cavallaro et al. 1993). This can result in the contamination of soils, and of the produce grown on them (Thornton 1991). In response to this problem, the quality of sewage sludge must now be assessed before it can be applied to agricultural land.

In addition to land-based disposal, aerial dispersion of pollutants, and subsequent contamination of soils, is a major problem in many urban and industrial areas. Such pollution is not necessarily confined to areas immediately adjacent to centres of urban and industrial development, but may be transferred several hundreds of kilometres, often traversing national boundaries. Sources of atmospheric pollution which may lead to soil contamination include acid emissions derived from the burning of fossil fuels, and radionuclide emissions derived from nuclear accidents, such as that at Chernobyl in Ukraine in 1986 (Livens and Loveland 1988). Other airborne contaminants often found in soils include organic compounds such as dioxins. A major problem with regard to the management of contaminated land is the absence of a consistent set of guidelines for threshold levels of the many and varied pollutants (Hortensius and Nortcliff 1991) (section 8.3.4).

Another area in which soils are relevant to urban and industrial landscapes is that of land reclamation and restoration. This applies particularly to sites which have been used for opencast mineral extraction. Prior to excavation at such sites, topsoil and subsoil are usually removed and stored temporarily. Storage often takes the form of banks or baffles which serve to hide the excavations, and to reduce levels of noise and dust. Once extraction is complete, the site is infilled with overburden, and subsoil and topsoil are then replaced (e.g. McRae 1989). In some instances, however, there is insufficient bulk to fill the excavations, therefore landfill may be a viable option. The main problems experienced with soil restoration at opencast sites include lack of aeration in stored soil mounds, and compaction by machinery on replacement of the soil (Harris *et al.* 1989). Erosion may also occur in the absence of a protective vegetation cover, particularly if the topsoil is poorly reinstated (McIntosh and Barnhisel 1993). Once the subsoil and topsoil have been replaced, applications of lime and fertilisers are often required, in combination with sensible tillage and land husbandry practices, and drainage systems need to be installed before conditions become suitable for effective plant growth (sections 8.2.2 and 8.2.3). In a study of the cropping potential of restored soils at a site in northern England, Younger (1989) concluded that the installation of an effective drainage scheme was particularly important. Here, yields of winter wheat and spring barley were found to be 30–40 percent lower on undrained land than on drained land.

7.4.2 Recreation and leisure

In the developed world in particular, the amount of time available for outdoor recreational and leisure activities has increased enormously in recent decades. To a greater or lesser extent, all of these activities put pressure on soils and ecosystems, and if they are poorly managed, land degradation may result. This can occur, for example, in the form of footpath erosion, which is a problem on many long-distance footpaths (e.g. Stewart and Cameron 1992). Also, in many upland and mountain areas, increased pressure on ski slopes has caused severe damage to rare and sensitive alpine plant communities. In Scotland, for example, damage to alpine vegetation communities has been particularly severe in parts of the Cairngorm area, and access has been restricted at a number of sites to encourage recolonisation (e.g. Bayfield and Aitken 1992).

Expansion of tourism in parts of the developing world has had an adverse impact on soils and ecosystems. In the Himalayan foothills of Nepal, for example, the growing number of trekkers has put pressure on already dwindling woodfuel resources through their demand for lighting, hot water, and other luxuries which they take for granted in their home countries. A similar argument may be applied to safari tourism in east Africa. As indicated earlier, exploitation of woodfuel has been associated with increased rates of soil degradation in many developing countries (section 7.3.3).

Another, more local aspect of recreation and leisure in which soils are important is that of sportsfield design and construction. In recent years, a considerable amount of research has emerged regarding the construction of artificial soil profiles on sports fields (e.g. Stewart and Scullion 1989) (Figure 7.17). Important requirements for such profiles include a durable turf

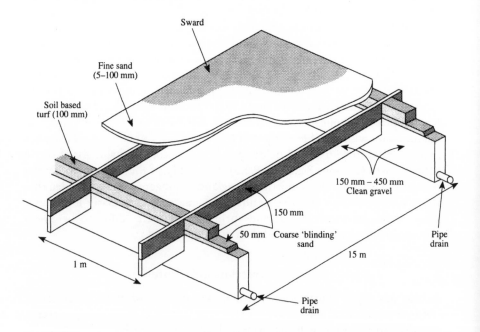

Figure 7.17 Sportsfield soil and turf design (from Stewart and Scullion 1989)

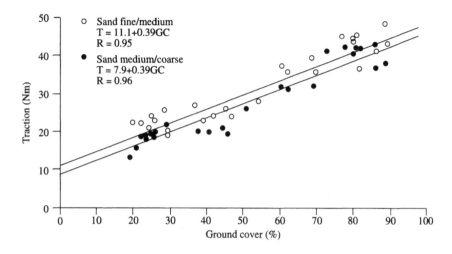

Figure 7.18 Effect of ground cover and texture on the traction properties of sportsfield soils (from Baker 1989)

surface, a subsoil with high compressive and shear strength, and adequate drainage (e.g. Adams 1986, Baker 1989) (Figure 7.18).

7.5 SUMMARY

This chapter has focused on the interactions between soils and landuse systems. Landuse systems are open systems characterised by inputs, outputs, transfers and transformations of material and energy. They may be classified as agricultural or non-agricultural, agricultural systems being further sub-divided into arable, pastoral and forestry practices, and non-agricultural systems into urban/industrial and recreational/leisure activities. Landuse systems vary dramatically in their details and scales of operation. They are also the product of numerous changes and developments which have taken place particularly over the last few decades. These include the development of improved crop, grass and tree varieties, more effective rotation systems, improved agrochemical treatments, more effective tillage practices, new breeds of livestock, and improved drainage and irrigation techniques.

Soil plays a central role in the effective operation of landuse systems. Important physical soil properties include texture, structure, porosity and pore size distribution, hydrological characteristics, and mechanical strength and stability. Similarly, important biochemical properties include organic content, nutrient status and acidity. In agricultural systems, these properties have an influence on the germination and emergence of crops, root network development, the nature and timing of mechanical operations (trafficability), animal stocking rates, the behaviour of pests, diseases and agrochemicals, and drainage and irrigation practices. In non-agricultural systems, they have an influence on construction and engineering practices, and on waste disposal and land restoration techniques.

Soil type, which is defined by specific combinations of the above properties, also has an influence on landuse potential. This is observed most clearly in geographical studies of soil and landuse patterns at the regional or national scale where wide variations in soil type can be found. Under appropriate management regimes and environmental conditions, many soil types are potentially productive, although some are more flexible than others in terms of the variety of landuse practices they can support. The least flexible soil types include Entisols, which are shallow and immature, and Spodosols which are highly leached and acidic, while the more flexible soil types include Alfisols, Mollisols, Histosols and Andisols. Other soil types may be highly productive for short periods of time but require long periods of recovery between phases of cultivation (e.g. Oxisols and Ultisols). In addition to soil properties and soil types, a number of other factors have an influence on landuse systems. These include climate (precipitation and temperature), which affects the degree of SMD and length of the growing season, and relief (altitude, aspect and slope angle), which controls many landuse operations, particularly the use of agricultural machinery. Important human factors include the rate of population growth, socio-economic characteristics, the nature of political and infrastructural support systems (e.g. transport, communications, education and incentive schemes), and the nature of land tenure and farm policies; these differ markedly between the developed and the developing world.

With regard to agricultural land use, a selection of arable, pastoral and forestry practices have been examined in terms of their influence on the soil. Arable practices include shifting cultivation in humid tropical environments which is based on three phases of activity – clearance, cultivation and bush-fallow. Temperate crop rotation systems have evolved from arable–ley rotations and mixed livestock farming, with plentiful supplies of organic matter, towards a cereal monoculture where traditional organic nutrient amendments have been replaced by inorganic fertilisers. Low external input agriculture is becoming increasingly important in the developing world where farmers can rarely afford to invest in costly agricultural machinery and agrochemicals. Pastoral activities focus on the impact of grazing on natural and semi-natural grasslands and on pasture improvement techniques. Pasture improvement is achieved either by replacement of existing herbage in conjunction with fertiliser and lime treatments, or by improvement of existing grasses. Forestry practices in low latitude regions include agroforestry, where trees and crops are combined, and social forestry where farmers are encouraged to transplant saplings from locally managed tree nurseries onto their own land. In temperate regions, afforestation with needleleaf species is widely practised in upland environments where the soils are of limited agricultural value. Specific urban and industrial activities that relate to soil include construction and associated engineering programmes, waste disposal and land restoration. Soil-related recreational activities include long-distance walking and associated footpath management, skiing, tourism in the developing world and related woodfuel demand, and soil profile construction in sportsfield design.

Clearly, the above list of landuse activities is by no means exhaustive and only a selection has been examined here. This chapter has focused on the role of soils in landuse systems and on the impact of selected landuse systems on the soil itself. Many of these impacts are manifested in a number of environmental problems – these are examined further in the following chapter. In terms of management of landuse systems, emphasis should be placed on the maintenance of soil quality through sustainable landuse practices, rather than on soil degradation through exploitative landuse activities.

8

SOILS AND
ENVIRONMENTAL PROBLEMS

8.1 INTRODUCTION

All landuse activities, particularly those which are poorly managed, involve destruction or disturbance, to a greater or lesser extent, of natural and semi-natural ecosystems. Almost invariably, however, it is these ecosystems, in equilibrium with their environment, which offer most effective protection to the soil which supports them. A major consequence of ecosystem destruction and disturbance is that of soil degradation (e.g. Barrow 1991). This has been defined as the decline in soil quality caused through its misuse by human activity. More specifically, it refers to the decline in soil productivity through adverse changes in nutrient status, organic matter, structural stability and concentrations of electrolytes and toxic chemicals (Lal and Stewart 1990a). Soil degradation incorporates a number of environmental problems, some of which are interrelated, including erosion, compaction, water excess and deficit, acidification, salinisation and sodification, and toxic accumulation of agricultural chemicals and urban/industrial pollutants. In many instances, these have led to a serious decline in soil quality and productivity, and it is only in recent decades that the finite nature of soil as a resource has become widely recognised.

Soil degradation is not a new phenomenon. Archaeological evidence suggests that it has been on-going since the beginning of settled agriculture several thousand years ago (section 5.2.2). The decline of many ancient civilisations, including the Mesopotamians of the Tigris and Euphrates valleys in Iraq, the Harappans of the Indus valley in Pakistan and the Mayans of Central America, was due in part to soil degradation (Olson 1981b). More recently, an event of major significance was the dustbowl which occurred in the Great Plains of the American midwest during the 1930s. At this time, intensive agricultural practices, employed in the eastern states, were transferred to the drier midwest where the soils are lighter textured and more susceptible to erosion. A number of years of drought, combined with crop failure and destruction of the protective organic-rich topsoil, resulted in severe wind erosion.

According to the Global Assessment of Soil Degradation project, about 15 percent of the global land area between 72°N and 57°S is degraded (Ayoub 1991). Of this, an area slightly less than that of India (c. 300 million hectares) is strongly degraded, largely as a result of deforestation (113 million hectares), inappropriate management of cropped land (83 million hectares) and overgrazing (75 million hectares). In recent decades the global rate of soil degradation has increased dramatically, and is likely to increase further as we approach the twenty-first century; in 1983 it was estimated at 5–7 million ha a^{-1} and is set to rise beyond 10 million ha a^{-1} by the year 2000 (Lal and Stewart 1990a).

The effects of soil degradation are not restricted to the soil alone, but have a number of off-site implications. Soil erosion, for example, is often associated with increased incidence of flooding, siltation of rivers, lakes and reservoirs, and deposition of material in low-lying areas (e.g. Morgan 1986). These problems may be compounded in areas where infiltration capacity is reduced due to compaction, hardsetting or induration of soils. Salinisation and sodification of soils are often associated with poor quality irrigation water (Szabolcs 1986), while soil acidification is commonly linked with acidification and aluminium contamination of surface waters (e.g. Adams et al. 1990). Leaching of fertilisers and pesticides from agricultural soils may also lead to contamination of surface and shallow ground waters (e.g. Addiscott et al. 1991). In addition, contamination of soils by urban and industrial pollutants, such as heavy metals and radionuclides, may lead to toxic accumulation in arable produce and in herbage for grazing animals, thus having important implications for human health (Thornton 1991).

The extent of soil degradation is influenced by a number of factors, many of which are interrelated, namely soil characteristics, relief, climate, land use, and socio-economic and political controls (Figure 8.1). In many studies of soil degradation and its wider environmental implications, the socio-economic and political controls are often overlooked, or at least not examined in any detail, perhaps because of the difficulties associated with the collection of reliable and comparable data. Increasingly, however, these controls on landuse systems are being viewed as central to the issue of soil degradation, particularly in the developing world (Blaikie 1985, Millington et al. 1989, Thapa and Weber 1991, Hudson 1992, Moyo et al. 1993).

Management of soil degradation, whether at a global, regional or local scale, is clearly a complex issue and represents one of our most challenging environmental problems. Emphasis should be placed on sustainable rather than exploitative landuse practices; this theme was highlighted by the World Soil Charter (FAO 1981) which called for a commitment by governments, agencies and land users to 'manage the land for long term advantage rather than short term expediency' (FAO 1983, Stocking 1992). The problem requires a holistic, multidisciplinary approach involving the collaborative and co-ordinated efforts of ecologists, agronomists, soil scientists, hydrologists, engineers, sociologists and economists. Moreover, the

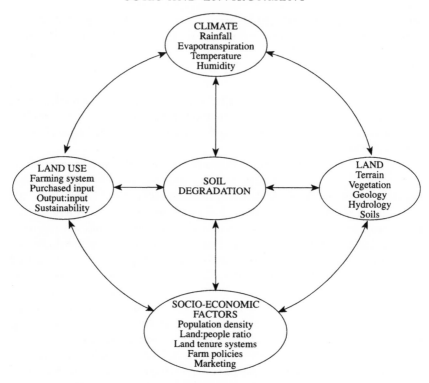

Figure 8.1 The causal factors of soil degradation and their interactions (from Lal and Stewart 1990b)

involvement of government and non-government organisations, aid agencies and the farmers themselves is essential to the success of research and development in this area. Such involvement should facilitate the implementation of education, training and incentive programmes. Imposition from above of high-technology, high-cost solutions by technical experts from developed countries is certainly not the answer in the developing world (Hudson 1992, Stocking 1992). Inevitably such solutions are not economically viable and low-technology, low-cost options, such as low external input agriculture, agroforestry and social forestry (section 7.3), are often the only answer (Sanchez and Benites 1987, Trudgill 1990). Hence, the approach to soil conservation has shifted in recent years from a rather technocentric standpoint to a more ecocentric position. Central to this approach are the concepts of land husbandry and sustainable development, which place emphasis on the land users themselves rather than on the technical experts and advisors (Hudson 1992, Moyo *et al.* 1993).

This chapter aims to examine a selection of the most pressing soil degradation problems, and in each case the causal factors, on- and off-site effects, and management strategies will be considered.

8.2 PHYSICAL PROBLEMS

8.2.1 Erosion

Soil erosion occurs when the rate of removal of soil by water and/or wind exceeds the rate of soil formation. Generally, rates of soil formation are very low, with profiles developing at a rate of about 1 cm every 100–400 years (FAO 1983); assuming an average bulk density of 1.33 g cm^{-3}, this equates to about 0.3–1.3 t ha^{-1} a^{-1}. It is important to differentiate between natural or background erosion, and erosion which has been accelerated largely as a result of human activity. Background erosion rates are often similar to rates of soil formation at < 1.0 t ha^{-1} a^{-1}, although in mountainous areas they may be considerably higher (Morgan 1986). In contrast, rates of accelerated erosion commonly exceed 10 t ha^{-1} a^{-1} and sometimes exceed 100 t ha^{-1} a^{-1}. Some of the highest soil erosion rates have been observed in the loess plateau area of China and in the Himalayan foothills of Nepal, where values in excess of 200 t ha^{-1} a^{-1} have been recorded (Bojie Fu 1989, Pearce *et al.* 1990). Similarly, in India, gully erosion results in a loss of about 8,000 ha of land per year (FAO 1983). The processes of erosion have already been discussed in relation to soil losses in section 3.2.4, in terms of soil hydrology in section 6.2.3, and in sections 6.4.2 and 6.4.3 with reference to the influence of soils on the geosphere.

The extent of soil erosion is governed by a number of factors. Those of particular importance include erosivity of the eroding agent, erodibility of the soil, slope steepness and length, landuse practices and conservation strategies. These factors are summarised in the Universal Soil Loss Equation which has been used widely in the modelling and prediction of soil erosion (e.g. Colby-Saliba 1985):

$$E = R.K.L.S.C.P$$

where E = mean annual soil loss, R = rainfall erosivity index, K = soil erodibility index, L = slope length, S = slope steepness, C = cropping factor which represents the ratio of soil loss under a given crop to that from bare soil, and P = conservation practice factor which represents the ratio of soil loss where contouring and strip-cropping are practised to that where they are not (Morgan 1986). Although widely used, this model has been the subject of extensive criticism. For example, it assumes that a vegetation cover is always protective which is not necessarily the case; erosion on land with a good cover of crops planted in rows can be greater than on land which is sparsely vegetated. It is also water erosion based and cannot be used in areas affected extensively by wind erosion. More specifically, it focuses on rill and

241

inter-rill erosion and is not easily applied to areas where gully and stream bank erosion are widespread. Its universal nature has also been questioned, particularly in terms of its application to tropical soils. Furthermore, it should be emphasised that this model does not consider the wide range of socio-economic and political factors which play a crucial role in terms of their influence on the degree of soil erosion (e.g. Millington *et al.* 1989, Thapa and Weber 1991); these will be examined later. Alternative models include SLEMSA (soil loss estimator for southern Africa) and CREAMS (chemicals runoff and erosion arising from agricultural management systems) (Morgan 1986).

Erosivity is a measure of the potential of the eroding agent to erode and is commonly expressed in terms of kinetic energy. The erosivity of rainfall relates to the detaching power of raindrops, which increases with increasing rainfall intensity and is largely a function of drop size. This relationship is non-linear, however, and at intensities in excess of 50 mm hr^{-1} the increase in kinetic energy is minimal as turbulence prevents further increase in drop size. Rainfall erosivity tends to be greatest in areas where there is a marked seasonal concentration of rainfall. There are a number of rainfall erosivity indices, although one of the most widely used is the EI_{30} index which is a compound index of kinetic energy and the maximum 30-minute rainfall intensity (Figure 7.11). Indices of wind erosivity are based largely on the velocity and duration of the wind (Morgan 1986).

Erodibility of the soil is a measure of its resistance to detachment and transport, and depends on a number of soil characteristics, particularly texture, organic content, structure and permeability. In terms of both water and wind erosion, the most erodible soils tend to be characterised by low clay and organic contents, and poor structural stability. Similarly, soils with an intermediate texture (fine sand to coarse loam) are more erodible than both coarse-textured sandy soils, where particle size and mass are large, and fine-textured clayey soils, which are more cohesive (section 3.2.4). The erodibility factor (K) in the Universal Soil Loss Equation can be determined from soil texture, organic content, structure and permeability using the nomograph shown in Figure 8.2.

Land use is perhaps the most significant factor influencing soil erosion, for two main reasons. First, many landuse practices leave the soil devoid of a protective vegetation cover, or with only a partial cover, for significant periods of time and second, they involve mechanical disturbance of the soil. Specific aspects of land use often associated with accelerated soil erosion include expansion and intensification of arable cultivation, overgrazing, deforestation, certain forestry practices, site clearance in preparation for urban and industrial construction, and a number of recreational activities such as walking and skiing.

Arable cultivation has expanded and intensified dramatically in recent decades. Relatively steep slopes, formerly covered by grass or trees, have

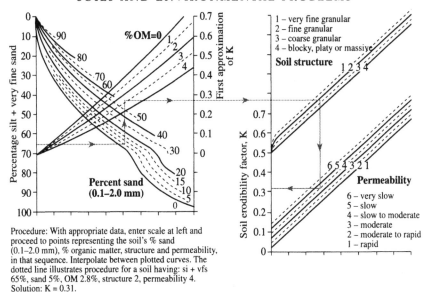

Procedure: With appropriate data, enter scale at left and proceed to points representing the soil's % sand (0.1–2.0 mm), % organic matter, structure and permeability, in that sequence. Interpolate between plotted curves. The dotted line illustrates procedure for a soil having: si + vfs 65%, sand 5%, OM 2.8%, structure 2, permeability 4. Solution: K = 0.31.

Figure 8.2 Nomograph for the calculation of soil erodibility (K), for use in the Universal Soil Loss Equation (from Wischmeier *et al.* 1971)

been converted to arable cropping, while an increased use of heavy agricultural machinery has resulted in compaction of the soil. This in turn has led to reduced infiltration capacity, particularly along wheel tracks, thus resulting in increased surface runoff and erosion (e.g. Robinson and Naghizadeh 1992, Hill 1993). Similarly, increased reliance on tillage activities, throughout the cropping cycle, has rendered soils more susceptible to erosion. This problem has been compounded by the decline in levels of soil organic matter, and hence structural stability, largely in response to increased use of inorganic fertilisers. In addition, the tendency to increase field sizes on arable land has meant that there are fewer physical breaks and barriers in the landscape, such as tree lines, hedgerows and walls, to restrict erosion. Susceptibility to erosion is further increased if land is cultivated with the slope rather than parallel to the contours. These problems are well illustrated by Boardman (1990a, 1991a) in studies of soil erosion in southeast England (Figure 8.3).

Overgrazing is particularly common in drought-affected parts of the developing world, such as the Sahelian region of sub-Saharan Africa, and the rangelands and communal lands of eastern and southern Africa. In a study of the impact of grazing on soils of the savanna region of Nigeria, for example, Aweto and Adejumbobi (1991) attribute enhanced surface runoff and erosion to compaction of the soil and destruction of the protective vegetation cover by grazing animals, and to the adoption of inappropriate

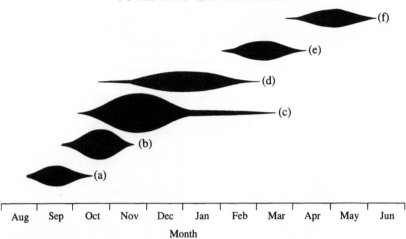

Figure 8.3 Erosion risk on the South Downs, southeast England. The timing and degree of risk is indicated for different crops or management practices. The thickness of the envelope denotes erosion risk: (a) oil seed rape and grass ley, (b) ploughed/cultivated in preparation for winter cereals, (c) winter cereals, (d) ploughed/cultivated in preparation for spring cereals, (e) spring cereals, (f) late spring and early summer crops e.g. maize (from Boardman 1991a)

burning strategies. Deforestation, largely for logging and woodfuel purposes, is also common in many parts of the developing world. Trees are well known for their ability to protect soils from erosion, particularly on steeply sloping terrain. Their root systems, and the organic material which they supply, help to stabilise the soil, while water uptake and canopy interception serve to reduce the frequency and intensity of surface runoff.

In addition to deforestation, many forestry practices are associated with accelerated soil erosion, including the needleleaf forestry programmes which have become widespread in many areas of upland Britain. Here, erosion is most serious during the pre-planting stages of land preparation and drainage, and after harvesting (Soutar 1989). In relation to urban and industrial land use, construction and associated disturbance of land may lead to increased soil erosion (e.g. Wolman 1967) (Figure 8.4). Even certain recreational activities have been implicated in this problem, including walking and skiing (section 7.4.2).

A number of socio-economic and political factors have been associated with accelerated soil erosion, particularly in the developing world. These include population pressure, skewed land resource distribution, poverty and marginalisation, increasing demand for woodfuel, innappropriate land tenure and farm policies, small size of land-holdings and poor infrastructure (McGregor and Barker 1991, Thapa and Weber 1991). In many developing countries, population growth is rapid and the demand for agricultural land

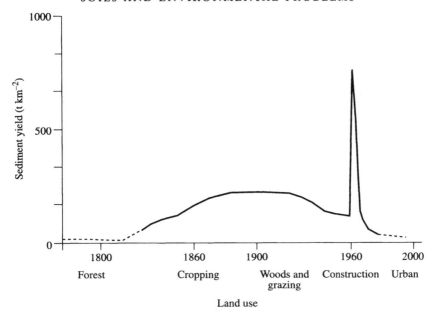

Figure 8.4 Relationships between sediment yield and changing land use, Maryland, USA (from Wolman 1967)

and woodfuel is ever increasing (Table 8.1). Furthermore, agricultural systems are characterised by a skewed land resource distribution where a minority of affluent and powerful landowners control a majority of the land area. The poorest farmers are thus forced onto marginal land, which is particularly susceptible to erosion, and often end up in a vicious spiral of

Table 8.1 Woodfuel as a proportion of energy use in selected countries

Country	Woodfuel as a percentage of total energy consumption
Nepal	98
India	36
Thailand	63
Angola	74
Tanzania	94
Brazil	33
Chile	16

Source: Based on Thapa and Weber 1991

245

debt. Rural–urban migration, abandonment of land and increased soil erosion are often responses to this poverty trap situation (Millington *et al.* 1989) (Figure 8.5).

In many parts of the developing world, large areas of land are utilised for mono-cultivation of cash crops, which are not necessarily best suited to soil conditions, rather than for indigenous mixed food cropping. Such commercial pressure on agricultural systems, as well as contributing to the problem of marginalisation discussed above, has a detrimental effect on soil quality and is unlikely to be sustainable in the long term. There is also little political support in terms of education, training and incentive schemes to encourage farmers to adopt more sustainable landuse practices. The establishment of appropriate and comprehensive soil conservation and land husbandry programmes is further hindered by the small size of land-holdings and the large numbers of farmers involved (McGregor and Barker 1991, Thapa and Weber 1991) (Table 8.2).

The on- and off-site effects of soil erosion are considerable. At the global scale, it is estimated that unless soil conservation measures are introduced on

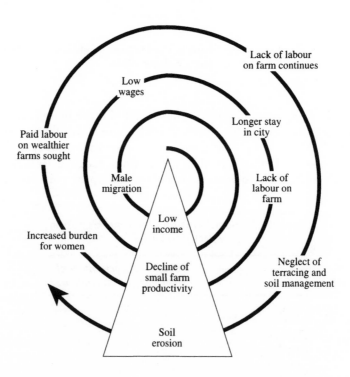

Figure 8.5 The spiral of poverty, land abandonment and soil erosion in the developing world (from Millington *et al.* 1989)

Table 8.2 Distribution of agricultural land-holdings in selected developing countries by size of holdings and proportionate area

Country	Size of holdings (%)			Area (%)		
	< 5 ha	5–50 ha	> 50 ha	< 5 ha	5–50 ha	> 50 ha
Asia						
India	91	9	0	51	46	3
Philippines	86	14	0	51	37	12
Thailand	72	28	0	39	61	0
Africa						
Algeria	69	30	1	14	63	23
Zaïre	99	1	0	60	5	35
Zambia	94	6	0	34	19	47
Latin America						
Argentina	15	39	47	0	2	98
Brazil	37	45	16	1	12	87
Columbia	60	32	8	4	19	78

Source: Thapa and Weber 1991

all cultivated land, 544 million ha of potentially productive rain-fed crop land will be lost, and agricultural production will decrease by almost 20 percent, by the year 2000 (FAO 1983). Undoubtedly, these effects will be felt most severely in those developing countries which are least able to cope with the problem. It should be noted that the deterioration in soil productivity is disproportionate to the amount of soil eroded, as it is the nutrient-rich and structure-supporting constituents in the topsoil which are lost most readily.

The off-site impacts of soil erosion are just as serious as the on-site effects, and are often more dramatic. Numerous incidents of disruption to communication systems, flooding, siltation of water supplies and damage to property have been reported (e.g. Ives and Messerli 1989, Boardman 1990a, Robinson and Blackman 1990, Ellis *et al.* 1993). In financial terms, the Conservation Foundation of the USA estimates the annual on-site and off-site costs of soil erosion, on a national basis, to be about $40 million and $3 billion respectively (Boardman 1990b). Allocation of responsibility and accountability for off-site incidents, and the implications for government policy and legislation, are currently the subject of debate in Britain (e.g. Boardman 1990a); this situation is in marked contrast to that in many other countries, such as Canada and New Zealand, where government policies on soil conservation and farmer education are well established and relatively successful (e.g. Boardman 1991b, Jakobson and Dragun 1991).

In terms of the management and remediation of soil erosion, Morgan (1986) identifies three main strategies – agronomic, soil management and mechanical (Table 8.3). Agronomic practices aim to protect the soil through sensible cropping programmes and are based on the encouragement of a dense vegetation cover and plant root network. Under these conditions, the time period over which the soil is left bare, and thus susceptible to erosion, is minimised. An example of an agronomic practice is strip-cropping, in which alternate rows of different crops are arranged, usually across the slope, so that they grow and are harvested at different times, thus ensuring at least a partial cover of vegetation for much of the year. Using this approach, arable crops may be combined with strips of grass or trees (agroforestry) (section 7.3.3). Retention of crop stubble in the soil, rather than burning, and the planting of trees or shrubs around the heads of gullies to reduce the rate of headward recession, are other examples of agronomic practices.

Soil management techniques aim to increase the resistance of soil to erosion, and focus mainly on the improvement and maintenance of soil structure. Such practices include mulching, where organic residues are added to the soil (e.g. Zuzel and Pikul 1993), and reduced or zero-tillage options, including direct drilling, where mechanical disturbance of the soil is minimised (e.g. Moran *et al.* 1988). Soil erodibility may also be reduced by

Table 8.3 The effectiveness of various soil conservation strategies in the control of soil erosion

Practice	Rainsplash D	Rainsplash T	Runoff D	Runoff T	Wind D	Wind T
Agronomic measures						
Improving vegetation cover	*	*	*	*	*	*
Increasing surface roughness	–	–	*	*	*	*
Increasing surface depression storage	+	+	*	*	–	–
Increasing infiltration	–	–	+	*	–	–
Soil management						
Fertilisers, manures, soil conditioners, bio-engineering techniques	+	+	+	*	+	*
Subsoiling, drainage	–	–	+	*	–	–
Mechanical measures						
Contouring, ridging	–	+	+	*	+	*
Terraces	–	+	+	*	–	–
Shelterbelts	–	–	–	–	*	*
Waterways	–	–	–	*	–	–

Source: Morgan 1980
Note: – = no control; + = moderate control; * = strong control; D = detachment phase; T = transport phase

using synthetic soil conditioners (e.g. Nadler 1993). These consist largely of organic polymers such as PVA (polyvinyl alcohol), PAM (polyacrylamide) and PEG (polyethyleneglycol), and are applied as a surface film or are incorporated into the topsoil. Treatment is costly, however, and tends to be restricted to specialised applications such as the stabilisation of sand dune systems, and protection of the fine tilth of seedbeds (White 1987). Bio-engineering techniques, including the use of biodegradable geotextiles, have also been used to reduce the susceptibility of soils to erosion. At the Cairngorm ski slopes in Scotland, for example, these have been used successfully to facilitate the recovery of sensitive alpine plant communities (Bayfield and Aitken 1992).

Mechanical techniques aim to reduce the energy of the eroding agent and often involve the modification of surface topography. Examples include terracing (Figure 8.6), contour bunding and the construction of diversionary spillways to direct water away from areas which are particularly susceptible to erosion. Shelter belts may also be planted to reduce wind erosion, although these will compete with crops for soil moisture, which may be problematic in drier areas. At the more local scale, check dams may be constructed in gully systems in an attempt to reduce erosion risk, while

Figure 8.6 Terraced agricultural land in Nepal (courtesy of T. Douglas)

gabions (wire baskets filled with stones) may be installed to reduce bank erosion and headward recession. In addition to their use on agricultural land, mechanical techniques are also used to control footpath erosion. These include the installation of drainage channels to direct water flow away from damaged stretches of footpath, and the construction of elevated sections to avoid damage to poorly drained soils. In extreme circumstances, footpaths are temporarily or permanently re-routed to facilitate the recovery of damaged stretches.

In terms of the relative effectiveness of the above measures, agronomic techniques tend to offer greatest protection to the soil, as they are able to control both the detachment and transport phases of soil erosion (Table 8.3). Soil management and mechanical techniques tend to be less effective, particularly in their control of the detachment phase of erosion. For this reason, it is usual to combine soil management and mechanical techniques with agronomic measures. Agronomic practices are often used in preference to others, not only because they are effective in the control of erosion, but also because they are relatively cheap to adopt, can be adapted to fit in with traditional cultivation methods and are acceptable to local people.

In recent decades, there has been a change in focus away from a largely technocentric approach to soil conservation, towards a broader and more ecocentric view of land management. This places emphasis not only on the soils themselves, but also on the landuse system, the socio-economic and political controls, and the wider off-site implications of soil erosion (e.g. Hudson 1992, Stocking 1992). More important, however, is that the ecocentric approach focuses on the land users themselves and on their role in the determination of their own future. This is in marked contrast to the imposition of decisions from above by technical experts, which perhaps explains, at least in part, the failure of many early soil conservation projects (Hudson 1992). Specific practices which have been effective in the control of soil erosion, particularly though not exclusively in the developing world, include agroforestry, social forestry and low external input agriculture (section 7.3).

8.2.2 Compaction

Soil compaction involves the compression of a mass of soil into a smaller volume and is usually expressed in terms of dry bulk density, porosity and resistance to penetration. The bulk density of soils which have been compacted by agricultural machinery, for example, may exceed 1.5 g cm^{-3}, while that of comparable uncompacted soils is usually between 1.0 and 1.5 g cm^{-3}. The ease with which soils are compacted depends on a number of characteristics, particularly texture, occurring most readily in soils which contain appreciable quantities of clay. The compacting force, which usually acts in a vertical direction, causes alignment of the clay platelets in a direction which is more or less parallel to the ground surface. Alignment of the clay particles in this way often leads to the formation of compaction or cultivation pans which may be a few centimetres in thickness (Figure 8.7). Such pans tend to form at a depth of about 20–30 cm, are often characterised by a well developed platy structure, and are commonly associated with impeded drainage and restricted development of plant root networks. In silty soils, raindrop impact may lead to surface crusting which is another form of soil compaction. Surface crusts may be several millimetres in thickness, and are commonly associated with reduced infiltration, increased surface runoff and accelerated soil erosion, and restricted germination and emergence of crops (Le Bissonnais and Singer 1993). In addition to textural controls, susceptibility to compaction also increases with increasing organic matter and water contents.

Closely related to soil compaction is the process of hardsetting which involves an increase in bulk density but without the application of an external load. Hardsetting soils are characterised by their low structural stability. During and after wetting, slaking and collapse of aggregates lead to uniaxial shrinkage and dispersal of clay and silt and, on drying, the soils harden without restructuring. Acceptance of hardsetting soils as a distinctive group has been hindered by the difficulties experienced in distinguishing hardsetting characteristics from other forms of soil behaviour (Mullins *et al.* 1987) (Figure 8.8).

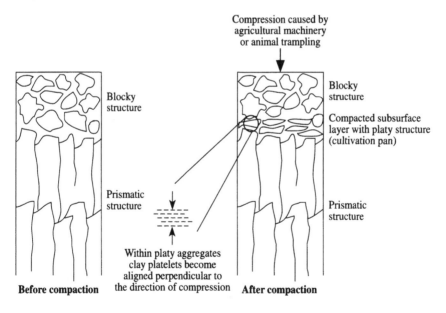

Figure 8.7 Soil compaction and the formation of cultivation pans

251

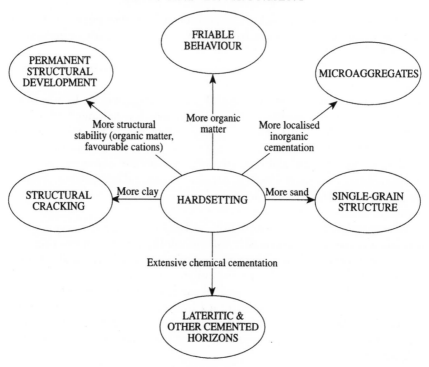

Figure 8.8 Hardsetting in relation to other types of soil behaviour (from Mullins *et al.* 1987)

One of the main causes of soil compaction is the excessive use of heavy agricultural machinery, and compaction is particularly severe along wheelings left by these vehicles (e.g. Liebig *et al.* 1993). For example, bulk density values of 2.2 g cm^{-3} and 1.3 g cm^{-3} have been recorded in soils of wheeled and inter-wheel areas respectively (McRae 1989). The intensification of arable cultivation in recent decades has been facilitated to a large extent by increased mechanisation. Most procedures in the cropping cycle, from tillage and seedbed preparation, through drilling, weeding and agrochemical applications to harvesting, are now largely mechanised, particularly in the developed world. The timing of cultivation in relation to precipitation and soil water levels is known to be critical with respect to soil compaction. If soils are cultivated when they are near or above field capacity, or their plastic limit (sections 2.3.4 and 2.3.6), then severe structural damage and compaction are likely to occur (Briggs and Courtney 1989) (Figure 8.9).

Soil compaction is not restricted to arable land but is also common on land used for grazing and forestry. Grazing animals are well known for their ability to cause compaction, or poaching, especially during wet conditions.

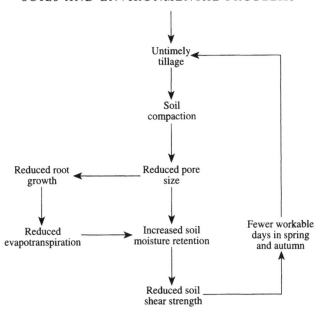

Figure 8.9 Feedback effects of soil compaction due to untimely tillage (from Briggs and Courtney 1989)

It is particularly common to find poorly drained or puddled areas in the vicinity of food troughs or gateways where animals tend to congregate. In a study of the impact of grazing animals on soils of the savanna region of Nigeria, Aweto and Adejumbobi (1991) reported significant compaction and loss of protective vegetation cover in grazed plots. In forestry plantations in upland Britain, heavy agricultural machinery is often used during the preparation and drainage of land prior to planting, and during harvesting operations; such mechanisation is often associated with the disturbance and compaction of soils (Soutar 1989). Similarly, soil compaction has been reported in association with the clearance of tropical rainforest, particularly where mechanical methods are used in preference to the traditional slash and burn approach (e.g. Nortcliff and Dias 1988) (Figure 8.10).

Soil restoration following mineral extraction, quarrying and opencast coal mining has also been implicated in the soil compaction problem. Compacted horizons result largely from the passage of heavy machinery during soil stripping and replacement, and from the shear forces produced during the lifting process (Harris *et al.* 1989). In clay and loam topsoils which had been stockpiled in southeast England, Abdul-Kareem and McRae (1984) reported bulk density values that were 20 percent higher than in similar topsoils at undisturbed control sites nearby. Elevated values are not restricted to

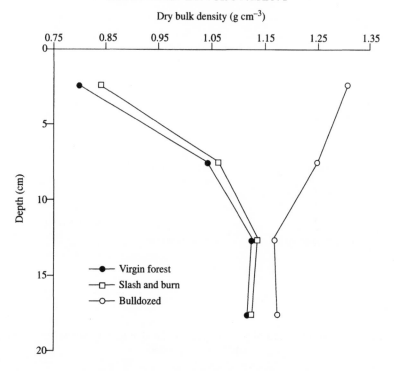

Figure 8.10 Dry bulk densities of soils measured six months after forest clearance
(based on Nortcliff and Dias 1988)

topsoils, however. King (1988), for example, observed dramatic increases in bulk density at the top of the subsoil, at about 0.3 m in depth, in restored soils in northeast England; values of 1.64 g cm^{-3} and 1.46 g cm^{-3} were recorded in the restored and unmined control sites respectively.

Compaction has adverse effects on a number of soil characteristics. Increased bulk density and the consequent decrease in porosity are associated with both increased waterlogging and poor aeration; these changes in turn have a detrimental effect on the thermal characteristics of soils (section 6.3.1). Under these conditions, plant growth may be restricted, particularly during the early stages of germination and emergence, and during the main phase of root network development (Figure 8.11). Similarly, the increased incidence of root rot in soils with compacted cultivation pans has been widely documented (e.g. Raghavan *et al.* 1990). Other characteristics that may be adversely affected by compaction include the size and diversity of soil organism populations, and the incidence of certain crop pests and diseases may also increase.

Soil compaction has been associated with declining crop yields in many areas. In Britain, for example, Briggs and Courtney (1989) present yield data

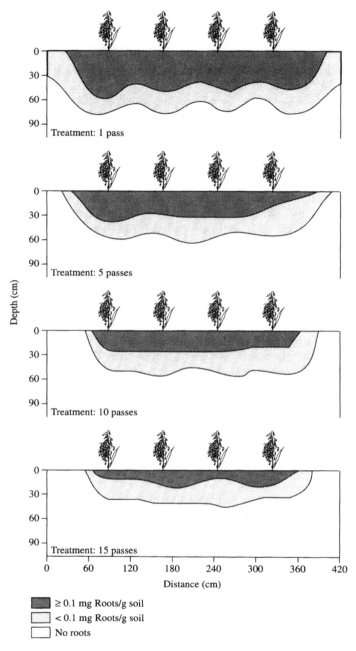

Figure 8.11 Compaction and restricted root network development: root density distributions in clay soils subjected to 1, 5, 10 and 15 wheel passes at a contact pressure of 62 kPa (from Raghavan *et al.* 1979)

for a number of crops grown in soils with a variety of textures, ranging from sand to clay. Yields from compacted sites were found to be 25–74 percent lower than those from uncompacted control sites. Similarly, yield reductions of up to 50 percent were observed for corn on compacted clay soils in Quebec (Raghavan *et al.* 1978).

The financial impact of soil compaction is difficult to assess due to the influence of a number of interrelated factors. In addition to yield reductions, damaged soil structure may have to be restored, surface runoff and soil erosion may increase, fertiliser usage becomes less efficient due to increased leaching losses, water loss by evaporation may increase, together with the operational costs of irrigation, and in soils which are susceptible to waterlogging, expensive drainage systems may have to be installed. It is estimated that the cost of fuel required for tillage of compacted soils may be up to 35 percent greater than that for uncompacted soils. In the USA alone, during the early 1970s, on-farm losses resulting from yield reductions on compacted soils exceeded $1 billion annually (Raghavan *et al.* 1990).

Successful management and remediation of soil compaction centres on two main approaches – improvement and maintenance of soil structure, and appropriate tillage practices. Good soil structure provides the basis for increased mechanical strength and stability, improved water retention and aeration, and protection of plant nutrients from leaching. Soil structure can be improved and maintained through the promotion of sensible crop rotation practices, particularly those which include significant periods of ley pasture, mulching and stubble retention (section 7.3.1). Unlike continuous cropping, such practices encourage the accumulation of soil organic matter and the development of strong plant root networks, both of which are essential ingredients in the development of good soil structure. Soil structure also benefits from the addition of certain nutrients, particularly the multivalent base cations which play a crucial role in the aggregation process (section 2.3.2).

In addition to soil structural improvements, good tillage practices are an essential component of effective management and remediation of soil compaction. Tillage of agricultural land is practised for a number of reasons – first, to create a fine tilth which facilitates effective germination and emergence of seedlings, second, to remove weeds which compete with crops for nutrients, water and light, third, to incorporate crop residues into the soil with the aim of improving soil structure and nutrient content, and fourth, to improve drainage and aeration in the root zone (rhizosphere) (Briggs and Courtney 1989). Although tillage is therefore of benefit to the soil, when mechanised it can lead to soil compaction. The timing of tillage is particularly important in this respect, and it should be avoided in wet conditions, particularly if soil moisture content is likely to exceed the field capacity or plastic limit. In an attempt to alleviate compaction resulting from the excessive use of heavy agricultural machinery, reduced or zero-tillage practices have been widely adopted (section 7.3.1). Such practices are inappropriate for poorly drained soils, however, where conventional tillage helps to improve drainage, aeration and the thermal characteristics of the topsoil (Pidgeon and Thorogood 1985). They should also be avoided on excessively drained soils, where increased nitrate leaching and denitrification (section 8.3.3) may lead to nitrogen deficiency, and on steep slopes where seed slots may be subjected to increased erosion. Thus, it appears that the benefits of reduced and zero-tillage are felt most strongly on medium-textured soils which are well drained and which possess a well developed structure. In some circumstances, compaction and associated poor drainage cannot be improved through good structural management and appropriate tillage practices alone, and installation of artificial drainage may be necessary (section 8.2.3).

8.2.3 Water excess and deficit

Excess water may be present in soils for a number of reasons. Compaction, for example, in the form of surface crusting, subsurface cultivation pans and hardsetting, often leads to reduced infiltration and restricted permeability (section 8.2.2). Excess water is also common in heavily textured clay soils, particularly in high rainfall areas, and in areas which are flat, low lying and prone to flooding. If excess water is found within about 40–50 cm of the surface, then the adverse effects on soil aeration and temperature are likely to lead to restricted root growth and crop performance (section 6.5.2).

In many cases, excess water may be removed from the soil through both sensible management of soil structure and appropriate tillage practices (section 8.2.2). Frequently, however, installation of artificial drainage is necessary. The aim of any artificial drainage programme is to lower the water table so that the amount of plant-available water in the rhizosphere is maximised. If the water table is lowered too much, the soil may become drought-susceptible. Conversely, if the water table remains too high, waterlogging and poor aeration are likely to continue. The efficiency of drainage depends largely on the lateral spacing, depth and size of drains (Figure 8.12). Generally, efficiency improves with decreased lateral spacing, although at spacings of greater than around 6 m the rate of inflow to the drains is independent of spacing. Drain depth controls the height of the water table and usually occurs at about 50 cm to 2 m. In terms of size, the ability of a drain to carry water increases with increasing cross-sectional area. The choice of drainage system depends on a number of factors, the most important being economic controls (installation and maintenance costs, expected yield improvements and availability of subsidies), management considerations (crop and tillage requirements), soil characteristics (texture, hydraulic conductivity and type of drainage problem) and climatic controls (rainfall amount, intensity and seasonality) (Bailey et al. 1980, Briggs and Courtney 1989).

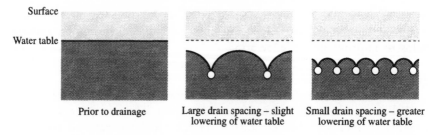

Figure 8.12 Impact of drain spacing on level of water table (based on Courtney and Trudgill 1984)

There are a number of specific drainage techniques which can be employed, either singly or in combination, depending on the type and extent of the drainage problem. Those most commonly used include subsoiling, mole drainage, pipe or tile drainage, open ditches or dykes, and regional or arterial drainage (Briggs and Courtney 1989). Subsoiling is a form of deep ploughing and is often used to break up compacted cultivation pans and indurated layers within the soil (Figure 8.13). The critical depth for this practice is about 20–50 cm; if it is shallower, then lifting of the soil may occur and if deeper, consolidation of the soil may result. Subsoiling is most effective when the soil is relatively dry and heavy; in wet conditions, plastic deformation and smearing may occur during passage of the subsoiling

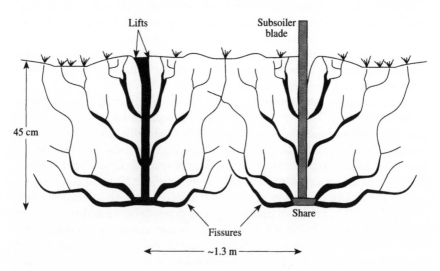

Figure 8.13 Fissure development in response to effective subsoiling (based on MAFF 1981)

implement, and this is likely to have an adverse effect on soil structure.

Mole drainage is produced by drawing a mole plough, which is a bullet-shaped instrument usually less than 20 cm in diameter, through the soil to produce a continuous passage. Mole drains are established at a depth of about 40–60 cm, with a lateral spacing of about 2–5 m. They are not permanent features and are susceptible to collapse and sediment blockage (Armstrong 1986). Consequently, they have to be renewed on a regular basis, usually after a period of about 5–7 years. During the installation of mole drainage, the moisture status of the soil is of crucial importance. If conditions are too wet, smearing and consolidation may result, while shattering may occur if the soil is too dry. In spite of the temporary nature of mole drainage, it is often adopted in preference to other drainage practices because it is cheap and easy to install. It is also commonly used as a form of secondary drainage in combination with primary tile or ditch drains.

Pipe or tile drainage consists of plastic or ceramic pipes of about 10–30 cm in diameter, and is installed at a depth of 0.5–2.0 m. The pipes usually contain perforations or slits in order to speed up the transfer of water from the soil. Once the pipes have been installed, the ditches are often back-filled with gravel, polystyrene or fuel-ash (Figure 8.14). This helps to improve the flow of water to the drains, while acting as a filter to remove fine sediments which may otherwise lead to drain blockage (e.g. Castle 1986, Chow et al. 1993). Open ditches or dykes are usually constructed at field margins and are relatively cheap and easy to maintain. This type of drainage has been used since settled agriculture began, and in Britain dates back to the Roman period. In some circumstances, drainage may need to be improved at the regional scale by arterial drainage. This can be achieved in a number of ways, including straightening of river channels to increase gradient and improve flow, deepening of river channels by dredging, and pumping of water from agricultural land. Flood prevention is also an important aspect of arterial drainage. In coastal areas, this is achieved through the construction of barrier systems which form part of sea defence strategies. Similarly, in low-lying floodplain areas, it may involve the construction of raised river banks or levées. Such regional or arterial drainage also has a long history, and large areas of land in many coastal, estuarine and low-lying riverine areas have been reclaimed in this way, such as the Fens, Somerset Levels and Romney Marsh in England (Curtis et al. 1976).

One of the best-known examples of this type of large-scale reclamation is the Polder scheme in the Netherlands, where large areas were reclaimed from the sea in the low-lying coastal fringes. Since the Zuyder Zee (Reclamation) Act of 1918, four polders have been reclaimed, creating about 165,000 ha of new land. This involved the construction of large dykes which were fed by lateral drains spaced at 1.6–2.0 km intervals. These drains were supplemented by main drains spaced at 300 m intervals and then a coarse network of field drains spaced at 8–24 m intervals (Briggs and Courtney 1989). Although such

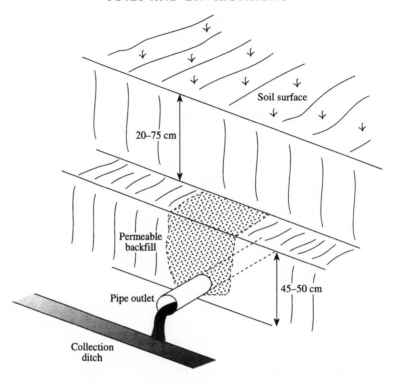

Figure 8.14 Pipe drain embedded in permeable backfill (from White 1987)

large-scale drainage programmes have produced some of the world's most fertile and productive land, through improvements to a range of soil physical and chemical characteristics, they have also created problems. Perhaps the most notable of these is soil shrinkage, which is particularly common in areas dominated by marine clays and peats. In parts of the Fens of eastern England, for example, it is estimated that the surface has been lowered by as much as 5 m following the long history of reclamation and drainage (Curtis *et al.* 1976). In an area which is already low lying, surface lowering places even greater pressure on existing flood prevention and drainage schemes, many of which need to be upgraded. Also of concern in areas where large-scale reclamation and drainage are widespread, is the destruction of important wetland habitats, such as those in the Somerset Levels of southwest England (Briggs and Courtney 1989).

Soil water deficit occurs most commonly in areas where rainfall is low in comparison with potential evapotranspiration. It is also common in coarse-textured, sandy soils which are excessively drained and low in organic matter, and in fine-textured, clay soils which contain large vertical cracks or which are poorly structured and therefore of low permeability; such soils have a low plant-available water capacity. Soil water deficit does not necessarily occur throughout the year but is most common during warmer and drier periods. In Britain, for example, it rarely occurs in the October–March period but is common in the April–September period, especially in southern and eastern areas where it may occur in three to five years out of ten. In the USA it is particularly common in the central and southwestern states, but is relatively rare in the southern and eastern states where rainfall is greater and more reliable. Soil water deficit has the greatest impact where it occurs during the main period of crop growth, when it is demonstrated by excessive wilting of plants. In some circumstances, intense evaporation may lead to salinisation and sodification of soils (section 8.3.2).

The main way in which soil water deficit is managed is by irrigation. This has been practised since the beginning of settled agriculture, not only in areas where rainfall is low throughout the year, but also where seasonal drought is a problem, in spite of respectable annual rainfall totals. It is used to replenish the soil moisture store and to leach toxic salts out of the rhizosphere, and as such it is a fundamental component of agriculture in many dryland areas. The effectiveness of irrigation depends on the method and timing of water application, and on the amount and quality of water applied. There are three main methods – surface application, overhead application and subsurface application (Briggs and Courtney 1989).

Surface application usually involves flood and furrow irrigation where water is diverted from existing river channels or canals directly onto the ridge and furrow systems of agricultural fields (Figure 8.15a). This low cost and low technology approach is often preferred in developing countries, although it is also widely used in the developed world (e.g. Mitchell and Van Genuchten 1993). One of the main problems with this method, however, is that capillary action may lead to redistribution of salts from the wet furrow areas to the drier ridges where the crops are often planted (section 8.3.2).

Overhead irrigation methods include spray or sprinkler, and trickle or drip systems. Spray irrigation systems involve the high pressure application of water through upright nozzles which may be attached to fixed or moveable pipes in a variety of ways, such as self-propelled water guns and centre-pivot systems. Although water can be applied uniformly and efficiently using these methods, equipment costs may be prohibitive and evaporative losses are high, particularly on windy days. Moreover, soil erosion may occur as a result of overhead application, particularly in the early stages of crop growth when vegetation cover is low. Trickle irrigation involves the application of water through flow regulating emitters which are placed at regular intervals in close proximity to the plants (Figure 8.15b). In this way water is supplied more directly to the plant and evaporative losses are minimised. Perhaps the most effective methods of irrigation involve subsurface applications of water. Here, ditches and underground drains or

(a)

(b)

Figure 8.15 Examples of irrigation practices: (a) flood and furrow irrigation in Nepal using the *shaduf* system to transfer water, (b) trickle irrigation of citrus fruits in southern Spain

pipes are used and, although their installation can be costly, evaporation, soil structural damage and erosion losses are minimised.

In addition to irrigation, soil water deficit may also be managed through the control of evaporation from the soil surface. This is most commonly achieved through the addition of organic mulches, or through the establishment of a vegetation cover which can offer some degree of shade. These strategies are usually combined with sensible irrigation practices in order to increase water use efficiency in dry environments.

8.3 CHEMICAL PROBLEMS

8.3.1 Acidification

Acidification has a number of natural and anthropogenic causes (Rowell and Wild 1985) (Table 8.4). The main natural causes are long term leaching (section 3.2.3) and microbial respiration. The acids found in rainwater (carbonic acid) and in decomposing organic material (humic and fulvic acids) can stimulate leaching by dissociating into H^+ ions and their component anions which then displace or attract base cations from the soil exchange complex (section 3.2.3). Leaching of bases is most common where precipitation exceeds evapotranspiration, and includes many eastern parts of North America, and northwestern parts of Europe, where soils have been subjected to leaching throughout most of the 10,000–15,000 years of post-glacial time (Catt 1985b). Microbial respiration also leads to soil acidification through the production of CO_2 (carbon dioxide), which is dissolved in soil water to form carbonic acid. Other natural processes associated with soil acidification are plant growth and nitrification. During plant growth, nutrient base cations are obtained through root systems in exchange for H^+ ions, thus leading to increased soil acidity. Nitrification is an oxidative process of organic decomposition whereby NH_4^+ (ammonium) ions are converted to NO_3^- (nitrate) ions by nitrifying bacteria, with H^+ ions as a by-product:

$$NH_4^+ + 1.5O_2 \Rightarrow NO_3^- + 4H^+$$

These ions are then available for displacing and attracting base cations from the soil exchange complex, as mentioned above, thus leading to soil acidification.

The main anthropogenic causes of acidification include certain landuse practices, such as needleleaf afforestation, excessive use of inorganic nitrogen fertilisers, land drainage, and acid deposition resulting from urban and industrial pollution. Needleleaf afforestation has been associated with the acidification of soils and surface waters for a number of reasons (Hornung 1985, Miller 1985). First, needleleaf trees produce litter which is very acidic in comparison with most broadleaf species. Second, because of their high canopy surface area, needleleaf trees are able to 'scavenge' acid pollutants

Table 8.4 Causes of soil acidification

Source		H^+ addition or equivalent ($kg\ H^+\ ha^{-1}\ a^{-1}$)
Natural	CO_2 in soil pH > 6.5 (calcareous soils)	7.2–12.8
	Organic acids in acid soils and from vegetation	0.1–0.7
Acid 'rain'	Wet deposition	0.3–> 1.0
	Dry deposition	> 0.3–> 2.4
	NH_3 and NH_4^+ oxidation	0.7
Land use	Cation excess in vegetation	0.2–2
	NH_4^+ oxidation (agricultural soils) and leaching	4–6
	Oxidation of N and S from organic matter and leaching	0–10

Source: Rowell and Wild 1985

from the atmosphere, later releasing them into the soil via throughfall and stemflow (Table 8.5). Third, due to modifications of the surface and soil hydrology by drainage channels and shallow root networks, water transfer is rapid and is concentrated either at the surface or in the uppermost layers of the soil. Under these conditions, the residence time of the water in the soil is limited, as is the depth to which it can percolate. Thus, the contributions of weathering and ion exchange reactions to the buffering process are limited (Bache 1983, Miller 1985).

Excessive use of inorganic nitrogen fertilisers in agricultural systems has also been associated with soil acidification, partly through the process of nitrification (see previous equation). If levels of NO_3^- ions in the soil are in excess of plant requirements, they will behave as mobile anions, thus encouraging the leaching process. The acidifying effect of nitrogen fertiliser

Table 8.5 Weighted mean concentration ($\mu mol\ l^{-1}$) of selected solutes in bulk precipitation, throughfall and stemflow at Llyn Brianne, Wales (Oct. 1988–Sept. 1989)

	Precipitation	Throughfall	Stemflow
pH	4.98	4.37	4.05
H^+ ($\mu mol\ l^{-1}$)	11	43	89
Cl^- ($\mu mol\ l^{-1}$)	89	344	413
SO_4^{2-} ($\mu mol\ l^{-1}$)	53	181	248

Source: Soulsby and Reynolds 1992

Table 8.6 Soil pH in response to different rates of nitrogen fertiliser application, Gleadthorpe, England

Total annual N rate (kg ha^{-1})	0	150	300	450	600	750
Mean pH after 4 years	6.9	6.4	6.1	6.0	5.6	5.4

Source: Chalmers 1985

was demonstrated in a four-year experiment during which different loadings of NH_4NO_3 (ammonium nitrate) fertiliser were applied to grassland plots (Chalmers 1985); soil pH declined markedly with increased fertiliser loading (Table 8.6). Land drainage is another important cause of soil acidification, particularly in soils which contain appreciable quantities of sulphide minerals. Once the land has been drained, soil aeration improves and sulphide minerals are oxidised to form sulphate compounds. Sulphate ions can then combine with H^+ ions in the soil to produce H_2SO_4 (sulphuric acid). Under extreme circumstances, soil acidity may fall below pH 3.0 as in the case of the acid sulphate 'cat-clay' soils which are common in mangrove swamp environments (Foth 1990).

Another important anthropogenic source of acidity in soils and surface waters is atmospheric deposition. Here, gases derived from industrial and motor vehicle emissions, particularly SO_2 (sulphur dioxide) and NO_x (oxides of nitrogen), are either dissolved in precipitation (*wet deposition*) or deposited directly (*dry deposition*) (Fowler *et al.* 1985). Essentially, acidity is derived from H_2SO_4 (sulphuric acid) and HNO_3 (nitric acid) which undergo dissociation in rain and soil water. Values of pH for acid deposition in industrial areas of Europe and North America are often less than 4.0. Some of the lowest values, however, (pH < 3.0) have been observed in acid mists and fogs, a phenomenon known as *occult deposition*. This type of acidity is common in upland areas of Britain where it may be scavenged by the canopies of needleleaf forestry plantations and then transferred to the soil, as mentioned above.

The acidification of soils has an impact on a number of soil characteristics. Hallbacken and Tamm (1986), for example, report pH reductions of up to 1.5 units over a 55-year period in southern Sweden; these reductions were attributed to high levels of acid deposition derived from other areas of northern and western Europe. Nutrient loss through leaching is another effect of soil acidification. Nilsson (1986), for example, suggests that the supply of available calcium and magnesium in topsoils will be halved within twenty to thirty years, due to continued acidification and leaching. In addition to nutrient losses, the mobility of aluminium in soils increases as they acidify, especially when soil pH falls below about 5.0. Levels of aluminium are often particularly high in the B (spodic) horizons of podzolic soils (Wilson 1986). If organic acids predominate in the soil, aluminium is mobilised in the form of soluble organo-metallic complexes (section 3.2.2). If mineral acids predominate, however, then aluminium is mobilised in its ionic, labile-monomeric form (Al^{3+}) (e.g. Wilson 1986, Adams et al. 1990); this form of aluminium is particularly toxic to many freshwater organisms, including fish (e.g. Ormerod et al. 1989). In a similar way to aluminium, heavy metals, such as lead, zinc and cadmium, are more readily mobilised in acidic soils.

Acidification of soils, and associated nutrient leaching, has also been implicated in damage to trees in forested areas, particularly in central Europe and Scandinavia (e.g. Rehfuess 1985). It is more likely, however, that nutrient leaching acts synergistically with other factors including ozone pollution, acid deposition, NH_4^+ (ammonium) uptake, drought and frost to produce stress in the tree (Figure 8.16) (e.g. Rose 1985). Signs of tree damage include needle discoloration, crown defoliation and deformed branch structures. Increased soil acidity and associated changes in aluminium mobility also have serious implications for the quality of surface waters which receive drainage from acidified soils (e.g. Hornung 1985, Miller 1985).

Soils vary dramatically in their ability to buffer acidity, a characteristic known as the *buffering* or *acid neutralising capacity*. The buffering capacity of a soil is defined as the amount of acid that needs to be added to cause a reduction in pH of one unit (Trudgill 1988). Soils which contain significant quantities of base-rich, weatherable minerals have a high buffering capacity, whereas those which are dominated by quartz and similarly resistant minerals have a low buffering capacity (Figure 8.17). In Britain, for example, soils with the highest buffering capacities are found in eastern and southern areas where calcareous parent materials are particularly common. Soils with the lowest buffering capacities are found in areas with base-poor, often siliceous, parent materials which predominate in upland areas in the west and north (Catt 1985b).

Surface water acidification is widespread in areas where the geology and soils are base-poor, and levels of acid deposition are high (e.g. Jones et al. 1986). The hydrological pathway taken by drainage waters and their residence time in the soil are particularly important factors in this respect. Water flowing rapidly over the surface or through the uppermost soil horizons undergoes little buffering and may carry organic acids from the soil, therefore contributing significantly to surface water acidification. If water is allowed to percolate slowly through the deeper mineral horizons, however, its acidifying effect is likely to be limited due to increased buffering (Bache 1983). Surface water acidification is exacerbated in areas where needleleaf afforestation is widespread, as is the case in many upland parts of northern and western Britain. Hornung et al. (1987), for example, examined the chemistry of streams under a needleleaf forest plantation and adjacent moorland in North Wales. The streams under forest were more acidic, and contained higher levels of aluminium, SO_4^{2-} (sulphate) and Cl^-

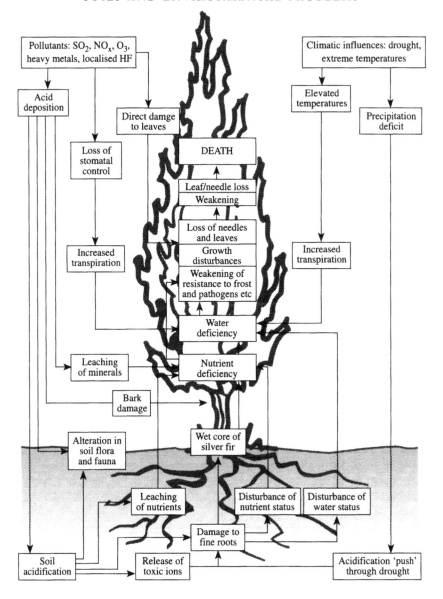

Figure 8.16 Tree damage and stress interactions (from Longhurst *et al.* 1987)

(chloride) (Table 8.7). Similarly, in a study of water quality in streams draining land used for rough grazing and forestry in Wales, Adams *et al.* (1990) found levels of ionic aluminium to be particularly high in the forest streams. These findings are also consistent with observations in many

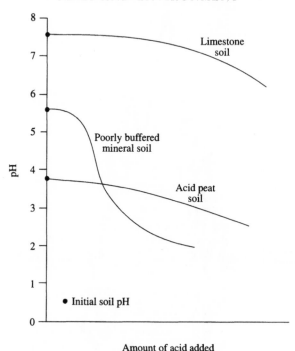

Figure 8.17 Variations in buffering capacity for three contrasting soil types
(from Trudgill 1988)

forested areas of Canada and the USA (e.g. Arthur and Fahey 1993).
Elevated levels of aluminium and heavy metals in potable water supplies
have been implicated in certain mental disorders in humans, notably
Alzheimer's disease, although the medical evidence is not conclusive (e.g.
Maugh 1984).

In addition to the effects of land use, stream flow conditions have an

Table 8.7 A comparison of the chemistry of a forested and a moorland stream in
North Wales

	Forested stream	*Moorland stream*
pH	4.6	5.5
Al (μmol l^{-1})	62	22
SO_4^{2-} (μmol l^{-1})	149	89
Cl^- (μmol l^{-1})	336	186

Source: Hornung *et al.* 1987

268

important influence on surface water acidity. Storm runoff is known to be more acidic than base-flow because it flows rapidly over the surface of the soil, or through the uppermost horizons, and is unable to percolate into the deeper mineral horizons where it might be buffered (e.g. Soulsby 1992). Reynolds *et al.* (1986), for example, report pH values of 4.4 and 6.4 for storm-flow and base-flow respectively in central Wales. Similarly, acid surges associated with storm runoff and snowmelt have been reported in eastern Scotland (Morris and Thomas 1986).

There are a number of approaches to the management and remediation of soil and surface water acidity, including liming, sensible forestry management and reduction of acid emissions into the atmosphere. Liming has long been practised on agricultural land to remedy the problems of acidity, slow organic matter turnover, poor nodulation in some legumes, calcium and molybdenum deficiency, and aluminium and manganese toxicity (White 1987). Liming materials most commonly used include ground limestone, chalk, marl and basic slag, the main active constituent being $CaCO_3$ (calcium carbonate); other minor components include quicklime (CaO), slaked lime ($Ca(OH)_2$) and $MgCO_3$ (magnesium carbonate). The lime requirement of a soil varies depending on its buffering capacity and is usually expressed as the amount of $CaCO_3$ (t ha^{-1}) required to raise the pH of the top 15 cm of soil to the desired value. In temperate areas, the ideal soil pH is about 6.5 for arable crops and about 6.0 for grassland. In tropical areas, however, pH values of about 5.5 are often preferred, particularly in soils with high exchangeable aluminium contents, where phosphorus availability may be restricted under more alkaline conditions (White 1987).

Liming is also used to combat surface water acidity. In many parts of Scandinavia, for example, particularly in Sweden, lakes are often limed directly (e.g. Nyberg and Thornelof 1988). This practice is rather costly, however, as the time period over which the lime is effective (*hydraulic retention time*) is relatively short, mostly ranging from several weeks to two years (e.g. Porcella 1988). Furthermore, fish populations do not benefit fully from such treatments as they tend to breed and spawn in the feeder streams which are not limed. In an attempt to overcome these problems, catchment liming is sometimes practised. Here, the whole catchment benefits from treatment, and the time-scale over which the treatment is effective is usually much greater than when only the lakes are treated (e.g. Howells and Brown 1986, Brown 1988, Dalziel *et al.* 1988). The cost of treatment may be reduced substantially by strategic applications of lime to specific sites within the catchment, such as spring and seepage areas, and stream headwaters. In some catchments, flow-activated lime wells have been used to supply lime to streams during periods of high flow when acidification is known to be a major problem.

Acidification of surface waters in afforested areas may be controlled through sensible forestry management practices, rather than by chemical means (Miller 1985, Forestry Commission 1988). For example, the termination of drainage ditches before they enter water courses, sometimes in deep check drains which run across the slope, will help to increase the residence time of water in the soil, thus facilitating the buffering process. Other useful practices include not planting in the riparian buffer zones adjacent to water courses, and increasing the proportion of broadleaf trees in plantations.

Most of the acidity mitigation strategies outlined above are curative in nature rather than preventative. In the long term, perhaps the most effective strategy is to reduce acid emissions, particularly of SO_2 (sulphur dioxide), into the atmosphere. European emissions of SO_2 from coal- and oil-fired

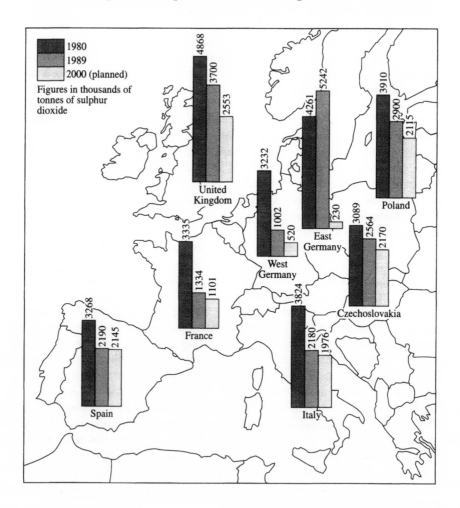

Figure 8.18 Cuts in SO_2 emissions in Europe since 1980 (from Pearce 1993)

power stations have decreased dramatically since 1980, partly as a result of installing flue-gas desulphurisation equipment and partly by switching to other energy sources such as gas, nuclear power or wind power (Figure 8.18). During the mid-1980s, in an attempt to form a more objective and scientific basis to the issue of cuts, and how large they need to be so that long-term ecosystem recovery is ensured, the 'critical loads' approach was devised (Pearce 1993). The critical load is the highest loading of acid that can be taken by an ecosystem before long-term damage occurs. In some of the most sensitive areas of Britain, it was found that the reduction required for ecosystems to recover is about 90 percent. This is considerably greater than the 60 percent reduction, by the year 2005, to which Britain is currently committed.

8.3.2 Salinisation and sodification

Salinisation describes the accumulation of salts in the soil, whereas sodification or alkalisation refers to the dominance of the soil exchange complex by Na^+ ions. These processes occur largely in semi-arid and arid environments where evaporation exceeds precipitation, and where soil parent materials and ground waters are rich in sodium salts. In such dry environments, soil water movement is driven largely by evaporation and occurs in an upward direction by capillary action (section 6.2.2). As water evaporates, salts precipitate out to form saline (solonchak or white alkali) soils (section 3.3.2). Slight dissolution of these salts and the consequent release of their constituent ions into solution may lead to dominance of the soil exchange complex by Na^+ ions, resulting in the formation of sodic (solonetz) soils.

Saline soils occur most commonly in low-lying floodplain areas where the water table is relatively high, thus leading to intense capillary action. In contrast, sodic soils are found most commonly on slopes immediately above valley floors and flood plains where the water table is lower and drainage potential is greater. On higher slopes in such areas, a leached and more acidic variety of the solonetz soil, known as a solod, is often found (section 3.3.2). This catenary association of solonchak, solonetz and solod soil types is widely observed in semi-arid areas, including the Indus valley of Pakistan, the Nile valley of Egypt and Sudan, and the Tigris and Euphrates valleys in Iraq and Syria. They are also found in many parts of central and southern Australia and around the Caspian and Aral seas in central Asia. Even in England and Wales, where the climate is relatively wet, salt-affected soils occupy about 6 percent of the agricultural land area (Loveland *et al.* 1986), being especially common in low-lying coastal and estuarine areas.

Soil salinity is measured in terms of the electrical conductivity of a saturated extract (EC_e), whereas soil sodicity is established from the exchangeable sodium percentage (ESP) or sodium adsorption ratio (SAR) (section 2.4.2). Soils are classified as saline, if the EC_e value is 4.0 mmho cm^{-1} or more, pH is 8.5 or less and ESP is less than 15 percent. Sodic soils are non-saline, have a pH of more than 8.5 (usually less than 10.0) and an ESP value of 15 percent or above. Soils are classified as saline-sodic if they are both saline and have an ESP value of 15 percent or above. The salinity of ground waters and other water supplies, which may be used for irrigation purposes, is also assessed on the basis of electrical conductivity (EC). Most water supplies have EC values of 0.15–1.5 mmho cm^{-1}, and are said to be of high salinity if the EC value exceeds 0.75 mmho cm^{-1} (White 1987).

Although salinisation and sodification occur naturally in semi-arid and arid environments, they are often exacerbated as a result of human activity. In parts of southwest Australia, for example, removal of indigenous eucalyptus forest has resulted in extensive salinisation and sodification of

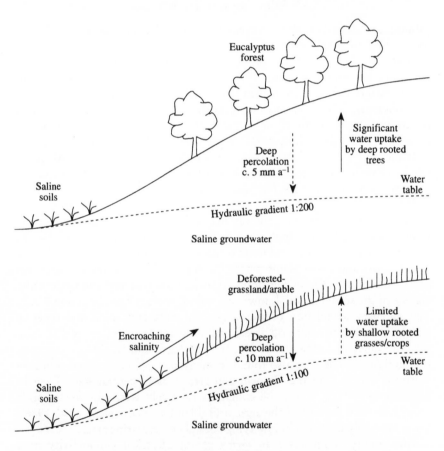

Figure 8.19 Impact of eucalyptus deforestation on level of water table and soil salinity in southwest Australia (based on White 1987)

soils. This has occurred because the deeply rooted trees have been replaced by shallow-rooted grasses and crops, which are less effective at lowering the ground-water level (Figure 8.19). Capillary action is most intense, and salinity and sodicity are greatest, in soils where the water table is within about 2 m of the surface. Another important cause of soil salinity and sodicity is poor irrigation practice. Overwatering leads to a rise in the water table which in turn causes enhanced capillary action. Similarly, poor maintenance of irrigation channels and canals results in leakage of water onto adjacent agricultural land. This has contributed to increased soil salinity and sodicity in parts of the Indus valley in Pakistan, for example.

In the Nile valley in Egypt, river impoundment resulting from construction of the Aswan dam has been associated with a deterioration in soil quality in what were once very fertile floodplain areas. Prior to this, flooding would occur annually in response to heavy seasonal rains in the mountainous source regions of the river. The seasonal flood waters were particularly important to agriculture in the Nile valley for three main reasons – they were an important source of irrigation water, they brought with them fertile silt deposits which were incorporated into the soils, and they helped to flush out salts from the soil. Since river impoundment, the benefits of seasonal flooding have been removed, and the resulting decline in soil quality has been compounded by agricultural intensification, in order to meet both the needs of a rapidly growing population and the demands of overseas markets for cash crops.

Salinisation and sodification of soils are not only associated with arable land, but may also occur in pastoral landuse systems. Lavado and Taboada (1987), for example, examined the salt regimes of soils under grazed and enclosed (ungrazed) natural grassland in the Flooding Pampa of Argentina. Here, soils are affected to varying degrees by salts and Na^+ ions, the water table is high and flooding is common during the wet season. The salt concentration in the topsoil of the grazed land increased dramatically and episodically after flooding, whereas on the ungrazed land it did not (Figure 8.20). This difference was attributed to compaction and impeded drainage, and to reduced vegetation cover on the grazed land, which led to increased evaporation and enhanced capillary action in the soils.

Soil salinity and sodicity have a detrimental effect on both chemical and physical aspects of soil quality. In a chemical context, high salinity and sodicity are associated with elevated soil pH, and under these conditions the availability of certain plant nutrients is reduced (section 6.5.3), resulting in severe disturbance of the plant nutrient balance as a whole. Elevated salt and Na^+ concentrations in soils are also highly toxic to many plants, although tolerance levels vary between different species. Crops such as barley, cotton and sugar beet, for example, have a relatively high tolerance level, whereas sugar cane, onion and lettuce have a very low tolerance level (Table 8.8). As with arable crops, grasses vary in their tolerance to soil salinity and sodicity. Bermuda grass, for example, has a high tolerance level whereas Guinea grass is much less tolerant.

Of particular significance in sodic and saline-sodic soils is the destabilisation and breakdown of soil structure. Deterioration of soil structure is not usually a

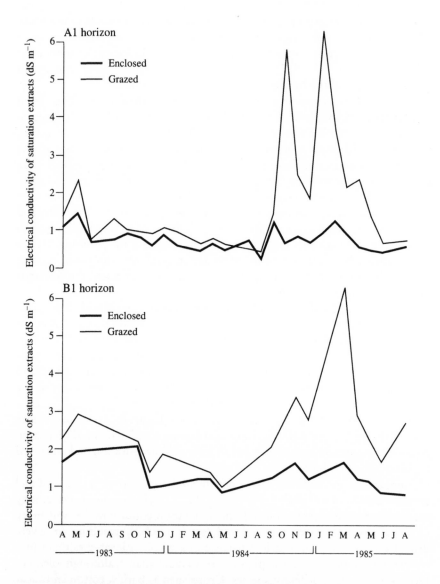

Figure 8.20 Soil salinity in the A1 and B1 horizons, depth and electrical conductivity of ground water, and flood periods in the Flooding Pampa of Argentina (from Lavado and Taboada 1987)

274

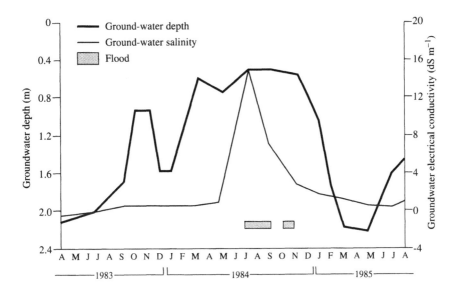

Figure 8.20 (continued) Soil salinity in the A1 and B1 horizons, depth and electrical conductivity of ground water, and flood periods in the Flooding Pampa of Argentina (from Lavado and Taboada 1987)

problem in saline soils, provided that the concentration of soluble salts remains high. However, if salts are depleted by leaching, and Na^+ ions begin to dominate the soil exchange complex, then swelling and deflocculation of clays may occur, thus resulting in aggregate dispersion. These changes, together with translocation of the dispersed clay, may lead to reduced macroporosity and permeability. Although clay swelling may be reversed through increases in salt concentration, deflocculation and translocation are not reversible and may cause permanent deterioration of soil structure (White 1987).

Soil salinity and sodicity management and remediation are complex issues, and may involve a number of approaches. Appropriate irrigation and drainage techniques are critical in the effective management of saline, saline-sodic and sodic soils (section 8.2.3). In addition to the amount of water applied, the quality of the irrigation water is important. Use of water which contains high concentrations of salts can lead to increased salinity and sodicity in soils. As the water evaporates, ionic concentrations increase and ions in solution (particularly Na^+) are exchanged with those held on the soil exchange complex (notably Ca^{2+}) until a state of equilibrium is achieved (section 2.4.2). Saline-sodic soils in particular require careful management in

Table 8.8 Salt tolerance of selected crops, grasses and trees, measured in terms of electrical conductivity of the saturated soil extract

Crop	Salt tolerance threshold (dS m⁻¹)	50% yield (dS m⁻¹)	50% emergence (dS m⁻¹)
Barley	8.0	18.0	16–24
Cotton	7.7	17.0	15
Sugar beet	7.0	15.0	6–12
Sorghum	6.8	15.0	13
Wheat	6.0	13.0	14–16
Soybean	5.0	7.5	–
Beet, red	4.0	9.6	14
Date palm	4.0	16.0	–
Spinach	2.0	8.5	–
Peanut	3.2	5.0	–
Sugar cane	1.7	9.8	–
Tomato	0.5	7.6	–
Safflower	–	14.0	–
Cowpea	1.3	9.1	16
Corn	1.7	5.9	21–24
Lettuce	1.3	5.2	16
Onion	1.2	4.2	5.6–7.5
Rice	3.0	7.2	18
Bermuda grass	6.9	14.8	–
Rye grass	5.6	12.1	–
Asparagus	4.1	29.0	–
Alfalfa	2.0	9.0	–
Sesbania	2.3	9.3	–
Berseem	1.5	9.5	–
Squash	4.7	9.9	–

Source: Based on Gupta and Abrol 1990

this respect because, although structural stability is maintained if the concentration of salts in the soil solution is high, if soluble salts are depleted by leaching, structural deterioration will occur as Na^+ ions begin to dominate the soil exchange complex. Reclamation of such soils requires leaching with water of a low enough SAR to facilitate Ca^{2+} exchange for Na^+, but a sufficiently high total salt concentration to maintain soil structure and permeability. As the soil ESP is gradually reduced, the salinity of the leaching water may also be reduced until reclamation is complete. Intermittent leaching is more effective than continuous ponding as it allows time for salts to diffuse into the depleted outer regions of aggregates from where they are removed during the next leaching phase. During continuous ponding, water flows preferentially through the larger inter-aggregate channels and does not remove salts trapped within the aggregates.

Soil salinity and sodicity can also be alleviated using chemical amendments. Sodic soils, for example, can be improved by treatment with gypsum

$(CaSO_4.2H_2O)$ (Frenkel *et al.* 1989, Armstrong and Tanton 1992). About 3 tonnes of gypsum are required to reduce by one unit the ESP of a volume of soil which is 1 ha by 15 cm deep (White 1987). During treatment, Ca^{2+} ions displace Na^+ ions from the soil exchange complex. This benefits the soil in a number of ways – nutrient status is improved, the degree of alkalinity is reduced and structural stability is increased. If $CaCO_3$ (calcium carbonate) is present in the soil, Ca^{2+} ions may be released through the addition of sulphur to the soil. The acidity generated during the oxidation of sulphur to SO_4^{2-} (sulphate) leads to dissolution of the $CaCO_3$; H_2SO_4 (sulphuric acid) may also be used for this purpose.

In addition, salinisation and sodification of soils may be controlled by manipulation of surface microrelief and by reducing evaporation from the soil surface; this helps to reduce the intensity of capillary action. On land which has been irrigated using the flood and furrow method, capillary action may lead to redistribution of salts from the wet furrow areas to the drier ridges where the crops are often planted; to some extent, this problem can be alleviated by modifying the shape of the ridges. Salt concentrations tend to be particularly high on the ridge tops, so planting in this position should be avoided. Instead, planting should take place on the sides of ridges where salt concentrations are relatively low (Figure 8.21). Evaporation may be reduced

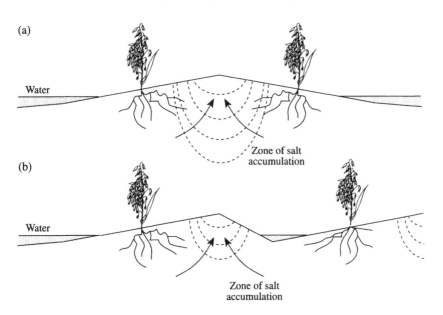

Figure 8.21 Microtopography of ridge and furrow systems designed to avoid salinity damage to crops: (a) paired crop rows on broadly sloping ridges, (b) single crop rows on asymmetric ridges (from White 1987)

through the addition of organic mulches to the soil surface. Mulching plays a crucial role in the conservation of soil moisture in soils of both warm and dry environments. If levels of soil salinity and sodicity are not too high, it may be possible to utilise salt-tolerant crops and grasses with minimal reductions in yield, thus avoiding the need for ameliorative treatments. The replanting of trees can also reduce the level of the water table and thus the extent of capillary rise of salts towards the surface (Schofield and Bari 1991).

8.3.3 Agrochemical pollution

In recent decades, the use of inorganic fertilisers has increased dramatically at the expense of more traditional organic nutrient treatments. Between 1950 and 1985, the global use of fertilisers increased from 14 million tonnes to 125 million tonnes, an increase of almost 900 percent (Saull 1990). Inorganic fertilisers are used in preference to organic treatments because the nutrients are in a more readily available form and are released rapidly after application. Organic material releases its nutrients slowly, through decomposition processes, and only when conditions are suitable (warm and moist), not necessarily when crops need them.

Fertilisers are applied in a variety of forms – solution, suspension, emulsion and solid. The solid forms vary in particle size from fine powders to coarse granules, and are either spread evenly (broadcast) over the soil surface or mechanically placed, by drilling, into the rhizosphere; generally, the rate of nutrient release decreases with increasing particle size. Fertilisers are based on compounds of plant macronutrients (e.g. nitrogen, phosphorus and potassium) and micronutrients (e.g. zinc, copper, boron and molybdenum), and a variety of nutrient combinations are available depending on the nature of the nutrient problem (Table 8.9).

There are five main fates of fertilisers applied to the soil – plant and animal uptake (immobilisation), adsorption and exchange in the soil (fixation), leaching and loss in soluble form through drainage, volatilisation and gaseous losses to the atmosphere (e.g. denitrification), and surface loss in solid form by runoff and erosion (Ross 1989) (Figure 8.22).

In terms of plant uptake, the percentage recovery of nutrients varies markedly between the different types of fertiliser. During the first year of application, the recovery of nitrogen from inorganic nitrogen fertilisers is about 50–65 percent, whereas from organic manures it is only about 20–30 percent (Fink 1982). Similarly, the recovery of phosphorus and potassium from inorganic fertilisers is about 5–15 percent and 75 percent respectively. In terms of fixation in the soil, many of the phosphorus and potassium fertilisers are of relatively low solubility and the nutrients released are often strongly adsorbed. In contrast, many of the nitrogen fertilisers are highly soluble, nutrient release is rapid, and in most soils adsorption is limited. The differences in degree of immobilisation by plants and animals, and in the

Table 8.9 Composition of some commonly used fertilisers

Nutrient	Compound		Composition
(a) *Nitrogen*			(% N)
solids	Ammonium sulphate	$(NH_4)_2SO_4$	21
	Potassium nitrate	KNO_3	13.8
	Ammonium chloride	NH_4Cl	26
	Ammonium nitrate (Nitro Chalk, if mixed with $CaCO_3$)	NH_4NO_3	35
	Urea (Carbamide)	$CO(NH_2)_2$	46
	Calcium cyanamide	$CaCN_2$	21–22
	Gold-N (sulphur-coated, slow-release fertiliser)		32
liquids	Ammonia gas liquors (dilute NH_3 solutions)		1–4
	Aqueous ammonia gas	NH_3	21–29
	Anhydrous ammonia gas	NH_3	82
(b) *Phosphorus*			(% P)
	Single superphosphate	$Ca(H_2PO_4)_2$; $CaSO_4$	8–9.5
	Triple superphosphate	$Ca(H_2PO_4)_2$	20
	Monoammonium phosphate	$(NH_4)H_2PO_4$	20–30
	Diammonium phosphate	$(NH_4)_2HPO_4$	21
	Basic slag		5–10
	Ground mineral phosphate (GMP)/crushed rock phosphate (CRP)		12.5
(c) *Potassium*			(% K)
	Muriate of potash (potassium chloride)	KCl	50
	Sulphate of potash (potassium sulphate)	K_2SO_4	40–42
	Saltpetre (potassium nitrate)	KNO_3	36

Source: Ross 1989

extent of fixation in the soil, help to explain why leaching losses of nitrogen are usually far greater than those of phosphorus and potassium. Levels of nitrogen in drainage waters, for example, are often in the range 15–150 kg ha^{-1} a^{-1}, whereas levels of phosphorus are generally < 1 kg ha^{-1} a^{-1}; even in exceptional circumstances in eroded areas, phosphorus levels rarely exceed 10 kg ha^{-1} a^{-1} (White 1987). Nitrogen losses of up to 30–35 percent of that applied have been recorded for both leaching and denitrification. Leaching is most common in coarse-textured and well drained soils, whereas denitrification losses are greatest in fine-textured, waterlogged and poorly aerated soils. Similarly, surface losses are greatest at

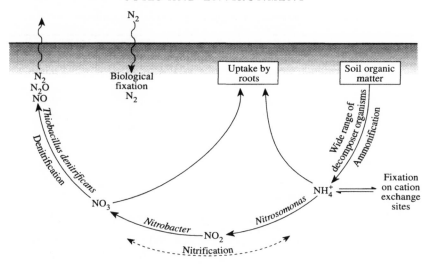

Figure 8.22 The soil nitrogen cycle and the fate of nitrogen fertilisers applied to soils (from Ross 1989)

sites which are most susceptible to surface runoff and erosion (section 6.4.2). In terms of the environmental problems associated with fertiliser use, perhaps the area which has received most attention in recent years, by way of research, is that of nitrate leaching (e.g. Addiscott *et al.* 1991).

In the last few decades, levels of nitrate in water supplies have increased dramatically, particularly in intensively cultivated areas where inputs of nitrogen fertiliser have been high. This association is not necessarily causal, however, and may be indirect. Nitrogen fertilisers are not the only source of leached nitrate, as illustrated by the Broadbalk experiments carried out over a 150-year period at Rothamsted Experimental Station in southeast England. It was found that here much of the leached nitrate is derived from the vast reserves of organic nitrogen in the soil, rather than from fertilisers (Saull 1990, Addiscott *et al.* 1991).

The main factors which influence the extent of nitrate leaching include land use, soil characteristics and climate. In terms of land use, nitrate leaching tends to be greatest when there is no crop cover to utilise the nitrate released from fertilisers or organic reserves. With spring planted cereals, for example, the leaching risk is greatest during the wet autumn and winter period when the soil is left bare. In order to maximise uptake and minimise leaching, fertilisers should be applied just before and during the period of maximum crop growth, and large applications at any one time should be avoided. Tillage practices also have an effect on nitrogen losses through leaching and denitrification (e.g. Drury *et al.* 1993). Tillage improves topsoil drainage and aeration and, as a result, rates of organic matter decomposition increase. Colbourn (1985) compared nitrogen losses by leaching and denitrification on conventionally cultivated and direct drilled land in southern England. Leaching losses were considerably greater from the conventionally cultivated land, whereas denitrification losses were smaller. For both tillage regimes, however, maximum nitrogen losses occurred during the autumn period after the harvest, and during the spring following fertiliser application.

Nitrate leaching is not restricted to arable land. Old clover-rich grasslands in particular are able to fix large quantities of nitrogen, and this is released when the grassland is ploughed up (e.g. Roberts 1987). This problem is not as serious in short-term arable leys where nitrogen fixation is considerably less. Even land under needleleaf forestry is not immune to the nitrate leaching problem, although rates of nitrogen release are relatively low in the often acidic and waterlogged soils (e.g. Adamson et al. 1987). Leaching occurs most frequently during the land preparation stages, prior to planting, where improved drainage and aeration lead to increased rates of organic mineralisation, and after felling when the source of nitrogen uptake has been removed (e.g. Stevens and Hornung 1988).

With respect to soil characteristics, coarse-textured soils are particularly susceptible to leaching due to the often poor structural development and relatively large pore sizes. Rapid drainage in such soils often leads to nitrate contamination of ground-water supplies, particularly in areas which are intensively cultivated. In contrast, clay soils are often characterised by good structural development, small pore sizes and restricted drainage. Under these conditions, surface runoff may occur and leached nitrate is more likely to be transferred into surface waters than into ground waters (Saull 1990). Waterlogging and resulting anaerobic conditions, often found in clay soils, may lead to significant nitrogen losses through denitrification. In soils of tropical environments, which often possess significant anion exchange capacity (AEC), nitrate may be retained in the soil to some extent, thus restricting the leaching process (e.g. Wong et al. 1990), but this situation does not apply in soils of temperate environments where AEC is usually negligible or non-existent.

In terms of climatic influences on nitrate leaching, rates of mineralisation of organic nitrogen increase with increasing temperature and are particularly high in warm, moist conditions (section 4.3). The timing of individual rainfall events, in relation to mineralisation and fertiliser application, is also important. If a significant amount of rain falls within a month or so of a phase of active mineralisation, or fertiliser application, then nitrate leaching is likely to occur, particularly if there is no crop to take up the released nitrogen (e.g. Colbourn 1985), or if uptake is restricted due to poor crop growth (e.g. Campbell et al. 1993). Smith and Stewart (1989) found that levels of nitrate in rivers draining into Lough Neagh, Northern Ireland, were positively associated with nitrogen fertiliser application rates and December to May river flow, and negatively associated with April to September rainfall and June to September air temperature.

Increased levels of nitrate in surface and ground waters have been implicated, together with phosphate, in a number of environmental and health problems. In surface waters, increased nutrient levels lead to excessive algal growth and oxygen depletion, a process known as *eutrophication*. Stagnation, and shading of the bottom environment by algal blooms, causes increased fish mortality and restricted aquatic plant growth. Eutrophication is not restricted to fresh water, but may also occur in marine environments. In the North Sea, for example, algal blooms have had an adverse effect on salmon and trout farms off the coast of Norway; this damage has incurred costs in excess of $200 million. In spite of this problem, however, countries of the European Union continue to dump 1.5 million tonnes N a^{-1} into adjacent marine areas, more than 60 percent of which is derived from agricultural runoff (Saull 1990). The nitrate in potable water supplies has important implications for human health. The incidence of 'blue-baby syndrome' (methaemoglobinaemia), for example, is often greatest where nitrate concentrations in water supplies are high. Similarly, the incidence of disorders of the digestive tract, including stomach cancer, are also associated with elevated nitrate levels, although the medical evidence is by no means conclusive (Addiscott *et al.* 1991).

In response to these environmental and health issues, legislation has been established in an attempt to control levels of nitrate in potable water supplies. The 1980 European Community 'Drinking Water Directive' (80/778), which actually came into force in 1985, set a limit of 50 mg l^{-1} of nitrate for drinking water; this is equivalent to 11.3 mg l^{-1} of NO_3-N (nitrate nitrogen). The latest European Union Directive aims to reduce this value by about 50 percent to 5.7 mg l^{-1} NO_3-N. In Britain, nitrate levels in drinking water are greatest in intensively cultivated areas where water supplies are derived from deep aquifer sources, particularly in the east and south of the country; these sources contribute about 40 percent of the potable water in England and Wales. In the USA, the Drinking Water Standard sets the legal threshold for NO_3-N at 44 mg l^{-1}, although this value is commonly exceeded in the more intensively cultivated areas of the mid-west. In Kansas, for example, 25 percent of rural drinking water wells exceed this limit (Jones and Schwab 1993).

Management and remediation of the nitrate leaching problem is a complex issue which depends largely on the adoption of sensible landuse strategies. In terms of nitrogen fertiliser application, slow-release varieties should be used in preference to more soluble types. Similarly, the leaching problem may be reduced if NH_4-N (ammonium nitrogen) compounds are used instead of NO_3-N compounds, although increased nitrification may lead to soil acidification (section 8.3.1). This may be controlled, however, using the specific inhibitor of nitrification, 'N-Serve', in conjunction with the NH_4-N fertilisers. It may also be controlled using a mixed fertiliser-lime compound such as 'Nitrochalk'. In landuse regimes where the soil is bare for significant periods of time and the risk of nitrate leaching is high, a 'catch' crop may be grown such as white mustard, forage rape or rye-grass. Such crops take up the excess nitrogen and are then ploughed into the soil just before the next crop is planted; the nitrogen retained by the catch crop then becomes available to the new crop. The creation of riparian buffer zones adjacent to water courses can also contribute to the immobilisation of excess nitrate, thus reducing the risk of surface water contamination.

In a broader landuse context, the 'Nitrate Sensitive Areas' (NSA) and 'Nitrate Advisory Areas' (NAA) schemes, established in central and eastern areas of England where nitrate concentrations in drinking water regularly exceed the European Community 50 mg l^{-1} threshold, provide a useful set of guidelines for management of the nitrate leaching problem. The NSA scheme relies on the willingness of farmers to improve agricultural practices on a voluntary basis, with a view to reducing the amount of nitrate leached from their land. Guidelines are issued and farmers are encouraged, through incentive payments, to comply with them. The scheme has two levels – basic and premium – with current rates of compensation payment of £55-95 $ha^{-1} a^{-1}$ and up to £380 $ha^{-1} a^{-1}$ respectively (Burt and Hanwell 1992). The options for these levels, in terms of agricultural practices, are indicated in Table 8.10. The NAA scheme aims to promote good agricultural practice but is not supported financially. Its methods are to avoid exceeding the recommended optimum application levels of fertiliser for each crop, to avoid nitrogen fertiliser applications in the autumn, to avoid large nitrogen fertiliser applications at any one time, to minimise autumn ploughing of grassland, and to avoid application of nitrogen fertiliser to hedge bottoms, field margins and water courses.

In addition to the environmental problems associated with fertiliser

Table 8.10 Options of the Nitrate Sensitive Areas (NSA) scheme

---

NSA basic scheme:
- Limit levels of organic and inorganic N fertiliser at or below the economic optimum and place constraints on the timing of fertiliser applications.
- Sow a winter crop or catch crop to avoid bare land in winter.
- Limit grassland ploughing to include only leys in arable rotation.

NSA premium scheme:
- Conversion of arable land to grassland, unfertilised and ungrazed.
- Conversion of arable land to grassland, unfertilised with grazing.
- Conversion of arable land to grassland with limited fertiliser use and optional grazing.
- Conversion of arable land to grassland with woodland.

Source: Based on Burt and Hanwell 1992

application, a number of problems arise from the use of pesticides. A wide range of pesticides has been developed (more than 450 compounds), the types most commonly used being insecticides, fungicides and herbicides (Table 8.11); other varieties include nematicides, miticides, rodenticides and molluscicides. Pesticides behave in a variety of ways following application. They may be degraded either biologically or photochemically, adsorbed by organic matter, clay and oxides/hydroxides of iron and aluminium, washed into water courses by leaching (especially compounds with solubilities > 10 mg l^{-1} such as simazine, bromacil and aldicarb) and surface runoff, or they may undergo volatilisation into the atmosphere (especially surface applied compounds, and those of low solubility and vapour pressure such as organochlorines) (McEwen and Stephenson 1979, Ross 1989) (Figure 8.23).

Ideally, pesticides should control only the target organism and persist for long enough to achieve this before degrading into harmless products. This is not always the case, however, and a number of environmental concerns arise from pesticide use. These include persistence in the environment, toxicity in soil, vegetation and water supplies, and impact beyond the target organism, including bioaccumulation and its implications for human health

Table 8.11 The main insecticide and herbicide groups

Pesticide groups	Specific pesticide types
Insecticides	
Organochlorines	DDT
	Aldrin
	Heptachlor
Organophosphates	Parathion
	Malathion
Carbamates	Carbaryl
	Carbofuran
Herbicides	
Phenoxyacetic acids	2, 4-D
	2, 4, 5-T
Toluidines	Trifluralin
Triazines	Atrazine
	Simazine
Phenylureas	Fenuron
Bipyridyls	Diquat
	Paraquat
Glycines	Glyphosate

Source: Based on Ross 1989

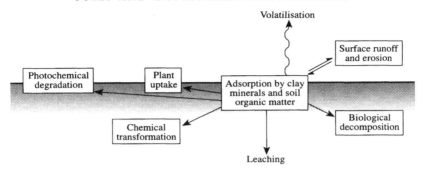

Figure 8.23 The fate of pesticides applied to soils (from Ross 1989)

(McEwen and Stephenson 1979, Kuhnt and Franzle 1993). In fact, the degradation products of some pesticides may be as toxic as the original source chemical but are not often measured by standard assays. The persistence and toxicity of many pesticide compounds and their degradation products are also dependent on a number of soil characteristics, notably clay and organic content, and pH.

The environmental impacts of pesticide use were felt most strongly with the early generation of pesticides, particularly the organochlorine compounds and those containing heavy metals. Generally, the organochlorine compounds are the most persistent of the pesticides and may survive for a number of years before they are degraded (Figure 8.24). Their residues have been found widely in soils, freshwater sediments, fish and cows' milk. Due to their high lipid solubility, they have also been found in the fatty tissues of animals. All of these findings have important implications for the ecology of food chains and thus for human health. An additional problem was that the effectiveness of organochlorine pesticides decreased as target organisms became resistant to them. The more recently developed organophosphate and carbamate pesticides are much less persistent than the organochlorines, with a half life of about six months, but are often more toxic. The phenoxyacetic acids, 2,4-D and 2,4,5-T, are also rapidly degraded but have been associated with animal and human growth and reproductive abnormalities.

In terms of management and remediation of the environmental problems associated with pesticide use, continued research and monitoring, with the aim of minimising effective persistence and toxicity, and maximising specificity, are essential. Attempts have been made to model the behaviour of pesticides in soils under different management regimes, with a view to reducing the risk of ground and surface water contamination (e.g. Johnson and Sims 1993, Sigua *et al.* 1993). In addition, the use of granular slow-release pesticides is being explored, together with ultra-low volume application

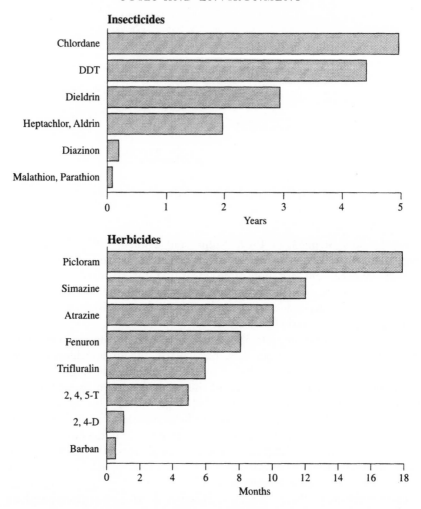

Figure 8.24 Variations in the time-scale of persistence for a selection of pesticides
(from Ross 1989)

techniques. The environmental problems may also be alleviated by adopting
alternative, non-chemical strategies of pest, disease and weed control. These
include direct approaches, biological and cultural methods and habitat
removal (Briggs and Courtney 1989). Direct approaches are aimed at clearly
identified animals and plants, and specific practices include hunting and
hand-weeding. Biological methods involve the use of predatory species,
preferably with a wide environmental tolerance range, to control the specific
target organism. Ideally, as the pest population grows, the predator popula-

tion should follow, and vice versa; this is known as a *density–dependence relationship*. If such a relationship does not exist, however, the predator itself may become a pest. Cultural methods involve the use of tillage, crop rotation, fertiliser application, liming and drainage techniques. These help to disturb the cycle of pests, diseases and weeds, and to improve the resistance of crops to them.

Habitat removal is perhaps the least acceptable approach to pest, disease and weed control. The basis of this approach is that semi-natural vegetation acts as a refuge for many pests, diseases and weeds, and that control may therefore be achieved through the removal of such habitats. However, this approach fails to recognise both the conservation value of these habitats and the fact that they may be an important source of predators as well as pests. Woodland and hedgerows in lowland arable areas, for example, are an important habitat for ladybirds, essential in the control of greenfly populations, and for bees, important in plant pollination. They also act as shelter belts, thus helping to control soil erosion (Briggs and Courtney 1989).

8.3.4 Urban and industrial pollution

Urban and industrial development has been associated with both physical degradation and chemical contamination of soils. Problems of physical degradation include erosion, compaction and structural damage resulting from construction activities and opencast mineral extraction (sections 8.2.1 and 8.2.2). Similarly, chemical problems result from waste disposal activities, discharge and spillage of liquid effluents, and atmospheric emissions, including acid deposition (section 8.3.1). Soils of urban and industrial environments are every bit as complex and variable in their characteristics as those in rural areas and merit classification in their own right. Hollis (1991), for example, suggests that the England and Wales soil classification system should be modified to include an 'anthropogenic' major soil group which could be subdivided on the basis of the types of contaminant materials present.

Physical disturbance and chemical contamination of soils in urban and industrial environments are not only recent phenomena. Archaeological investigations have revealed the extensive accumulation of building and domestic waste, although much of this was relatively harmless. Since the Industrial Revolution of the eighteenth and nineteenth centuries, however, the amount and variety of waste materials have increased dramatically. Furthermore, much of this waste is considerably more harmful and often less biodegradable than its early historical counterparts. It is particularly important that any survey of contaminated sites should include an assessment of this historical legacy, since the response of soils to stored pollutants may be delayed for long periods of time. This type of delayed response is potentially very serious and is often referred to as the 'chemical time-bomb' effect (Hekstra 1993).

Bridges (1991a) identifies four major sources of soil contamination in urban and industrial environments – construction and demolition waste, metalliferous materials, power generation emissions, and chemical and organic wastes. During construction and demolition a range of materials are disposed of in the soil environment, including broken bricks, tiles, glass, timber, piping, wiring and cables, insulation materials, mortar, concrete and plaster. Once in the soil, these materials undergo a number of chemical changes. Plaster, for example, contains large amounts of gypsum and at sites where the water table is high, the gypsum is dissolved and capillary action may bring it into contact with new concrete structures, thus leading to serious corrosion problems. Similarly, although the use of asbestos and its removal from existing buildings is now carefully controlled, significant quantities can be present in soils and overburden at sites with a long history of construction and demolition.

Metalliferous wastes, particularly heavy metals (e.g. lead, zinc, cadmium, copper and nickel), are commonly found in soils of areas where ore extraction and smelting have occurred (e.g. Macklin 1986, Rautengarten 1993). More locally, metal contamination occurs on land used for scrap metal dealing and munitions factories. In an attempt to provide systematic data on the extent of heavy metal contamination in urban soils in Britain, a national survey was commissioned by the Department of the Environment in 1981 (e.g. Culbard *et al.* 1988); 100 household gardens were sampled in each of 53 cities, towns and villages. In most of these localities, total lead levels in the soil exceeded 200 mg kg^{-1}, with values in excess of 500 mg kg^{-1} in industrial areas in the south and in former lead mining areas in the north (Figure 8.25). These values are considerably greater than the background levels of lead (usually < 100 mg kg^{-1}) normally found in uncontaminated soils.

Toxic metals may exist in the soil in a number of forms including adsorbed cations, attached to clay and humus colloids, and organo-metallic chelates (Bridges 1989). Their availability to plants depends on a number of soil characteristics, particularly cation exchange capacity (CEC), pH and the interdependence effects of other metals. In soils with a low CEC, the metals are not retained effectively and are likely to be either leached from the soil or taken up by plants, while in soils with a high CEC, they are likely to be fixed in the soil through adsorption processes. Similarly, the mobility and availability of heavy metals is considerably greater in acidic soils (pH < 5.5) than in near neutral or alkaline soils (Willett *et al.* 1994). Once mobilised, the metals may enter the food chain either through water supplies and aquatic organisms, or through arable produce and grazing animals. The effects of toxic metals in soils on human health are unclear, and it is therefore difficult to establish threshold concentrations above which toxicity problems are likely to occur. However, a number of countries have devised such threshold values, although these vary markedly in response to differences in environmental policy and legislation. The Dutch government, for example, has a

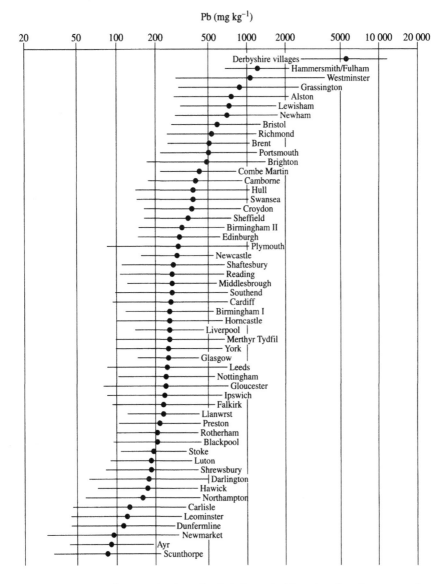

Figure 8.25 Lead contamination of soils in household gardens in urban areas of Britain. Values presented are geometric means ± 1 standard deviation for 100 households in each location (from Culbard *et al.* 1988)

system with three levels – an acceptable reference or background value, an indicative value for further investigation and an indicative value for clean-up (Table 8.12).

Table 8.12 Dutch guidelines for several soil metal pollutants

	Threshold concentration in soil (mg kg^{-1} dry weight)		
	A	B	C
Cr	100	250	800
Co	20	50	300
Ni	50	100	500
Cu	50	100	500
Zn	200	500	3,000
As	20	30	50
Mo	10	40	200
Cd	1	5	20
Sn	20	50	300
Ba	200	400	2,000
Hg	0.5	2	10
Pb	50	150	600

Source: Based on Thornton 1991
Note: A = reference (background) value; B = indicative value for further investigation; C = indicative value for cleaning up

A number of soil contaminants are derived from the power generation industry, including SO_2 (sulphur dioxide) from coal-fired power stations (section 8.3.1), and radionuclides from nuclear power stations and weapons testing. The anthropogenic radionuclides most commonly found in soils are those of caesium (^{137}Cs and ^{134}Cs). Much of the research in this field has occurred since the accident at Chernobyl in Ukraine in 1986, which resulted in large quantities of radionuclides being deposited across wide areas of Europe, including Britain (Figure 8.26). The behaviour of radionuclides in soils depends on a number of soil characteristics, particularly clay content and mineralogy, organic content, CEC, pH, NH_4^+ (ammonium) content and nutrient status (Livens and Loveland 1988). The radionuclides are immobilised most strongly in soils with a high CEC and near neutral pH values, where they are adsorbed onto clays, especially micaceous varieties, and humic materials. They are least well retained in acidic soils with a low CEC, where they may be available for plant uptake. Such conditions are widespread in upland areas of northern and western Britain where the milk and meat of grazing cattle and sheep were contaminated following the Chernobyl accident. In northern Scandinavia, reindeer herds grazing on slow growing lichen heath were similarly affected. The problem of radionuclide entry into the food chain from acid soils in Britain was not anticipated because much of the early research here had focused on clay soils with near neutral pH values where radionuclides are relatively immobile (Milne 1987).

A wide range of chemical waste materials have been implicated in the

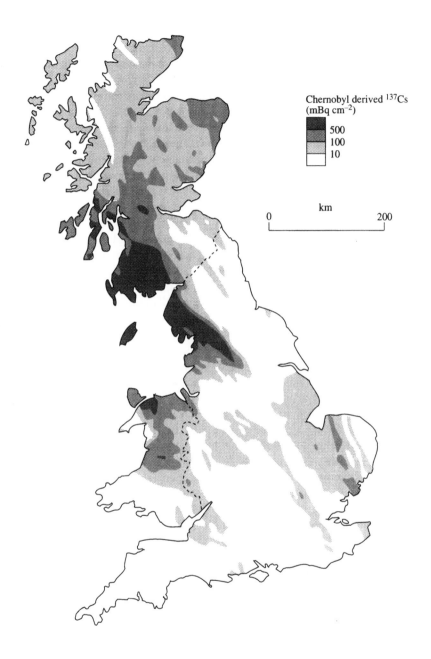

Figure 8.26 Deposition of ^{137}Cs over mainland Britain resulting from the Chernobyl accident (from Walling and Quine 1991)

contamination of soils. These include derivatives of detergents, fertilisers and pesticides (section 8.3.3), paints, dyestuffs, battery chemicals and leather tanning agents, to name but a few. Even silicon chip manufacture involves the use of chemical components, and leakage of solvents used in this process led to the contamination of drinking water in Silicon Valley, California (Anon. 1983). In recent years, particular concern has arisen regarding the production of dioxins; these are a particularly toxic group of chemicals and are associated with the manufacture of several organic pesticides. In the Italian town of Seveso, for example, an area of 1,800 ha was affected by dioxin contamination in 1976 and 700 people were evacuated; subsequently, the contaminated soil had to be excavated and removed. A major event which publicised the issue of chemical contamination of soils was the Love Canal disaster near Niagara, USA. Here, extensive dumping of chemicals over a 30-year period led to severe soil contamination by migrating leachates and gasses, with serious health implications for local people. This persuaded the US government to release funds to implement the Comprehensive Environmental Response Compensation and Liability Act of 1980 (CERCLA) (Bridges 1991a).

A number of organic wastes may lead to soil contamination. At Times Beach, Missouri, for example, the land was so badly contaminated by dioxins from waste oils, derived from the manufacture of the defoliant 'agent orange', that it was purchased by the Federal government (Bridges 1991a). Spillage of oils and related wastes is common on land used for storage and maintenance of motor vehicles, and for fuel storage. Sewage sludge, often applied to agricultural land adjacent to urban and industrial areas, often contains high concentrations of heavy metals (Table 8.13). The PCBs (polychlorinated biphenyls) are a group of organic solvents commonly implicated in soil contamination. These are used as dielectric fluids in transformers and are often released during the break up of electrical equipment.

Management of contaminated soils is a difficult prospect as the guidelines on threshold toxicity levels of contaminant materials, and on the vulnerability of soils to pollution, are still being developed and refined (e.g. Hortensius and Nortcliff 1991, Batjes and Bridges 1993). Management options depend to a large extent on the geological and hydrological characteristics, and size, of the contaminated area. Such areas vary dramatically from large designated landfill sites with records of waste disposal activities, to former industrial sites with a long and unclear history of pollution. Management options also depend on the nature of the contamination problem and whether it can be contained at the affected site, or whether leachates and gases threaten areas beyond it. Furthermore, it may be possible to treat the contamination on site, or it may be necessary to remove and relocate the contaminated soil.

Bridges (1991b) outlines a number of management approaches to the problem of contaminated soil. These include the use of physical stabilisation,

Table 8.13 Trace element concentrations in dry digested municipal sewage sludges

| Element | Reported range (mg kg^{-1} dry sludge) | | |
	Minimum	Maximum	Median
As	1.1	230	10
Cd	1	3,410	10
Co	11.3	2,490	30
Cu	84	17,000	800
Cr	10	99,000	500
F	80	33,500	260
Fe	1,000	154,000	17,000
Hg	0.6	56	6
Mn	32	9,870	260
Mo	0.1	214	4
Ni	2	5,300	80
Pb	13	26,000	500
Sn	2.6	329	14
Se	1.7	17.2	5
Zn	101	49,000	1,700

Source: Logan 1990

barrier, thermal and microbiological techniques. Stabilisation techniques involve treatment to reduce the solubility and mobility of the waste materials. This is achieved through the application of cement, lime, gypsum, silicate materials, epoxy-resins, polyesters or asphalt; these act as binding agents and help to stabilise the 'landform' within which the contaminated waste is stored. Barrier systems usually rely on physical containment using steel or concrete piling to prevent downward and lateral migration of toxic materials. Similarly, layered cover systems are often used to prevent upward migration of contaminants (Figure 8.27). Thermal techniques involve heating contaminated soil in rotary kilns or furnaces, whereas microbiological techniques involve the inoculation of waste materials with microbiological communities. Both of these techniques aim to convert the toxic waste materials into less harmful forms. Other approaches to the management of contaminated soil include chemical treatments which may be used to hydrolyse or oxidise contaminants into less dangerous products; similarly, acidic or alkaline wastes may be neutralised. In addition, physical methods may be employed to separate out contaminants according to particle size or density.

The number of sites officially designated as contaminated varies dramatically from country to country. In the UK there are 300 sites, covering an area of about 10,000 ha, where land is officially designated as contaminated. Unofficially, however, there may be between 50,000 and 100,000 sites, covering an area of over 100,000 ha (Bridges 1991b). In the USA 25,000 sites

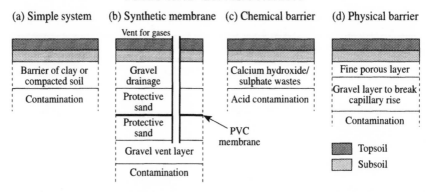

Figure 8.27 Examples of barrier cover systems used to prevent the migration of pollutants through leaching, capillary action and gaseous transfer (from Bridges 1991b)

have been designated as contaminated by the Environmental Pollution Agency. Similarly, in the Netherlands 6,000 sites are currently being reclaimed (Bridges 1991b). Yakowitz (1988) estimated that the cost of dealing with contaminated soils amounts to about US $8–12 per head in the industrialised countries. This figure does not include the former Eastern-bloc countries, however, as it is only in recent years that the rest of the world has become aware of the extent and severity of soil contamination here (e.g. Appelgren and Burchi 1993, Kozak *et al.* 1993). The problem of contaminated soils is often far too great to be dealt with by individual companies, hence governments are having to provide sizeable financial contributions towards the management of this problem. Estimates of national costs include £100 million in the UK, $20–100 billion in the USA, DK 410 billion in Denmark, DM 22 billion in the former West Germany, and Df 13 billion in the Netherlands (Bridges 1991b).

8.4 SUMMARY

This chapter has focused on the causal factors, on- and off-site effects, and management and remediation of soil degradation. This results from ecosystem disturbance and destruction, and from inappropriate landuse practices. It incorporates a number of environmental problems including soil erosion, compaction, water excess and deficit, acidification, salinisation and sodification, and toxic accumulation of agricultural chemicals and urban/industrial pollutants. In many cases these have led to a serious decline in soil quality and have a number of off-site impacts. The extent of soil degradation is influenced by a number of factors including soil characteristics, topography, climate, land use, and socio-economic and political controls. Management of soil degradation requires a holistic, multidisciplinary approach involving agronomists, soil scientists, hydrologists, engineers, sociologists, economists, politicians, and most importantly, the land users themselves. The approach to soil conservation has shifted in recent years from a rather technocentric standpoint to a more ecocentric position. Central to this approach are the concepts of land husbandry and sustainable development.

The extent of soil erosion is governed by a number of factors, in particular the erosivity of the eroding agent, erodibility of the soil, slope characteristics, landuse and conservation strategies, and socio-economic controls. Off-site effects of erosion include disruption of transport and communication systems, flooding, siltation of water supplies and damage to property. There are three main approaches to the management of soil erosion – agronomic methods (e.g. sensible cropping programmes), soil management strategies (e.g. mulching and tillage) and mechanical techniques (e.g. terracing). These approaches are often used in combination, although agronomic methods are often preferred as they are able to control both the detachment and transport phases of erosion, are relatively inexpensive and are easily incorporated into most landuse systems.

Compaction involves compression of a mass of soil into a smaller volume, thus leading to increased bulk density and resistance to compression, and decreased porosity. Closely related to compaction is hardsetting which involves an increase in bulk density without the application of an external load. The main causes of compaction are raindrop impact, which leads to surface crusting, and use of heavy agricultural machinery, which contributes to the formation of subsurface cultivation pans. Management of soil compaction focuses on the improvement of soil structure, appropriate tillage practices and, if necessary, installation of artificial drainage.

Excess water in soils may result from compaction, heavy texture and high rainfall, especially where sites are flat, low lying and prone to flooding. In many cases excess water may be removed through sensible management of soil structure and appropriate tillage practices. Frequently, however, artificial drainage is necessary in the form of subsoiling, mole drainage, pipe or tile drainage, open ditches or dykes, and regional or arterial drainage. Allied techniques include flood prevention through the construction of coastal defences or riverine levées. Soil water deficit occurs most commonly where rainfall is relatively low, particularly in coarse-textured soils which are excessively drained, poorly structured and low in organic matter. This problem is managed most commonly using irrigation. Specific irrigation techniques include surface (e.g. flood and furrow), overhead (e.g. spray or trickle) and subsurface applications. Soil water deficit may also be managed through the control of evaporation from the soil surface, often by mulching.

The main natural causes of soil acidification are long-term leaching, microbial respiration, plant nutrient uptake and nitrification. The main anthropogenic causes include excessive use of inorganic nitrogen fertilisers, needleleaf afforestation and acid deposition. Acidification of soils is commonly associated with surface water acidification, aluminium contamination and damage to freshwater ecosystems. Management approaches to soil and surface water acidification often centre on liming and sensible forestry management strategies; reduction of acid emissions into the atmosphere and assessment of critical pollution loads are also important in this respect.

Salinisation and sodification describe the accumulation in the soil of toxic salts and Na^+ ions respectively. These processes occur most commonly in soils of dry environments where intense evaporation leads to strong capillary action. The main causes of salinisation and sodification are the replacement of deep-rooted trees by shallow-rooted crops, and inappropriate irrigation practices. Overwatering, and the use of water which is high in Na^+ ions and low in total salts, may lead to destabilisation and breakdown of soil structure. Salinity and sodicity problems are managed most effectively through a combination of sensible irrigation practices, installation of drainage to permit the flushing away of salts, chemical treatments, evaporation control, and use of salt- and sodium-tolerant crops.

Agrochemical pollution has resulted from an increased use of inorganic fertilisers and pesticides in recent decades. The main concerns are nitrate leaching, and persistence and toxicity of pesticide residues. Nitrate leaching results not only from increased fertiliser usage, but also from mobilisation of organic nitrogen reserves within the soil. It is most serious in bare soils where there is no crop cover to take up the excess nitrogen, and in soils where old grassland has been ploughed up. Nitrate leaching is often associated with the eutrophication of surface waters, and is implicated in some conditions of poor human health. Management strategies include the use of slow-release varieties of fertiliser, keeping to recommended levels of application for each crop and avoiding large applications at any one time. Ploughing of grassland and application of fertilisers during the autumn period should also be avoided, as should fertiliser applications to hedge bottoms, field margins and water courses. Pesticide research and development should continue to focus on minimising effective persistence and toxicity, and maximising specificity. Granular and slow-release varieties and ultra-low volume application techniques are currently being explored. Alternatively, non-chemical strategies of pest, weed and disease control are also available; these include direct, biological and cultural methods, and habitat removal.

Urban and industrial development are associated with a number of soil problems including erosion, compaction and structural damage resulting from construction and opencast mineral extraction, and chemical contamination resulting from waste disposal, effluent spillage and atmospheric emissions. Specific sources of soil contamination include construction and demolition wastes, metalliferous materials, especially heavy metals, power generation emissions and chemical and organic wastes. Management of contaminated soils is a difficult prospect as the guidelines on threshold toxicity levels, and on the vulnerability of soils to pollution, are still being developed and refined. Furthermore, the long-term behaviour of contaminants in the soil environment, including the possibility of a chemical time-bomb effect, is not necessarily easy to predict. The main management approaches include stabilisation, barrier, thermal and microbiological techniques; others include chemical treatment and physical separation.

9

SOIL SURVEY AND LAND EVALUATION

9.1 INTRODUCTION

Soils are a complex, multivariate medium which play an important role in all environmental disciplines (section 1.2). Consequently it is necessary to understand the way in which they vary spatially and how their characteristics are suited to various forms of environmental investigation and utilisation. The spatial variation in soils has been recognised since the earliest times through its influence on agriculture, drainage and, therefore, human settlement. The earliest surveys were concerned primarily with basic agricultural characteristics such as texture and drainage (Avery 1990), but with the development of the environmental sciences during the twentieth century, soil survey became more sophisticated and aimed at a wider range of users, so that today there are now many types of survey, showing a variety of information at a range of different scales (Olson 1981a). Closely related to soil survey is land evaluation, whereby land is assessed on the basis of its suitability for particular purposes. In the first part of this chapter we will examine the methods of soil survey and land evaluation. This will then be followed by a discussion of the applications of the resulting information for both agricultural and non-agricultural purposes.

9.2 METHODS

9.2.1 Soil survey

Before embarking on a soil survey it is necessary to answer four main questions – what is the purpose of the survey and what information will therefore be recorded; how much detail is required and therefore at what scale will the survey operate; what level of accuracy is required; and how much time and what resources are available to conduct the survey? The information shown must be carefully related to the purpose of the survey; for example, a civil engineer is likely to be interested in soil depth and compaction, whereas an agriculturalist will wish to know its fertility and erodibility. The scale at which a survey operates will determine the amount

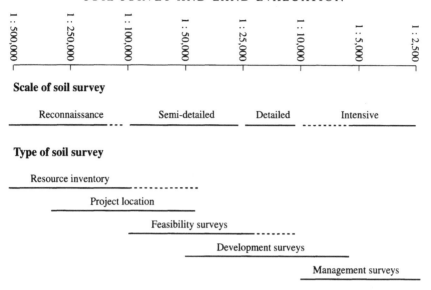

Figure 9.1 Scales and types of soil survey (from Young 1976)

of detail it is possible to show. For example, an area of 1 cm² on a map will represent an area on the ground of 0.1 × 0.1 km at a scale of 1:10,000, but will represent an area of 10 × 10 km at a scale of 1:1,000,000. In the former case, much more detail can therefore be shown. In general, surveys at scales larger than 1:100,000 are used for specific projects, while smaller scales are used for reconnaissance and compilation purposes (Figure 9.1). The accuracy of a survey will be determined by the size of the smallest area depicted on a map; this is known as the *fundamental mapping unit*. For example, for any given scale, a map with a fundamental mapping unit measuring 0.25 cm² will be more accurate than one with a unit of 1 cm². The time and resources available will determine the method of survey. For example, a small-scale reconnaissance survey might be made over a period of a few weeks, or even days, by a single person, while a more detailed survey may require a larger team working over many months or even several years. These four questions are often related; for example, the time and resources available will usually determine to some extent the degree of detail and accuracy provided, as will the purpose of the survey. The type of information recorded will also influence the choice of scale, because certain soil characteristics, such as texture or depth, can vary rapidly over a short distance, while others, such as soil type or drainage, may vary more slowly.

Having answered these questions, there are two principal types of survey method available – *ground survey* and *aerial survey*. Often these are used in combination, but for discussion purposes they will be considered separately.

Ground survey can take one of two forms – *grid survey* or *free survey*. Grid survey is an objective method that requires little prior knowledge of the area to be mapped. A grid is established over the area, and the grid intersections mark the positions of soil sampling locations in the field. From the data obtained, boundaries between soils of different characteristics can then be delineated (Figure 9.2a). The accuracy of the map is determined by the grid density; high densities produce more accurate maps, but this can be an extremely time-consuming and therefore costly method of survey. In contrast, free survey involves the location of soil boundaries on the basis of the surveyor's knowledge of soil-environmental relationships in the mapping area. Once the surveyor has determined the relationship between soil characteristics and slope, drainage and vegetation, it is possible to concentrate on establishing the location of the soil boundaries without spending much time on the intervening areas (Figure 9.2b). This technique is therefore more time-effective than grid survey, although it can only be used in areas of marked environmental contrasts; in areas of little relief or uniform vegetation, for example, it is usually necessary to use the grid survey method. The element of subjectivity in free survey also makes errors possible in that an important change in soil characteristics may be overlooked if it is not expressed at the surface by a change in environmental conditions. For

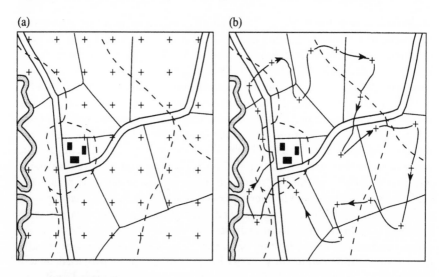

(a)   (b)

+   Inspection point

– – – Soil boundaries

—◄— Soil surveyor's traverse

Figure 9.2 Sample points for (a) grid survey and (b) free survey for an area of about 0.4 km^2 (from Bridges 1982)

example, soil conditions may change if the parent material changes from sandstone to shale, but this may not be manifested in the surface relief or vegetation cover. However, this problem can be overcome if a geological map of the area is available, and such maps often form an important part of a free survey strategy.

Field description and sampling of the soil can be conducted in a number of ways. The most rapid method is to use an auger, which can be operated either by hand or mechanically. However, this only provides information at a single point. In order to obtain a more detailed view of the soil, a pit must be excavated; this allows the soil to be viewed in three dimensions, and therefore for small-scale lateral and vertical variations to be recognised. For an even better view of local variability, a trench is required. The digging of pits or trenches is, however, time-consuming, so these are usually kept to a minimum and are often excavated mechanically. Methods which do not require any form of excavation are also available; for example, ground-penetrating radar and magnetic susceptibility measurements can allow the rapid identification of soil boundaries (e.g. Collins and Doolittle 1987, Williams and Cooper 1990). Description of a soil during survey is usually made on a systematic basis, describing a predetermined set of attributes according to a standardised system; this information can then be logged on a computer database for rapid retrieval and display.

Aerial survey is based on the use of images of the ground recorded from aircraft or satellites (known as *remote sensing*), relating the tones of these images to particular soil characteristics (Mulders 1987). In the case of photographs taken using light-sensitive film, tonal variations occur due to variations in light reflectivity, and these are controlled principally by surface type. For example, soil moisture can influence the reflectivity of the surface (section 6.3.1), which can be expressed either directly at the surface if a soil is unvegetated, or via the vegetation cover which derives its moisture from the soil; for any given type of vegetation, areas where it is parched are often more reflective and may therefore indicate areas of more freely draining soils with lower water-retention properties. Alternatively, different types of vegetation within the mapping area might indicate different soil drainage conditions, such as where marshland occurs in poorly drained areas. Variations in moisture may in turn relate to soil textural variations or to differences in drainage due to parent material or slope characteristics. Reflectivity can also be determined by the organic and carbonate contents of soils (Major *et al.* 1992).

Other parts of the electromagnetic spectrum can also be used to record soil conditions. For example, false colour infra-red imagery can show moisture and vegetation type (Plate 15), while far- and thermal infra-red parts of the spectrum can be used to detect thermal conditions (Landon 1991). The effectiveness of spectral differentiation of soil conditions can, however, be limited by atmospheric constituents such as moisture, carbon dioxide and dust particles (Coleman *et al.* 1993).

Aerial survey can be expensive, but has the advantage over ground survey of being less labour-intensive, since large areas can be mapped with only limited fieldwork. This is therefore particularly advantageous in surveying areas of remote or inaccessible terrain. However, some initial field observations must normally be undertaken in order to be able to relate the photographic information to the soil conditions, a procedure known as *ground truth*. Although it is possible to survey an area using only aerial techniques, especially if the survey is of a reconnaissance nature, more detailed surveys usually involve ground techniques or the use of ground and aerial methods in combination.

Soil survey data are usually shown on a map in one of two forms – as values at individual points or as units delineated by boundaries. Examples of the former are binary and Boolean maps, based on a grid sampling system (Rudeforth and Webster 1973) (Figure 9.3). Examples of units delineated by boundaries are seen in most national soil surveys, where the units are usually classified according to soil type, soil properties or environmental characteristics. However, because soils are complex media, drawing boundaries on a map to differentiate one area of soil from another will inevitably mean that the soil shown within a mapping unit is not totally homogeneous. The conventional approach to this problem adopted by traditional, pedologically based surveys has been to recognise different levels of purity of mapping units. For example, in the British Isles, the most detailed mapping units are recognised as being simple if they contain < 15 percent of soil which is different from the one represented by the unit; such units are known as soil *series*. In the USA the most detailed mapping unit is known as a *consociation*, in which about 75 percent of the unit comprises a single soil series. Units which are more variable are referred to as *complexes* (Avery 1990). These

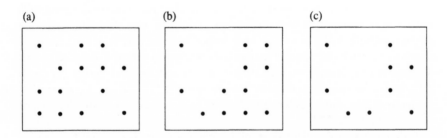

Figure 9.3 The principle of binary and Boolean soil mapping on a grid square base:
(a) a binary map showing soils at least 0.5 m deep, (b) a binary map showing freely drained soils in the same area as the previous map, (c) a Boolean map showing soils at least 0.5 m deep that are also freely drained, produced by combining the data from the previous two maps

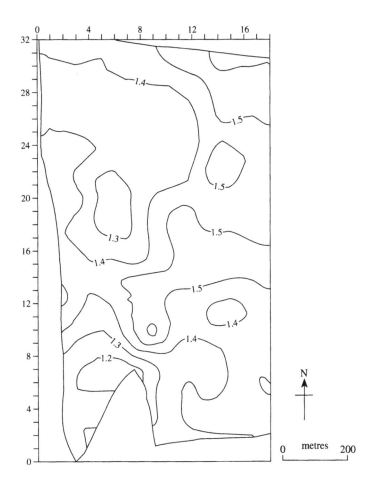

Figure 9.4 A map of exchangeable potassium produced by kriging (from Webster and Oliver 1990). The units are $\log_{10}$ (mg K/kg soil)

often contain a number of different series within them, whose spatial distribution is too variable to be able to map at the particular scale used. Where a group of series occurs in relation to a common environmental characteristic such as parent material, these are sometimes mapped as an *association*. In both the USA and British Isles, series are usually named after the localities in which they were first studied or where they are most extensively developed.

Although these concepts are useful, the location of boundaries on traditional soil maps remains subjective, being determined by the judgement and experience of the soil surveyor. However, numerical approaches have been developed as an alternative to this system, as in the case of soil classification (section 3.3.3), in order to increase the objectivity of boundary location. These involve the plotting of a continuous variable (Z) in the vertical plane relative to its geographic axes (X and Y), and joining points of inferred equal value by the use of mathematical interpolation techniques such as *kriging* (Webster and Oliver 1990) (Figure 9.4). Although these techniques can be effective in intensive, single-property surveys, they require a large quantity of data derived from systematic grid sampling, and complex computation, and are therefore of limited use for more general and smaller-scale surveys. In this respect, they suffer from the same limitations as the numerical methods used in soil classification (section 3.3.3).

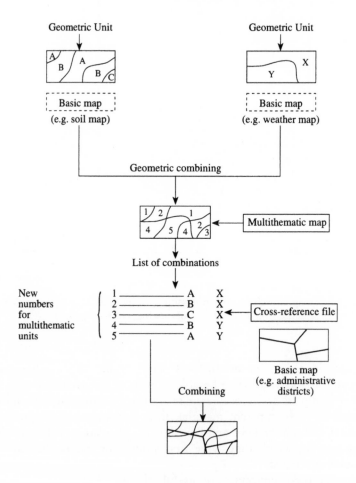

Figure 9.5 Example of GIS procedure (from King *et al.* 1986)

A recent development in soil survey methods has been made with the computerised storage, combination and display of data for mapping, known as *geographical information systems* (GIS). This has allowed conventional maps to be replaced by soil survey information compiled specifically to suit a particular user, and displayed by computer graphics (e.g. King *et al.* 1986) (Figure 9.5). This dispenses with the problem of 'fixed' boundaries and types of information on a conventionally printed map, since the boundaries and information can now be 'drawn to order'. Although at an early stage of development and application within soil survey, this approach clearly possesses great potential for the future.

9.2.2 Land evaluation

Land evaluation can be based directly on soil survey data, although it often uses this information in a more indirect way by producing an evaluation system designed specifically for a particular purpose, which can also include other types of environmental data. Such systems fall broadly into two categories – qualitative and quantitative – although the principles of their formulation are the same in each case; the aim is to produce a rating system by which land can be evaluated for a particular purpose. Qualitative systems involve assigning categories which may be descriptive, for example 'good', 'moderate', 'poor', or designated by a number or letter. In either case the criteria upon which each category is based need only be descriptive and not too closely defined. For example, an agricultural land evaluation may designate an area of light textured soils on gentle slopes to category 1, an area of heavier textured soils or those on moderately steep slopes to category 2, and an area of very heavy soils or those on very steep slopes to category 3. However, this type of system is obviously prone to subjectivity, which can lead to inconsistency when applied by more than one evaluator.

To overcome this problem, quantitative criteria can be used to differentiate between categories, as in the case of soil classification (section 3.3.3). For example, in the previous case, clay percentages could be assigned to textural characteristics and angular values to slopes. The approach can be extended by the use of indices based on a number of soil and other environmental characteristics. This is usually obtained by adding or multiplying together the values of the various criteria used; in the example given above, this could be achieved by multiplying the percentage clay by the slope angle. Land evaluation classes are then based on ranges of index values (Young 1976). An alternative to the use of classes is a continuous scale of assessment related to factors affecting the productivity of the land. These are multiplied together to produce an index of productivity, using factors such as soil moisture, depth and organic matter content (Nortcliff 1988). Spatial patterns of land evaluation can be shown using either a traditional map or a GIS, as in the case of soil survey (section 9.2.1).

9.3 APPLICATIONS

9.3.1 Agricultural applications

Soil surveys fall into two main groups – those produced, usually by pedologists, as part of a national survey programme, and those conducted for specialist purposes. The former tend to be genetically based, concerned principally with soil properties and types resulting from the operation of soil-forming processes and their associated environmental influences. Although this information is primarily pedological, it may also be useful for other purposes. Typical soil type, parent material, drainage and relief are usually shown, and the maps are often accompanied by a report which provides basic physical and chemical data for representative profiles of the various soil types recognised. Information which can be of use for agricultural purposes includes colour, texture, structure, depth, organic matter content, consistence, moisture and chemistry (sections 7.2.1 and 7.2.2) (Bridges and Davidson 1982). For example, colour can influence the reflectivity of the soil surface, and therefore the rate of warming, which can be important at the start of the growing season; dark-coloured soils may therefore warm more quickly than lighter-coloured ones, although dark colours may also be associated with high organic contents which can reduce warming rates because of low thermal diffusivities (section 6.3.1).

Texture can provide an agriculturalist with an indication of moisture characteristics; coarse-textured soils are often freely drained although they may also be prone to drought, while fine-textured soils often hold more nutrients although they can be prone to waterlogging or even drought if they experience vertical cracking; loamy soils therefore often possess the most favourable moisture characteristics for crop growth. Structure is an important property in terms of aeration and drainage. Soils with a poor structure may be prone to crusting, puddling, compaction or near-surface waterlogging, all of which inhibit crop growth (section 8.2). The structure therefore affects the workability of a soil; heavy agricultural machinery can only be used for limited periods of the year or under particular moisture conditions on soils with a poor structure. Surveys which include information on the workability of a soil are therefore useful in this respect. Soil depth can be important in terms of crop root development and moisture retention, which may be limited in shallow soils, although in some cases this appears not to be of any great significance (Burnham and Mutter 1993).

Organic matter data can provide an indirect indication of the structural, moisture retention and nutrient potential of a soil; these properties generally increase with organic content, although very high organic values may be associated with poor structure and low nutrient content, as in the case of certain peats. Consistence is related to soil structure, but indicates the way in which a soil responds to disturbance (Bridges and Davidson 1982). It can therefore indicate, for example, the likelihood of the soil developing clods or surface-sealing, both of which can inhibit seedling growth. Moisture data are often provided in soil survey reports, which can therefore give a direct indication as to the suitability of a soil for agriculture. Alternatively, this can be inferred from soil type information; for example, gleys and peats usually have high moisture contents which may inhibit crop growth unless the soils are artificially drained, while brown earths and rendzinas are generally freely drained. Chemical data provided by soil surveys are usually limited to basic properties such as pH and carbonate and exchangeable base contents, although this can clearly be of much use to an agriculturalist. Even where such information is absent, it is often possible to infer chemical character-istics from soil type distribution. For example, podzolic soils are acidic and therefore low in nutrients, while chernozems and brown earths will have moderate to high nutrient contents. Specialist surveys focus on one or more of a wide range of soil characteristics of specific interest to agriculturalists. Such surveys can be conducted at either a regional or more local level, and are often used in the planning of agricultural programmes in developing countries (Landon 1991).

The forms of information discussed above are not only used by agricultur-alists directly to assess land quality and plan cultivation and management programmes, but can also provide a basis for the development of specific land evaluation systems. A system which has become widely used, or has formed the basis for similar schemes elsewhere, is that developed in the USA (Klingebiel and Montgomery 1961). Eight classes of soil capability are recognised: Class I soils offer few limitations to cultivation, Class II have some limitations that reduce the choice of plants or they require moderate management to maintain productivity, and Class III have severe limitations to cultivation and/or require special management such as erosion control or artificial drainage. Class IV soils have very severe limitations that restrict the range of crops and/or require very careful management. Classes V to VIII are generally unsuited to arable cultivation, but may be used for other purposes. Class V soils are of the highest quality within this group, but cannot be cultivated because of factors such as wetness or climatic limitations, and they are therefore best suited to grazing, woodland or wildlife habitat. Class VI soils respond to management, but this is not possible for soils in Class VII. Class VIII soils pose severe limitations on grazing and woodland, and are best suited to recreation, wildlife, water supply or aesthetic purposes.

Related to this system is that of Bibby and Mackney (1969) for use in Britain. Here seven classes are distinguished; Classes 1 to 4 are suitable for arable farming, Classes 5 and 6 are most appropriate for grazing and forestry, and Class 7 is best left for wildlife or amenity purposes. The FAO (1976) system was developed for use worldwide, and is a multi-level hierarchical scheme. At the highest level it comprises only two orders – suitable, and not suitable because of technical, economic or environmental constraints. These

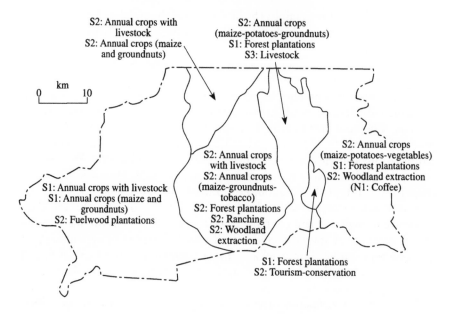

Figure 9.6 Proposed landuse alternatives for the soil landscape regions of the Dedza project area, Malawi (from Young and Goldsmith 1977)

are divided into classes, for example highly, moderately and marginally suitable, which are in turn divided into sub-classes based on the major limitation such as moisture or nutrient availability. Finally, the sub-classes are divided into units which are used to distinguish aspects such as particular management requirements. This system has been of particular use in developing countries, as in the case of Malawi, where it was used to propose landuse alternatives within soil landscape regions in the central part of the country (Young and Goldsmith 1977) (Figure 9.6). It has also been used more recently in Zambia by Chinene (1992) in relation to crop suitability. Geographical information systems have been used to develop land evaluation systems, for example in Spain (De la Rosa *et al.* 1992), Canada (McBride and Bober 1993) and Thailand (Liengsakul *et al.* 1993), which allows the production of computer-generated land suitability maps at a variety of scales based on a range of soil and environmental data. These methods are also being developed for use at a global scale (Oldeman and van Engelen 1993).

9.3.2 Non-agricultural applications

Although pedologists and agriculturalists are often the principal users of soil survey and land evaluation systems, there are many other users both academic and applied. Academic users include geomorphologists, ecologists and archaeologists, while applied fields include hydrology, civil engineering, recreation, waste disposal and landuse planning (Olson 1981a, Jarvis 1982). Geomorphological information can be obtained from data which relate to factors such as soil erodibility and slope stability. For example, dry, fine-textured soils may be susceptible to wind erosion in environments with little vegetation cover, while clayey, poorly drained soils are likely to be unstable on steep slopes. Some national surveys also provide information regarding geomorphological components of the landscape in which the soils occur. Soil survey data can be useful to ecologists in interpreting the distribution of vegetation types because soils usually play an important role in determining this distribution (section 6.5). For example, needleleaf forest or heathland is often associated with acidic soils such as podzols, while areas of marshland vegetation will generally coincide with poorly drained gleys or peats. The availability of physical and chemical data in a soil survey will allow a more detailed vegetation–soil association to be established. Archaeologists can use soil maps to locate potential occupation sites in the past and to explain past landuse practices. For example, the Maya civilisation in central America avoided Vertisols as construction sites for heavy buildings because these soils are prone to poor drainage and mechanical instability; shallow Mollisols developed over limestone were favoured for this purpose. The Mollisols were also initially very fertile, and were therefore cleared of forest and used for cultivation, but they then became vulnerable to erosion, leading not only to soil deterioration but also to silting of drainage channels and reservoirs, and in many locations this is associated with site abandonment and decline of the civilisation (Olson 1981a, b).

Hydrologists can obtain information from soil surveys with respect to the influence of soil physical conditions on surface water and ground-water movement (section 6.2). For example, fine-textured or poorly structured soils will usually have low infiltration rates, which can lead to rapid surface runoff and therefore to an increased risk of flooding. Shallow soils, or those with low-permeability pans, can have a similar effect. However, soil survey information may not always relate closely to hydrological conditions; for example, McKeague and Topp (1986) reported that saturated hydraulic conductivity values of soils were incompatible with a number of drainage groups assigned to soils on the basis of soil survey information. Specialised surveys can provide more specific hydrological information, for example in the case of the map of Winter Rain Acceptance Potential (Farquharson et al. 1978). Five classes are distinguished, ranging from very high (1) to very low (5), and these are determined on the basis of soil hydrological properties and slope angle. Such information can be used in planning river management schemes.

The use of soil survey information by civil engineers relates largely to the mechanical and hydrological properties of the soil (e.g. Westerveld and

Van Den Hurk 1973, Brink *et al.* 1982). Although the topsoil is usually removed prior to construction, subsoil properties such as shrinkage, compressibility and frost susceptibility can be important in determining the stability of foundations, while permeability can influence underground drainage installations. For example, the shrinkage of clayey soils can cause disturbance to foundations and above-ground structures; disturbance can also be caused by silty, frost-susceptible soils. Highly compacted fragipans can cause waterlogging, thus affecting drainage. Soil chemical characteristics may also be important in some cases, because steel and concrete can be attacked in acid soils or those containing sulphates or sulphides (Jarvis 1982). Soil depth can also be relevant, for example in terms of the quantity of soil requiring penetration or removal if bedrock foundations are necessary, and the availability of soil for earthwork constructions such as road or flood defence embankments, reservoir retaining walls, and banks to obscure quarry or opencast mining operations from view. Specialised land evaluation for civil engineering purposes is based on properties such as texture, plasticity, stoniness, water table depth and slope, for example the Soil Survey Staff (1971) system which expresses soil limitations for shallow excavations as 'slight', 'moderate' or 'severe', based on a combination of these factors.

The construction and management of recreational facilities such as playing fields, campsites and footpaths can be aided by soil survey and land evaluation data. For example, in the case of playing fields and campsites, good drainage and low stone content are important soil characteristics, while footpaths should also be well-drained and resistant to erosion (section 7.4.2). Waste disposal can be a problem in certain types of soil, so again soil survey and land evaluation data can be useful in planning disposal sites and management policies (Olson 1981a, Hodgson and Whitfield 1990). For example, the disposal of septic tank effluent is less likely to cause a direct health hazard in freely drained soils, where surface or near-surface accumulation of pollutants is minimised, and disposal sites may be located where these soils occur on land unsuitable for agriculture. However, the rapid drainage of waste can cause contamination of rivers and water supplies, so it is important that the soils are not too highly permeable. Lagoons for storing waste products require a low-permeability floor and containing walls; clayey soils are therefore better suited for these features than are coarser-textured soils. Specialist soil survey data can also be useful in land restoration programmes, such as data relating to soil volume and plastic limit, for planning soil removal, storage and restoration (McRae 1989).

The use of soil and land evaluation surveys in landuse planning relates to a combination of some of the factors already considered. For example, the mechanical and hydrological properties of soils are taken into consideration with respect to the construction of buildings and roads, and drainage and waste disposal systems; physical and chemical characteristics are also relevant to the type of vegetation required for recreational areas such as parks, playing fields, golf courses and natural wildlife habitats (Bauer 1973, Jarvis 1982). Related to landuse planning is the study of environmental degradation and restoration. For example, Raina *et al.* (1991) have used remote sensing for the mapping of land degradation by wind and water erosion in northwest India, and Evans (1990b) has used existing national soil maps as a basis for mapping soil erosion risk in England and Wales, based on landuse, soil type and landform criteria. At a more detailed level, Aspinall *et al.* (1988) have mapped heavy metal concentrations using a 1 km grid in order to determine the pattern of industrial pollution in Tyneside, England.

9.4 SUMMARY

Soil survey and land evaluation are important aspects of soil-environmental study from a variety of viewpoints, pure and applied, and agricultural and non-agricultural. Having decided the purpose of a soil survey, what information to record, the level of detail and accuracy required, and the time and resources available, the survey can use one or a combination of two methods – ground and aerial survey. Ground survey can be conducted either by a grid survey, which requires little prior knowledge of the area to be mapped, or by a free survey, which relies on the surveyor's knowledge of soil-environmental relationships. Aerial survey uses images of the ground recorded from aircraft or satellites, and is less labour intensive, although some limited field observations (ground truth) must also be made. Soil survey data can be recorded on a map either as point values or in areas. The problem of subjectivity in locating area boundaries has been addressed by the use of numerical techniques, although these methods require a large database and are therefore best suited to large-scale, specialist surveys. The recent development of geographic information systems has improved the flexibility of soil map compilation by the use of computerised data storage and mapping techniques. Land evaluation can be based directly on soil survey data, or on the use of these data in combination with other forms of environmental data. The evaluation may be qualitative or quantitative, the latter using individual criteria or combinations of these criteria in the form of indices.

National soil surveys can provide useful information to agriculturalists, in the form of environmental characteristics such as parent material, drainage and relief, along with soil types and their basic physical and chemical properties. Specialist surveys provide more specific information and usually operate at a larger scale. Soil survey data can also be used in the development of agriculturally based land evaluation systems, which categorise areas according to their limitations to cultivation, pastoral farming or forestry. Testing of these schemes in specific field situations has shown them to be successful in a variety of environments. Non-agricultural applications of soil surveys and land evaluations include geomorphology, ecology and archaeology, plus applied fields such as hydrology, civil engineering, recreation, waste disposal and landuse planning. Specialist surveys are also used in environmental assessment programmes involving land degradation and restoration.

10

CONCLUSIONS

In this book we have examined the ways in which soils are both influenced by, and themselves influence, the environment. During the course of this study a number of general points have emerged concerning soil-environmental interactions, and although many of these points have been summarised at the end of individual chapters, it is useful finally to draw together some of these aspects in order to provide an overview of our present state of understanding, and to consider the role of soils in the environment of the future.

10.1 THE PRESENT

Over the past century or so, much has been achieved in the broad understanding of soil constituents and their behaviour, and of the processes by which soils develop, although of all the soil constituents, the organic fraction remains the most elusive in terms of both its composition and transformation (Ross 1989, Schnitzer 1991). The development of classification systems has helped to categorise soil data in a systematic manner, with the systems of the FAO-Unesco (1974, 1989) and the Soil Survey Staff (1975, 1992) now being widely used. While numerical methods offer a useful alternative approach to classification (Webster and Oliver 1990), these require a large quantity of data and are yet to replace traditional hierarchical systems at a national level.

The influence of environmental factors on soil formation is now well known in general terms, although it remains difficult to quantify their individual or combined effects (Johnson and Watson-Stegner 1987). The role of the time factor in soil formation is more easily quantified, and studies synthesising data from a wide range of environments have been particularly useful in this respect (e.g. Bockheim 1980, Birkeland 1992, Schaetzl *et al.* 1994). Information concerning rates at which soil changes have occurred in the past may be useful in the prediction of future changes, which might be of relevance to environmental planning. The formation of soils at a global scale in response to prevailing environmental conditions is becoming increasingly well understood, although it is now clear that many soils also possess characteristics

313

formed under different environmental conditions in the past (Martini and Chesworth 1992). Indeed, some of the most exciting developments have been in the areas of pedogenic history, environmental reconstruction and dating, and the past decade has seen much interest in this subject (e.g. Boardman 1985, Wright 1986, Bronger and Catt 1989, Retallack 1990).

The importance of soils in natural environmental systems has been shown by studies in a number of disciplines, in particular, hydrology, geomorphology, climatology and biology. Hydrological studies have emphasised the role of the soil in surface and subsurface water movement and storage, an understanding of which relates largely to developments in soil physics (e.g. Hillel 1982, Ward and Robinson 1990, Jury *et al.* 1991). Soils play a vital role in geomorphology since they form the key to weathering and can participate in sediment transport processes, above and at the surface, below ground and *en masse*, the understanding of which is dependent to a large extent on soil physics and chemistry (e.g. Trudgill 1986, Selby 1993). The role of soils in determining microclimatic conditions near, at and below the ground surface, is again clarified by an understanding of soil physics, in relation to radiation, temperature, moisture and air movement (e.g. Oke 1987, Monteith and Unsworth 1990). There now exists a vast body of knowledge relating to the role of soils in biological systems, particularly with regard to vegetation. Much of this has been obtained through agricultural studies connected with the limiting factors to plant growth and the improvement of crop yields (e.g. Foth and Ellis 1988, Wild 1988).

Much has been written on the subject of soils and land use, most of which falls into two major categories – agricultural and non-agricultural. In an agricultural context, important research themes include the role of organic matter in the improvement of soil fertility and structural stability (e.g. Haynes and Swift 1990), the benefits of reduced and zero-tillage operations to soil erosion control and soil water conservation (e.g. Mbagwu and Bazzoffi 1989), and the behaviour and fate of agricultural chemicals with a view to reducing levels of fertiliser and pesticide contamination (e.g. Addiscott *et al.* 1991). Other developments include advances in drainage and irrigation technology, particularly in relation to the management and remediation of saline and sodic soils (e.g. Gupta and Abrol 1990), improvement of forestry strategies to ensure long-term sustainability of forest resources and maintenance of soil quality, particularly in the developing world (e.g. Szott *et al.* 1991), and more effective pasture improvement and management regimes (e.g. Newbould 1985). In a non-agricultural context, emphasis has been placed on the restoration of soils following opencast mineral extraction (e.g. McRae 1989), the contamination of soils by industrial wastes and airborne pollution (e.g. Bridges 1991a), and the response of soils to various recreational pressures (e.g. Stewart and Scullion 1989, Stewart and Cameron 1992).

For many aspects of soil-environmental study it is useful to record, or make reference to, the spatial distribution of soils. Techniques of ground and aerial survey have been applied at many scales and in many different environments, but the problem with traditional soil survey information is that it is restricted to a limited number of soil characteristics which are confined to a single map and which are therefore often of only general use to a particular discipline. With recent advances in computerised information systems, however, there now exists the possibility of providing much more user-specific data (King *et al.* 1986). Closely allied to soil survey is land evaluation, and in recent years this has been applied increasingly to landuse planning in both rural and urban areas (e.g. Hodgson and Whitfield 1990, Greenwood 1993).

10.2 THE FUTURE

At the close of the twentieth century we stand at a crossroads – before us lies the prospect of either allowing the continuing environmental deterioration of our planet, or actively participating in environmental improvement in an attempt to safeguard its future. In either case soils, as an essential condition-ally renewable resource, will continue to play a vital role (e.g. Lal and Stewart 1990a, b, Barrow 1991). Never before have global environmental problems been so severe. These include soil erosion in response to increasing demands for agricultural land and woodfuel (e.g. Thapa and Weber 1991), compaction of soils by heavy agricultural machinery (Raghavan *et al.* 1990), acidification of soils resulting from acid deposition, needleleaf afforestation and excessive use of nitrogen fertiliser (e.g. Rowell and Wild 1985), salinisation and sodification of soils due to poor cropping and irrigation practices (e.g. Gupta and Abrol 1990), agrochemical pollution resulting from increased reliance on fertilisers and pesticides (e.g. Jones and Schwab 1993), and the contamination of soils by materials such as heavy metals and radionuclides derived from industrial activity (e.g. Livens and Loveland 1988, Thornton 1991). Although most of these problems are of global significance, their impact is often felt most strongly in the developing world where population growth is most rapid and poverty is widespread.

In recent years, research into ways of addressing these problems has focused on the approach of land husbandry (e.g. Hudson 1992, Stocking 1992). This approach extends beyond the soil to consider the land users themselves, farm practices and systems of land tenure, together with broader aspects of socio-economic and cultural frameworks. The aim of land husbandry is to ensure that landuse systems are sustainable in the long term, and this depends ultimately on the maintenance of soil quality. Examples of sustainable landuse practices include agroforestry (e.g. Szott *et al.* 1991), social forestry (e.g. Kirkby and Sill 1991) and low external input agriculture (e.g. Sanchez and Benites 1987); as part of these sustainable landuse programmes, wood clearance is discouraged. In the developed world many of these environmental problems can be tackled effectively, although some are more persistent than others. An area of particular concern is the management of soils which have been contaminated as a result of industrial activity (e.g. Bridges 1991a, b).

Added to these problems there is the threat of global warming as a result of fossil fuel burning and depletion of the ozone layer, with the possibility of large-scale vegetation change (Woodward 1989). For example, the boreal forests of North America and Eurasia may expand northwards, thereby increasing the extent of acid podzolic soils. Similarly, at lower latitudes increased desertification could lead to an increase in the area of Aridisols. Changes in soil conditions would have important implications for agriculture in terms of factors such as moisture, nutrient content, structure and erodibility (Parry 1990). Recent research has attempted to predict the response of soils and crops to various global warming scenarios (e.g. Tate 1992), and this must continue in order to understand the consequences for both natural ecosystems and agricultural production. Associated with global warming is the melting of glaciers and ice caps, which may lead to a rise in sea level. Although there is much debate over the extent to which this will occur, even modest rises could have a significant effect on low-lying and often densely populated regions, such as Bangladesh, Egypt, Thailand and Indonesia. The resulting decrease in the area of habitable land would probably necessitate more intensive agricultural production, with attendant increases in soil degradation and related environmental problems.

In theory it is possible to tackle all of the major environmental problems which face us, and soil science will undoubtedly play a vital role. However, the profile of soil science within the scientific community, and awareness of its importance by the public at large, are not high relative to the traditional sciences and technological development, and consequently its teaching and research resources tend to be limited. In order to address this problem, soil scientists must not become isolated; they must work in collaboration with others, including agronomists, hydrologists, land economists, social scientists, politicians and, of course, the land users themselves (McCracken 1987, Gardner 1991, Greenland 1991, Simonson 1991, Greenwood 1993). It is within this holistic framework that the environmental problems outlined above can best be solved.

Environmental problems may be reduced in the long term by biotechnological developments which provide more efficient sources of food. Alternatively, pressure on the land could be reduced by the development of new food sources on a large scale which do not require conventional soil as a growth medium (Mannion 1991). In the shorter term, however, we must hope that the increasing awareness of environmental problems continues, and that these are not only tackled by the developed world but also that less developed areas receive the resources they require in order to be able to participate fully in what is undoubtedly a global challenge.

GLOSSARY

Accessory mineral A soil mineral which occurs in relatively minor quantities.

Acid A substance that releases hydrogen ions. Conditions in which hydrogen ions exceed hydroxyl ions.

Acidification An increase in soil acidity resulting from the removal of alkaline components.

Active layer The layer of thawed soil above permafrost.

Adhesion Attraction or sticking together of unlike materials; often used to describe the affinity of water molecules for colloidal surfaces (surface tension).

Adsorption The attraction of ions or compounds to the surface of a solid or liquid.

Aerobic Conditions in which a continuous supply of molecular oxygen is present.

Aggregate A discrete unit formed by the aggregation of soil constituents. Also known as a *ped*.

Aggregation The process(es) by which aggregates are formed.

Albedo A measure of the ability of a surface to reflect incoming solar radiation.

Alkaline Conditions in which hydroxyl ions exceed hydrogen ions.

Alluvium Material deposited by running water, usually by a river.

Anaerobic Conditions in which molecular oxygen is absent.

Anion A negatively charged ion.

Anion exchange capacity The total amount of anions that a soil can adsorb by anion exchange.

Argillan A depositional coating of clay.

Atmosphere The Earth's gasses, both above ground and in the soil.

Atom The smallest unit in which an element can occur naturally.

Azonal soils Soils whose development is restricted, mainly due to lack of time.

Base A substance that reacts with hydrogen ions or releases hydroxyl ions.

Biogeochemical cycling The cycling of elements in an ecosystem by both organic and inorganic processes.

317

Biomass The total amount of organic matter in a community or per unit area.

Biosequence A sequence of soils whose differences in development relate primarily to the different biotic conditions under which they have formed.

Biosphere That part of the Earth in which life is found.

Biota A group term for living organisms, including both plants and animals.

Bioturbation The mechanical mixing of soil by biota.

Brown earth A soil showing limited acidification, often associated with temperate broadleaf forest.

Bulk density Combined density of solids and pore space in an undisturbed sample of dry soil.

Calcification The concentration of calcium compounds, particularly carbonate, by illuviation.

Calcrete A material consisting mainly of calcium carbonate, usually associated with semi-arid environments.

Capillary fringe The zone in which most of the soil pores are filled with water.

Carbonation The dissolution of mineral compounds by the presence of carbon dioxide dissolved in water.

Catchment The area of a drainage basin delimited by the watershed.

Catena A sequence of soils whose characteristics are associated with their position on a slope.

Cation A positively charged ion.

Cation bridging Electrostatic attraction of colloidal particles by multivalent cations to form domains and microaggregates.

Cation exchange capacity The total amount of cations that a soil can adsorb by cation exchange.

Chelation The formation of multiple bonds between a polyvalent metal cation and a (usually organic) molecule.

Chernozem A soil showing moderate calcification, associated with temperate grasslands.

Chronofunction A mathematical description of the relationship between a soil property and time.

Chronosequence A sequence of soils whose differences in development relate primarily to the different periods of time over which they have been forming.

Clay A mineral particle < 2 μm equivalent diameter.

Clay-humus complex The intimate association of colloidal clay and humus particles which plays a central role in ion exchange and aggregation processes. Also known as the *soil exchange complex*.

Clay mineral A mineral particle < 2 μm in diameter and possessing an orderly arrangement of atoms in layers, giving it a large surface area per unit mass.

Climosequence A sequence of soils whose differences in development relate primarily to the different climatic conditions under which they have formed.

Cohesion Electrostatic attraction of like particles or materials; often used to describe the intermolecular attraction of water molecules by hydrogen bonding.

Colloid A negatively charged organic or inorganic particle of very small dimensions, usually < 2 µm, with a large surface area per unit mass.

Colluvium Unconsolidated material transported on a slope.

Compound A substance comprising two or more elements.

Condensation The process by which vapour is changed to liquid (as in the case of water vapour to water).

Consistence A combination of soil properties that determine its resistance to crushing and its ability to be moulded or changed in shape.

Coversand A deposit of wind-blown material of predominantly sand-sized particles.

Cryoturbation The disturbance to soil resulting from freezing and thawing.

Dew point The temperature at which air becomes saturated with water vapour.

Diagenesis The alteration of sediments and paleosols usually by high pressures or temperatures during or after burial.

Dissociation The separation of a molecule into its component atoms or ions.

Domains Submicroscopic aggregates (< 5 µm in diameter) of colloidal particles.

Duricrust A hardened crust at or near the surface, comprising materials such as calcrete or silcrete.

Duripan Indurated material cemented by various forms of secondary silica. Also known as *silcrete*.

Ecosystem A system which involves the interaction of organisms, causing energy flows and the exchange of materials between living and non-living parts of the system.

Eh A measure, in volts, of the energy change required in the addition or removal of electrons.

Electrical double layer The negative charge of colloidal surfaces together with the adjacent positive charge of adsorbed cations.

Electron Small negatively charged particle which forms part of the structure of atoms.

Element The simplest form in which a substance can occur.

Eluviation Removal of material from a soil layer, in solution or suspension.

Erosion Wearing away of the land surface by water, wind, ice or gravity.

Eutrophication Nutrient enrichment of surface waters often resulting in the growth of algal blooms and reduced oxygen levels in the water.

Evaporation The process by which a liquid is changed into a gas (as in the case of water into water vapour).

Evapotranspiration The addition of water vapour to the atmosphere by the combined processes of evaporation from the Earth's surface and transpiration from vegetation.

Exfoliation Weathering of rocks by peeling off their surface layers.

Exhumed paleosol A buried soil which has become exhumed by erosion of the overlying material.

Field capacity The water content of a soil after drainage under gravity is more or less complete.

Fine fraction (earth) The fine (< 2 mm) fraction of soils upon which most analytical procedures are based.

Flocculation The clustering of colloidal particles to form aggregates.

Fractionation The separation of a material (usually organic matter) into various component parts.

Gelifluction The slow downslope movement of material, often saturated with water, over a frozen subsurface layer.

Geosphere The upper part of the Earth's crust, in which pedological and geomorphological processes operate.

Gilgai Hummocky microrelief produced by soils susceptible to large amounts of expansion and contraction due to high contents of expansive clays.

Gley A soil formed under poorly drained conditions, characterised by the reduction of iron to its ferrous state.

Grus A zone of disintegration surrounding a rock, resulting from weathering.

Heat capacity A measure of the heat absorbed or released per unit volume of a material for a temperature rise or fall of 1 degree.

Heavy metals Metals with a specific gravity of $> 6 \text{ g cm}^{-3}$. Also referred to as *trace metals*.

Holocene The most recent period of geological time, spanning approximately the last 10,000 years.

Horizon A soil layer, approximately parallel with the soil surface, with characteristics produced by soil-forming processes.

Humification The process of humus formation via organic matter decomposition.

Humus The end product of organic matter decomposition.

Hydration The uptake of water by a material. In the case of minerals, this constitutes a weathering reaction.

Hydraulic conductivity A measure of the rate at which water can move through a soil.

Hydraulic head A measure of suction.

Hydrograph A plot of runoff against time.

Hydrolysis A reaction involving the dissociation products (H^+ or OH^-) of water. A weathering reaction in which H^+ is added to a mineral structure.

Hydrosphere The Earth's water, which can occur in liquid, solid or gaseous form.

Illuviation The deposition of material removed (eluviated) from an upper horizon.

Illuviation cutan A depositional coating.

Intrazonal soils Soils whose distribution is associated with local conditions, such as parent material or drainage, to a greater extent than that of zonal soils.

Ions Atoms, groups of atoms or compounds which are electrically charged as a result of electron gain (anion) or loss (cation).

Isomorphous substitution Replacement of ions in mineral structures by others of similar size but of different charge.

Latent heat The heat released or absorbed per unit mass by a system in a changing phase, such as condensation or evaporation.

Laterite Highly leached and weathered soil, found mainly in low latitude regions.

Lateritisation The process of laterite formation involving the mobilisation of iron and formation of plinthite.

Latosolisation The process of laterite formation in which silica and bases are leached from the upper profile, leaving a residual accumulation of sesquioxides.

Leaching Transport of material in solution by water draining through a soil.

Lessivage The translocation of clay in colloidal suspension.

Lithosequence A sequence of soils whose differences in development relate primarily to the different parent materials from which they have formed.

Lithosol A weakly developed soil over bedrock.

Litter The dead components or excrement of biota, added to a soil surface.

Loess Fine-textured material, mainly of silt size, transported and deposited by wind.

Macrofauna Soil organisms larger than 1 mm in size; a term usually used to describe the larger vertebrate animals which live in the soil.

Macronutrients Nutrients used in relatively large quantities by plants.

Mass movement The movement *en masse* of material on a slope. This can operate in a variety of ways and at a variety of speeds.

Matric suction The combined effect of capillary and adsorptive forces with respect to soil water.

Mesofauna Small animals living on or in the soil, but larger than the microscopic forms (micro-organisms).

Micelle A colloidal particle with negative surface charge which behaves like a giant anion.

Micromorphology The microscopic study of soil morphological characteristics.

Micronutrients Nutrients used in very small amounts by plants. Also sometimes referred to as *trace elements*.

Micro-organism Plant and animal organisms living on or in the soil which can only be identified clearly with a microscope.

Mineralisation The conversion of organic substances to inorganic forms during organic matter decomposition.

Moder The type of organic matter in a soil surface horizon which shows moderate decomposition and a pH of around 5.5.

Molecule A chemical structure comprising two or more atoms.

Mor The type of organic matter in a soil surface horizon which shows limited decomposition and a pH < 5.5.

Mull The type of organic matter in a soil surface horizon which shows extensive decomposition and a pH > 5.5.

Nitrification The biochemical oxidation of ammonium to nitrate.

Nutrient cycle The movement of nutrients within ecosystems.

Osmosis The movement from a lower concentration solution to one of higher concentration.

Oxidation The removal of electrons from an atom, ion or molecule during a reaction.

Paleopedology The study of soil formation under past environmental conditions.

Paleosol A soil formed under environmental conditions different from those of the present. Also applied to buried soils, although these may not necessarily have formed under different conditions.

Paludification Formation of organic soils under poorly drained conditions.

Palynology The study of pollen and spores, usually of the fossil variety in the context of soil study.

Pan A soil layer or horizon that is strongly compacted or indurated.

Parent material The material from which a soil develops by soil-forming processes.

Ped A soil aggregate.

Pedogenesis The process of soil formation.

Pedology The study of soil formation.

Pedon A three-dimensional unit sufficiently large to be able to represent the vertical and lateral variability of a soil.

Pedorelict A soil feature formed under past environmental conditions.

Pedoturbation The mechanical transfer of soil materials by both biological and inorganic processes.

Permafrost Ground which is permanently below 0°C.

Permeability A measure of the ease with which a liquid or gas can pass through a particular material.

pF The hydraulic head expressed on a $\log_{10}$ basis.

pH A measure of the hydrogen ion activity of a soil, expressed on a negative $\log_{10}$ basis.

Piping The formation of subsurface pipes in soils by natural processes.

Plaggen soils Soils which have been improved by the addition of seaweed or manure.

Pleistocene The period of geological time immediately before the Holocene, and which, along with the Holocene, comprises the Quaternary period.

Plinthite An iron-rich material formed by iron and silica illuviation and clay synthesis.

Poaching Soil compaction resulting from animal trampling.

Podzol A soil characterised by the eluviation of iron and aluminium compounds from its upper profile, often associated with needleleaf forest or heathland.

Podzolisation The process by which podzols are formed.

Polycyclic soil A soil which exhibits pedological characteristics formed under more than one set of environmental conditions.

Polymer A large molecule formed by the joining together of many similar small molecules (monomers).

Polymerisation The process whereby polymers are formed.

Pores The space within a soil that is not occupied by solid material. Also known as *voids*.

Porosity A measure of the percentage of a soil by volume which is not occupied by solid material.

Pressure cutan A zone of localised reorientation of clay adjacent to a pore or larger particle, produced by pressure, usually associated with wetting and drying. Also known as *stress cutan*.

Profile The vertical section of a pedon.

Quaternary The period of geological time comprising the Pleistocene and Holocene.

Radius of hydration The radius of an ion together with the associated water molecules surrounding it.

Ranker A weakly developed soil occurring on steep slopes.

Redox A term used to denote the combined operation of reduction and oxidation.

Reduction The addition of electrons to an atom, ion or molecule during a reaction.

Regolith The weathered mantle of rock debris at the Earth's surface.

Regosol A weakly developed soil formed in regolith.

Relict paleosol A polycyclic soil occurring at the Earth's surface.

Remontée The process of nutrient cycling involving uptake by plant roots and return to the soil via organic matter decomposition.

Remote sensing The recording of images of the ground from aircraft or satellites.

Rendzina A soil showing limited development, formed from a limestone parent material.

Rhizosphere The soil zone surrounding plant roots.

Rill A small ephemeral channel which conducts flowing water at the surface.

Rubification Reddening of a soil, associated with the formation of hematite.

Runoff Water that runs over the soil surface, laterally through the soil, or in stream channels.

Sand A coarse mineral particle, between 2.00 mm and 0.05 or 0.06 mm in diameter, depending on size classification used.

Saturation zone The zone in which a soil is saturated with water.

Sensible heat Heat energy that is able to be sensed (cf. latent heat, which cannot be directly sensed).

Sesquioxides The oxides and hydroxides of iron and aluminium.

Silcrete See *duripan*.

Silicate minerals Minerals whose basic framework comprises silicon-oxygen tetrahedra.

Silt A mineral particle between 0.002 mm and 0.05 or 0.06 mm diameter, depending on size classification used.

Slaking Destruction of soil aggregates due to wetting and the resulting compression of trapped air.

Solifluction The slow downslope movement of material saturated with water. Often used to refer to gelifluction.

Solod A salt-rich soil in a more advanced state of leaching than solonetz.

Solonchak A soil of dry environments in which salts accumulate at or near the surface.

Solonetz A salt-rich soil in the early stages of leaching, characterised by dispersed organic matter and hydrolysed salts, giving a very high pH.

Solum The A and B horizons of a soil profile.

Solution A solvent in which one or more materials are dissolved.

Sorting The size range or distribution of mineral particles in the soil; a narrow range indicates good sorting whereas a wide range indicates poor sorting.

Stress cutan See *pressure cutan*.

Structure The nature and arrangement of aggregates in a soil.

Surface texture The surface appearance of soil mineral grains normally established using scanning electron microscopy.

Suspension A fluid in which small particles are held, but not in a dissolved state.

Terra fusca A soil showing slight rubification, developed over limestone.

Terra rossa A soil showing marked rubification, developed over limestone.

Texture The relative proportions of sand, silt and clay in a soil.

Thermal conductivity A measure of the ability of a material to conduct heat by molecular motion.

Thermal diffusivity The ratio of the thermal conductivity to the heat capacity of a substance.

Thin section A thin slice of undisturbed soil mounted on a microscope slide.

Till The debris deposited by a glacier or an ice sheet.

Tilth Physical characteristics of the seedbed following tillage of agricultural soils to break up large structural aggregates.

Toposequence A sequence of soils whose differences in development relate primarily to the different relief conditions under which they have formed.

Trace elements Elements occurring in soils or plants in only very small quantities. Also sometimes used to refer to *micronutrients*.

Translocation The transport of material within a soil, either in solution or suspension.

Valency The number of hydrogen atoms which one atom of an element will combine with or displace.

Vertisol A soil with a high content of expansive clay minerals, and therefore subject to extensive cracking during drying.

Voids See *pores*.

Warping The process whereby poorly drained land adjacent to an estuary is improved by the addition of estuarine alluvium.

Water table The interface between the capillary fringe and the saturation zone.

Weathering The physical or chemical alteration of rocks and minerals by processes operating within the soil.

Wilting point The point at which movement of water in liquid form ceases, and beyond which movement occurs by vapour diffusion.

Zonal soils Soils whose distribution is related to major climatic and vegetational zones.

REFERENCES

Abdul-Kareem, A.W. and McRae, S.G. (1984). The effects on topsoil of long-term storage in stockpiles. *Plant and Soil*, 76: 357–63.

Adams, W.A. (1986). Practical aspects of sportsfield drainage. *Soil Use and Management*, 2: 51–4.

Adams, W.A. and Evans, G.M. (1989). Effects of lime applications to parts of an upland catchment on soil properties and the chemistry of drainage waters. *Journal of Soil Science*, 40: 585–97.

Adams, W.A., Ali, A.Y. and Lewis, P.J. (1990). Release of cationic aluminium from acidic soils into drainage water and relationships with land use. *Journal of Soil Science*, 41: 255–68.

Adamson, J.K., Hornung, M., Pyatt, D.G. and Anderson, A.R. (1987). Changes in solute chemistry of drainage waters following the clear-felling of a Sitka spruce plantation. *Forestry*, 60: 165–77.

Addiscott, T.M., Whitmore, A.P. and Powlson, D.S. (1991). *Farming, Fertilizers and the Nitrate Problem*. CAB International, Wallingford.

Alexander, E.B., Mallory, J.I. and Colwell, W.L. (1993). Soil-elevation relationships on a volcanic plateau in the southern Cascade Range, northern California, USA. *Catena*, 20: 113–28.

Amundson, R.G., Chadwick, O.A., Sowers, J.M. and Doner, H.E. (1989). Soil evolution along an altitudinal transect in the Eastern Mojave Desert of Nevada, USA. *Geoderma*, 43: 349–71.

Anderson, C.J., Campbell, D.J., Ritchie, R.M. and Smith, D.L.O. (1989). Soil shear strength measurements and their relevance to windthrow in Sitka spruce. *Soil Use and Management*, 5: 62–6.

Anon. (1983). Pollution is on tap in Silicon Valley. *New Scientist*, 21 July: 180.

Appelgren, B.G. and Burchi, S. (1993). Technical, policy and legal aspects of chemical time bombs with emphasis on the institutional action required in eastern Europe. *Land Degradation and Rehabilitation*, 4: 437–40.

Arkley, R.J. (1967). Climates of some Great Soil Groups of the western United States. *Soil Science*, 103: 389–400.

Armstrong, A.C. (1986). Mole drainage of a Hallsworth Series soil. *Soil Use and Management*, 2: 54–8.

Armstrong, A.S.B. and Tanton, T.W. (1992). Gypsum applications to aggregated saline-sodic clay topsoils. *Journal of Soil Science*, 43: 249–60.

Arthur, M.A. and Fahey, T.J. (1993). Controls on soil solution chemistry in a subalpine forest in north-central Colorado. *Soil Science Society of America Journal*, 57: 1122–30.

REFERENCES

Aspinall, R., Macklin, M. and Openshaw, S. (1988). Heavy metal contamination in soils of Tyneside: a geographically-based assessment of environmental quality in an urban area. In: J.M. Hooke (ed.) *Geomorphology in Environmental Planning.* Wiley, Chichester, 87–102.

Avery, B.W. (1985). Argillic horizons and their significance in England and Wales. In: J. Boardman (ed.) *Soils and Quaternary Landscape Evolution.* Wiley, Chichester, 69–86.

Avery, B.W. (1990). *Soils of the British Isles.* CAB International, Wallingford.

Avery, B.W., Clayden, B. and Ragg, J.M. (1977). Identification of podzolic soils (Spodosols) in upland Britain. *Soil Science,* 123: 306–18.

Aweto, A.O. and Adejumbobi, D.O. (1991). Impact of grazing on soil in the southern Guinea savannah zone of Nigeria. *The Environmentalist,* 11: 27–32.

Ayanaba, A. and Jenkinson, D.S. (1990). Decomposition of carbon-14 labelled ryegrass and maize under tropical conditions. *Soil Science Society of America Journal,* 54: 112–15.

Ayoub, A.T. (ed.) (1991). *World Map of the Status of Human-induced Soil Degradation: a brief explanatory note.* United Nations Environment Programme and International Soil Reference and Information Centre, Wageningen.

Babel, U. (1975). Micromorphology of soil organic matter. In: J.E. Gieseking (ed.) *Soil Components, Vol. 1.* Springer, New York, 369–473.

Bache, B.W. (1983). The role of buffering in determining surface water composition. *Water Science and Technology,* 15: 33–45.

Bailey, A.D., Dennis, C.W., Harris, G.L. and Horner, M.W. (1980). *Pipe Size Design and Field Drainage.* Agricultural Development and Advisory Service, Land Drainage Series, Report No. 5, Cambridge.

Baize, D. (1993). *Soil Science Analysis: a guide to current use.* Wiley, Chichester.

Baker, S.W. (1989). Soil physical conditions of the root zone layer and the performance of winter games pitches. *Soil Use and Management,* 5: 116–22.

Barron, V. and Torrent, J. (1986). Use of the Kubelka-Munk theory to study the influence of iron oxides on soil colour. *Journal of Soil Science,* 37: 499–510.

Barrow, C.J. (1991). *Land Degradation.* Cambridge University Press, Cambridge.

Barry, R.G. and Chorley, R.J. (1992). *Atmosphere, Weather and Climate.* Routledge, London.

Bascomb, C.L. (1968). Distribution of pyrophosphate-extractable iron and organic carbon in soils of various groups. *Journal of Soil Science,* 19: 251–68.

Batjes, N.H. and Bridges, E.M. (1993). Soil vulnerability to pollution in Europe. *Soil Use and Management,* 9: 25–9.

Bauer, K.W. (1973). The use of soils data in regional planning. *Geoderma,* 10: 1–26.

Baumler, R. and Zech, W. (1994). Soils of the high mountain region of Eastern Nepal: classification, distribution and soil forming processes. *Catena,* 22: 85–103.

Bayfield, N.G. and Aitken, R. (1992). *Managing the Impacts of Recreation on Vegetation and Soils: a review of techniques.* Institute of Terrestrial Ecology (NERC), Banchory.

Beckmann, G.G. (1983). Development of old landscapes and soils. In: CSIRO *Soils: an Australian viewpoint.* Academic Press, London, 51–72.

Bell, M. and Walker, M.J.C. (1992). *Late Quaternary Environmental Change: physical and human perspectives.* Longman, Harlow.

Bell, M.A. (1993). Organic matter, soil properties and wheat production in the high valley of Mexico. *Soil Science,* 156: 86–93.

Bendelow, V.C. and Hartnup, R. (1980). *Climatic Classification of England and Wales.* Soil Survey Technical Monograph No. 15, Harpenden.

Bernier, N., Ponge, J.F. and André, J. (1993). Comparative study of soil organic layers in two bilberry-spruce forest stands (*Vaccinio-Piceetea*). Relation to forest dynamics. *Geoderma*, 59: 89–108.

Besly, B.M. and Turner, P. (1983). Origin of red beds in a moist tropical climate (Etruria Formation, Upper Carboniferous, UK). In: R.C.L. Wilson (ed.) *Residual Deposits: surface related weathering processes and materials*. Blackwell, Oxford, 131–47.

Bhattacharyya, T., Pal, D.K. and Deshpande, S.B. (1993). Genesis and transformation of minerals in the formation of red (Alfisols) and black (Inceptisols and Vertisols) soils on Deccan basalt in the Western Ghats, India. *Journal of Soil Science*, 44: 159–71.

Bibby, J.S. and Mackney, D. (1969). *Land Use Capability Classification*. Soil Survey Technical Monograph No. 1, Harpenden.

Binkley, D. and Sollins, P. (1990). Factors determining differences in soil pH in adjacent conifer and alder-conifer stands. *Soil Science Society of America Journal*, 54: 1427–33.

Bintliff, J. (1992). Erosion in the Mediterranean lands: a reconsideration of pattern, process and methodology. In: M. Bell and J. Boardman (eds) *Past and Present Soil Erosion: archaeological and geographical perspectives*. Oxbow, Oxford, 125–31.

Birkeland, P.W. (1984a). *Soils and Geomorphology*. Oxford University Press, New York.

Birkeland, P.W. (1984b). Holocene soil chronofunctions, Southern Alps, New Zealand. *Geoderma*, 34: 115–34.

Birkeland, P.W. (1992). Quaternary soil chronosequences in various environments – extremely arid to humid tropical. In: I.P. Martini and W. Chesworth (eds) *Weathering, Soils and Paleosols*. Elsevier, Amsterdam, 261–81.

Blaikie, P. (1985). *The Political Economy of Soil Erosion in Developing Countries*. Longman, Harlow.

Blume, L.J., Perkins, H.F. and Hubbard, R.K. (1987). Subsurface water movement in an upland Coastal Plain soil as influenced by plinthite. *Soil Science Society of America Journal*, 51: 774–9.

Boardman, J. (ed.) (1985). *Soils and Quaternary Landscape Evolution*. Wiley, Chichester.

Boardman, J. (1990a). Soil erosion on the South Downs: a review. In: J. Boardman, I.D.L. Foster and J.A. Dearing (eds) *Soil Erosion on Agricultural Land*. Wiley, Chichester, 87–105.

Boardman, J. (1990b). *Soil Erosion in Britain: costs, attitudes and policies*. Social Audit Paper No. 1, Education Network for Environment and Development. University of Sussex, Brighton.

Boardman, J. (1991a). Land use, rainfall and erosion risk on the South Downs. *Soil Use and Management*, 7: 34–8.

Boardman, J. (1991b). The Canadian experience of soil conservation: a way forward for Britain ? *International Journal of Environmental Studies*, 37: 263–9.

Bockheim, J.G. (1978). A comparison of the morphology and genesis of Arctic Brown and Alpine Brown soils in North America. In: W.C. Mahaney (ed.) *Quaternary Soils*. Geo Abstracts, Norwich, 427–52.

Bockheim, J.G. (1980). Solution and use of chronofunctions in studying soil development. *Geoderma*, 24: 71–85.

Boettcher, S.E. and Kalisz, P.J. (1990). Single-tree influence on soil properties in the mountains of east Kentucky. *Ecology*, 71: 1365–72.

Boettcher, S.E. and Kalisz, P.J. (1991). Single-tree influence on earthworms in forest soils in eastern Kentucky. *Soil Science Society of America Journal*, 55: 862–5.

Bojie Fu, (1989). Soil erosion and its control in the loess plateau of China. *Soil Use and Management*, 5: 76–82.

Bonneau, M. and Souchier, P. (1982). *Constituents and Properties of Soils.* Academic Press, London.

Booltink, H.W.G. and Bouma, J. (1991). Physical and morphological characterization of bypass flow in a well-structured clay soil. *Soil Science Society of America Journal*, 55: 1249–54.

Bradley, R.S. (1985). *Quaternary Paleoclimatology: methods of paleoclimatic reconstruction.* Allen & Unwin, Boston.

Brady, N.C. (1990). *The Nature and Properties of Soils.* Macmillan, New York.

Brewer, R. (1968). Clay illuviation as a factor in particle-size differentiation in soil profiles. *Transactions of the 9th International Congress of Soil Science, Adelaide*, 4: 489–99.

Brewer, R. (1976). *Fabric and Mineral Analysis of Soils.* Krieger, New York.

Bridges, E.M. (1982). Techniques of modern soil survey. In: E.M. Bridges and D.A. Davidson (eds) *Principles and Applications of Soil Geography.* Longman, London, 28–57.

Bridges, E.M. (1989). Toxic metals in amenity soil. *Soil Use and Management*, 5: 91–100.

Bridges, E.M. (1991a). Waste materials in urban soils. In: P. Bullock and P.J. Gregory (eds) *Soils in the Urban Environment.* Blackwell, Oxford, 28–46.

Bridges, E.M. (1991b). Dealing with contaminated soils. *Soil Use and Management*, 7: 151–8.

Bridges, E.M. and Davidson, D.A. (1982). Agricultural uses of soil survey data. In: E.M. Bridges and D.A. Davidson (eds) *Principles and Applications of Soil Geography.* Longman, London, 171–215.

Briggs, D.J. (1977a). *Sources and Methods in Geography: sediments.* Butterworth, London.

Briggs, D.J. (1977b). *Sources and Methods in Geography: soils.* Butterworth, London.

Briggs, D.J. and Courtney, F.M. (1989). *Agriculture and Environment.* Longman, Harlow.

Brindley, G.W. and Brown, G. (eds) (1980). *Crystal Structures of Clay Minerals and Their X-ray Identification.* Mineralogical Society, London.

Brink, A.B.A., Partridge, T.C. and Williams, A.A.B. (1982). *Soil Survey for Engineering.* Oxford University Press, Oxford.

Bronger, A. and Catt, J.A. (eds) (1989). *Paleopedology: nature and application of paleosols*, Catena Supplement 16. Catena, Cremlingen-Destedt.

Bronger, A. and Heinkele, T. (1989). Micromorphology and genesis of paleosols in the Luochuan loess section, China: pedostratigraphic and environmental implications. *Geoderma*, 45: 123–43.

Brown, D.J.A. (1988). The Loch Fleet and other catchment liming programs. *Water, Air and Soil Pollution*, 41: 409–15.

Brunsden, D. (1979). Weathering. In: C. Embleton and J. Thornes (eds) *Process in Geomorphology.* Arnold, London, 73–129.

Buchan, G.D. and Grewal, K.S. (1990). The power function model for the soil moisture characteristic. *Journal of Soil Science*, 41: 111–17.

Bullock, P. (1985). The role of micromorphology in the study of Quaternary soil processes. In: J. Boardman (ed.) *Soils and Quaternary Landscape Evolution.* Wiley, Chichester, 45–68.

Buol, S.W., Hole, F.D. and McCracken, R.J. (1980). *Soil Genesis and Soil Classification.* Iowa State Press, Ames.

Burke, I.C., Yonker, C.M., Parton, W.J., Cole, C.V., Flach, K. and Schimel, D.S.

REFERENCES

(1989). Texture, climate, and cultivation effects on soil organic matter content in U.S. grassland soils. *Soil Science Society of America Journal*, 53: 800–5.

Burke, R.M. and Birkeland, P.W. (1979). Re-evaluation of multiparameter relative dating techniques and their application to the glacial sequence along the eastern escarpment of the Sierra Nevada. *Quaternary Research*, 11: 21–51.

Burnham, C.P. and Mutter, G.M. (1993). The depth and productivity of chalky soils. *Soil Use and Management*, 9: 1–8.

Burt, T. and Hanwell, J. (1992). Land of many uses. *Geography Review*, 6: 2–8.

Burt, T.P. and Trudgill, S.T. (1985). Soil properties, slope hydrology and spatial patterns of chemical denudation. In: K.S. Richards, R.R. Arnett and S. Ellis (eds) *Geomorphology and Soils*. George Allen & Unwin, London, 13–36.

Busacca, A.J. (1989). Long Quaternary record in eastern Washington, U.S.A., interpreted from multiple buried paleosols in loess. *Geoderma*, 45: 105–22.

Butler, B.E. (1959). *Periodic Phenomena in Landscapes as a Basis for Soil Studies*, CSIRO Soil Publication No. 14. CSIRO, Canberra.

Campbell, C.A., Zentner, R.P., Selles, F. and Akinremi, O.O. (1993). Nitrate leaching as influenced by fertilisation in the Brown Soil zone. *Canadian Journal of Soil Science*, 73: 387–97.

Campbell, I.B. and Claridge, G.G.C. (1992). Soils of cold climate regions. In: I.P. Martini and W. Chesworth (eds) *Weathering, Soils and Paleosols*. Elsevier, Amsterdam, 183–201.

Canada Soil Survey Committee (1978). *The Canadian System of Soil Classification*, Canada Department of Agriculture Publication 1646. Supply and Services Canada, Ottawa.

Carling, P.A. (1986). Peat slides in Teesdale and Weardale, northern Pennines, July 1983: description and failure mechanisms. *Earth Surface Processes and Landforms*, 11: 193–206.

Carson, M.A. and Kirkby, M.J. (1972). *Hillslope Form and Process*. Cambridge University Press, Cambridge.

Carter, B.J. and Ciolkosz, E.J. (1991). Slope gradient and aspect effects on soils developed from sandstone in Pennsylvania. *Geoderma*, 49: 199–213.

Caseldine, C.J. and Matthews, J.A. (1985). ^{14}C dating of palaeosols, pollen analysis and landscape change: studies from the low- and mid-alpine belts of southern Norway. In: J. Boardman (ed.) *Soils and Quaternary Landscape Evolution*. Wiley, Chichester, 87–116.

Castle, D.A. (1986). A rationale for permeable backfill placed over pipe drains. *Soil Use and Management*, 2: 58–61.

Catt, J.A. (1985a). Soil particle size distribution and mineralogy as indicators of pedogenic and geomorphic history: examples from the loessial soils of England and Wales. In: K.S. Richards, R.R. Arnett and S. Ellis (eds) *Geomorphology and Soils*. George Allen & Unwin, London, 202–18.

Catt, J.A. (1985b). Natural soil acidity. *Soil Use and Management*, 1: 8–10.

Catt, J.A. (1986a). *Soils and Quaternary Geology: a handbook for field scientists*. Oxford University Press, Oxford.

Catt, J.A. (1986b). The nature, origin and geomorphological significance of clay-with-flints. In: G. de G. Sieveking and M.B. Hart (eds) *The Scientific Study of Flint and Chert*. Cambridge University Press, Cambridge, 151–9.

Catt, J.A. (1989). Relict properties of soils of the central and north-west European temperate region. In: A. Bronger and J.A. Catt (eds) *Paleopedology: nature and application of paleosols*. Catena, Cremlingen-Destedt, 41–58.

Cavallaro, N., Padilla, N. and Villarrubia, J. (1993). Sewage sludge effects on chemical properties of acid soils. *Soil Science*, 156: 63–70.

Central Soil and Water Conservation Research and Training Institute (1977). *Annual Report*. Dehra Dun, India.

Chalmers, A.G. (1985). Review of information on lime loss and changes in soil pH gained from ADAS experiments. *Soil Use and Management*, 1: 17–19.

Chartres, C.J. (1980). A Quaternary soil sequence in the Kennet Valley, central southern England. *Geoderma*, 23: 125–46.

Chartres, C.J. and Pain, C.F. (1984). A climosequence of soils on late Quaternary volcanic ash highland, Papua New Guinea. *Geoderma*, 32: 131–55.

Chinene, V.R.N. (1992). Land evaluation using the FAO Framework: an example from Zambia. *Soil Use and Management*, 8: 130–9.

Chow, T.L., Rees, H.W., Webb, K.T. and Langille, D.R. (1993). Modification of subsoil characteristics resulting from drainage tile installation. *Soil Science*, 156: 346–57.

Christopherson, R.W. (1992). *Geosystems: an introduction to physical geography*. Macmillan, New York.

Churchman, G.J. and Tate, K.R. (1987). Stability of aggregates of different size grades in allophanic soils from volcanic ash in New Zealand. *Journal of Soil Science*, 38: 19–27.

Churchward, H.M. and Gunn, R.H. (1983). Stripping of deep weathered mantles and its significance to soil patterns. In: CSIRO *Soils: an Australian viewpoint*. Academic Press, London, 73–81.

Colbourn, P. (1985). Nitrogen losses from the field: denitrification and leaching in intensive winter cereal production in relation to tillage method of clay soil. *Soil Use and Management*, 1: 117–20.

Colby-Saliba, B. (1985). Comparative measures of effectiveness in farm-level soil conservation. *Soil Use and Management*, 1: 106–10.

Coleman, T.L., Agbu, P.A. and Montgomery, O.L. (1993). Spectral differentiation of surface soils and soil properties: is it possible from space platforms? *Soil Science*, 155: 283–93.

Collins, M.E. and Doolittle, J.A. (1987). Using ground-penetrating radar to study soil microvariability. *Soil Science Society of America Journal*, 51: 491–3.

Collins, M.E. and Shapiro, G. (1987). Comparisons of human-influenced and natural soils at the San Luis archaeological site, Florida. *Soil Science Society of America Journal*, 51: 171–6.

Colman, S.M. (1986). Levels of time information in weathering measurements, with examples from weathering rinds on volcanic clasts in the western United States. In: S.M. Colman and D.P. Dethier (eds) *Rates of Chemical Weathering of Rocks and Minerals*. Academic Press, Orlando, 379–93.

Conacher, A.J. and Dalrymple, J.B. (1977). The nine-unit landsurface model: an approach to pedogeomorphic research. *Geoderma*, 18: 1–154.

Cooke, G.W. (1984). Constraints on crop production – opportunities for the chemical industry. *Chemistry and Industry*, 20: 730–7.

Courtney, F.M. and Trudgill, S.T. (1984). *The Soil: an introduction to soil study*. Edward Arnold, London.

Crabtree, R.W. (1986). Spatial distribution of solutional erosion. In: S.T. Trudgill (ed.) *Solute Processes*. Wiley, Chichester, 329–61.

Cuanalo, H.E. and Webster, R. (1970). A comparative study of numerical classification and ordination of soil properties in a locality near Oxford. *Journal of Soil Science*, 21: 340–52.

Culbard, E.B., Thornton, I., Watt, J., Wheatley, M., Moorcroft, S. and Thompson, M. (1988). Metal contamination of British suburban dusts and soils. *Journal of Environmental Quality*, 17: 226–34.

331

Curi, N. and Franzmeier, D.P. (1987). Effect of parent rocks on chemical and mineralogical properties of some Oxisols in Brazil. *Soil Science Society of America Journal*, 51: 153–8.

Curtis, C.D. (1976a). Chemistry of rock weathering: fundamental reactions and controls. In: E. Derbyshire (ed.) *Geomorphology and Climate*. Wiley, Chichester, 25–57.

Curtis, C.D. (1976b). Stability of minerals in surface weathering reactions: a general thermochemical approach. *Earth Surface Processes and Landforms*, 1: 63–70.

Curtis, L.F., Courtney, F.M. and Trudgill, S.T. (1976). *Soils in the British Isles*. Longman, Harlow.

Cutforth, H.W., Shaykewich, C.F. and Cho, C.M. (1986). Effect of soil water and temperature on corn (*Zea mays* L.) root growth during emergence. *Canadian Journal of Soil Science*, 66: 51–8.

Dabney, S.M. and Selim, H.M. (1987). Anisotropy of a fragipan soil: vertical vs. horizontal hydraulic conductivity. *Soil Science Society of America Journal*, 51: 3–6.

Dahlgren, R.A., Ugolini, F.C., Shoji, S., Ito, T. and Sletten, R.S. (1991). Soil-forming processes in Alic Melanudands under Japanese pampas grass and oak. *Soil Science Society of America Journal*, 55: 1049–56.

Dalrymple, J.B., Blong, R.J. and Conacher, A.J. (1968). A hypothetical nine-unit landsurface model. *Zeitschrift für Geomorphologie*, 12: 60–76.

Dalziel, T.R.K., Proctor, M.V. and Dickson, A. (1988). Hydrochemical budget calculations for parts of the Loch Fleet catchment before and after watershed liming. *Water, Air and Soil Pollution*, 41: 417–34.

Daniels, R.B. and Hammer, R.D. (1992). *Soil Geomorphology*. Wiley, New York.

Davidson, D.A. (1980). *Soils and Land Use Planning*. Longman, Harlow.

Davis, B.N.K., Walker, N., Ball, D.F. and Fitter, A.H. (1992). *The Soil*. Harper Collins, London.

Dawson, A.G. (1992). *Ice Age Earth: Late Quaternary geology and climate*. Routledge, London.

Dearing, J.A., Maher, B.A. and Oldfield, F. (1985). Geomorphological linkages between soils and sediments: the role of magnetic measurements. In: K.S. Richards, R.R. Arnett and S. Ellis (eds) *Geomorphology and Soils*. George Allen & Unwin, London, 245–66.

De Coninck, F. and Righi, D. (1983). Podzolisation and the spodic horizon. In: P. Bullock and C.P. Murphy (eds) *Soil Micromorphology, Vol. 2 Soil Genesis*. AB Academic Publishers, Berkhamsted, 389–417.

De la Rosa, D., Moreno, J.A., Garcia, L.V. and Almorza, J. (1992). MicroLEIS: a microcomputer-based Mediterranean land evaluation information system. *Soil Use and Management*, 8: 89–96.

Denne, T., Bown, M.J.D. and Abel, J.A. (1986). *Forestry: Britain's growing resource*. UK Centre for Economic and Environmental Development, London.

Dimbleby, G.W. (1985). *The Palynology of Archaeological Sites*. Academic Press, London.

Donald, R.G., Anderson, D.W. and Stewart, J.W.B. (1993). The distribution of selected soil properties in relation to landscape morphology in forested Gray Luvisol soils. *Canadian Journal of Soil Science*, 73: 165–72.

Drury, C.F., McKenney, D.J., Findlay, W.I. and Gaynor, J.D. (1993). Influence of tillage on nitrate loss in surface runoff and tile drainage. *Soil Science Society of America Journal*, 57: 797–802.

Duchaufour, P. (1982). *Pedology*. George Allen & Unwin, London.

Edwards, C.A. and Lofty, J.R. (1977). *Biology of Earthworms*. Chapman and Hall, London.

Edwards, W.M., Shipitalo, M.J., Owens, L.B. and Norton, L.D. (1990). Effect of *Lumbricus terrestris* L. burrows on hydrology of continuous no-till corn fields. *Geoderma*, 46: 73–84.

Edwards, W.M., Triplett, G.B., Van Doren, D.M., Owens, L.B., Redmond, C.E. and Dick, W.A. (1993). Tillage studies with a corn–soybean rotation: hydrology and sediment loss. *Soil Science Society of America Journal*, 57: 1051–5.

Eghbal, M.K. and Southard, R.J. (1993). Micromorphological evidence of polygenesis of three Aridisols, western Mojave Desert, California. *Soil Science Society of America Journal*, 57: 1041–50.

Ekwue, E.I. (1990). Effect of organic matter on splash detachment and the processes involved. *Earth Surface Processes and Landforms*, 15: 175–81.

Ellis, S. (1980). Physical and chemical characteristics of a podzolic soil formed in Neoglacial till, Okstindan, northern Norway. *Arctic and Alpine Research*, 12: 65–72.

Ellis, S. (1983). Micromorphological aspects of arctic-alpine pedogenesis in the Okstindan Mountains, Norway. *Catena*, 10: 133–48.

Ellis, S. (1988). Pedogenesis on the basalt and associated deposits of Canna, western Scotland. *Catena*, 15: 281–7.

Ellis, S. and Matthews, J.A. (1984). Pedogenic implications of a ^{14}C-dated paleo-podzolic soil at Haugabreen, southern Norway. *Arctic and Alpine Research*, 16: 77–91.

Ellis, S. and Newsome, D. (1991). Chalkland soil formation and erosion on the Yorkshire Wolds, northern England. *Geoderma*, 48: 59–72.

Ellis, S. and Richards, K.S. (1985). Pedogenic and geotechnical aspects of Late Flandrian slope instability in Ulvådalen, west-central Norway. In: K.S. Richards, R.R. Arnett and S. Ellis (eds) *Geomorphology and Soils*. George Allen & Unwin, London, 328–47.

Ellis, S., Taylor, D.M. and Masood, K.R. (1993). Land degradation in northern Pakistan. *Geography*, 78: 84–7.

Esteoule-Choux, J. (1983). Kaolinitic weathering profiles in Brittany: genesis and economic importance. In: R.C.L. Wilson (ed.) *Residual Deposits: surface related weathering processes and materials*. Blackwell, Oxford, 33–8.

Evans, J.G. (1972). *Land Snails in Archaeology*. Seminar Press, London.

Evans, R. (1990a). Soil erosion: its impact on the English and Welsh landscape since woodland clearance. In: J. Boardman, I.D.L. Foster and J.A. Dearing (eds) *Soil Erosion on Agricultural Land*. Wiley, Chichester, 231–54.

Evans, R. (1990b). Soils at risk of accelerated erosion in England and Wales. *Soil Use and Management*, 6: 125–31.

Fahey, B.D. (1983). Frost action and hydration as rock weathering mechanisms on schist: a laboratory study. *Earth Surface Processes*, 8: 535–45.

Fanning, D.S. and Fanning, M.C.B. (1989). *Soil: morphology, genesis, and classification*. Wiley, New York.

FAO (1976). *A Framework for Land Evaluation*, FAO Soils Bulletin No. 32. UN Food and Agriculture Organization, Rome.

FAO (1981). *World Soil Charter*. UN Food and Agriculture Organization, Rome.

FAO (1983). *Protect and Produce: soil conservation for development*. UN Food and Agriculture Organization, Rome.

FAO-Unesco (1974). *Soil Map of the World: volume 1, legend*. Unesco, Paris.

FAO-Unesco (1989). *Soil Map of the World: revised legend*. International Soil Reference and Information Centre, Wageningen.

Farmer, V.C. (1982). Significance of the presence of allophane and imogolite in podzol Bs horizons for podzolisation mechanisms: a review. *Soil Science and Plant*

Nutrition, 28: 571–8.

Farmer, V.C., Russell, J.D. and Smith, B.F.L. (1983). Extraction of inorganic forms of translocated Al, Fe and Si from a podzol Bs horizon. *Journal of Soil Science*, 34: 571–6.

Farquharson, F.A.K., Mackney, D., Newson, M.D. and Thomasson, A.J. (1978). *Estimation of Runoff Potential of River Catchments from Soil Survey*. Soil Survey Special Survey 11, Harpenden.

Fenwick, I.M. (1985). Paleosols: problems of recognition and interpretation. In: J. Boardman (ed.) *Soils and Quaternary Landscape Evolution*. Wiley, Chichester, 3–21.

Fenwick, I.M. and Knapp, B.J. (1982). *Soils: process and response*. Duckworth, London.

Fernandez, I.J. and Kosian, P.A. (1987). Soil air carbon dioxide concentrations in a New England spruce-fir forest. *Soil Science Society of America Journal*, 51: 261–3.

Fernandez, I.J., Son, Y., Kraske, C.R., Rustad, L.E. and David, M.B. (1993). Soil carbon dioxide characteristics under different forest types and after harvest. *Soil Science Society of America Journal*, 57: 1115–21.

Fink, A. (1982). *Fertilizers and Fertilization: introduction and practical guide to crop fertilization*. Verlag Chemie.

Finlayson, B.L. (1985). Soil creep: a formidable fossil of misconception. In: K.S. Richards, R.R. Arnett and S. Ellis (eds) *Geomorphology and Soils*. George Allen & Unwin, London, 141–58.

Fisher, G. and Yan, O. (1984). Iron mobilisation by heathland plant extracts. *Geoderma*, 32: 339–45.

FitzPatrick, E.A. (1986). *An Introduction to Soil Science*. Longman, Harlow.

Follmer, L.R. (1978). The Sangamon Soil in its type area – a review. In: W.C. Mahaney (ed.) *Quaternary Soils*. Geo Abstracts, Norwich, 125–65.

Forestry Commission (1988). *Forests and Water Guidelines*. Forestry Commission, Edinburgh.

Förster, H. (1993). Properties, dynamics and association of soils on high elevations of the Bavarian Forest (Germany) as illustrated by a toposequence on gneiss debris. *Catena*, 20: 563–79.

Foth, H.D. (1990). *Fundamentals of Soil Science*. Wiley, New York.

Foth, H.D. and Ellis, B.G. (1988). *Soil Fertility*. Wiley, New York.

Fowler, D., Cape, J.N. and Leith, I.D. (1985). Acid inputs from the atmosphere in the United Kingdom. *Soil Use and Management*, 1: 3–5.

Francis, G.S. and Kemp, R.A. (1990). Morphological and hydraulic properties of a silt loam soil in New Zealand as affected by cropping history. *Soil Use and Management*, 6: 145–51.

Franco-Vizcaino, E., Graham, R.C. and Alexander, E.B. (1993). Plant species diversity and chemical properties of soils in the Central Desert of Baja California, Mexico. *Soil Science*, 155: 406–16.

French, H.M. (1988). Active layer processes. In: M.J. Clark (ed.) *Advances in Periglacial Geomorphology*. Wiley, Chichester, 151–77.

Frenkel, H., Gertsl, Z. and Alperovitch, N. (1989). Exchange-induced dissolution of gypsum and the reclamation of sodic soils. *Journal of Soil Science*, 40: 599–611.

Gafni, A. and Brooks, K.N. (1990). Hydraulic characteristics of four peatlands in Minnesota. *Canadian Journal of Soil Science*, 70: 239–53.

Gardner, W.R. (1991). Soil science as a basic science. *Soil Science*, 151: 2–6.

Garwood, E.A., Tyson, K.C. and Clement, C.R. (1977). *A Comparison of Yield and Soil Conditions During 20 Years of Grazed Grass and Arable Cropping*. Grassland

Research Institute, Technical Report No. 21. Hurley.

Gerrard, A.J. (1981). *Soils and Landforms*. George Allen & Unwin, London.

Gerrard, A.J. (1985). Soil erosion and landscape stability in southern Iceland: a tephrochronological approach. In: K.S. Richards, R.R. Arnett and S. Ellis (eds) *Geomorphology and Soils*. George Allen & Unwin, London, 78–95.

Gerrard, A.J. (1988). *Rocks and Landforms*. Unwin Hyman, London.

Ghuman, B.S. and Lal, R. (1985). Thermal conductivity, thermal diffusivity, and thermal capacity of some Nigerian soils. *Soil Science*, 139: 74–80.

Gile, L.H., Peterson, F.F. and Grossman, R.B. (1966). Morphological and genetic sequences of carbonate accumulation in desert soils. *Soil Science*, 101: 347–60.

Gilman, K. and Newson, M.D. (1980). *Soil Pipes and Pipeflow*, British Geomorphological Research Group Research Monograph 1. Geo Abstracts, Norwich.

Goudie, A.S. (1983). Calcrete. In: A.S. Goudie and K. Pye (eds) *Chemical Sediments and Geomorphology: precipitates and residua in the near-surface environment*. Academic Press, London, 93–131.

Goudie, A.S. (1985). Duricrusts and landforms. In: K.S. Richards, R.R. Arnett and S. Ellis (eds) *Geomorphology and Soils*. George Allen & Unwin, London, 37–57.

Goudie, A.S. (1989). Weathering processes. In: D.S.G. Thomas (ed.) *Arid Zone Geomorphology*. Belhaven, London, 11–24.

Goudie, A.S. (1992). *Environmental Change*. Clarendon Press, Oxford.

Goudie, A.S., Cooke, R.U. and Evans, I.S. (1970). Experimental investigation of rock weathering by salts. *Area*, 4: 42–8.

Goulding, K.W.T. and Stevens, P.A. (1988). Potassium reserves in a forested, acid upland soil and the effect on them of clear-felling versus whole-tree harvesting. *Soil Use and Management*, 4: 45–51.

Graham, R.C. and Buol, S.W. (1990). Soil–geomorphic relations on the Blue Ridge Front: II. Soil characteristics and pedogenesis. *Soil Science Society of America Journal*, 54: 1367–77.

Greenland, D.J. (1979). Structural organization of soils and crop production. In: R. Lal and D.J. Greenland (eds) *Soil Physical Properties and Crop Production in the Tropics*. Wiley, Chichester, 47–56.

Greenland, D.J. (1991). The contributions of soil science to society – past, present, and future. *Soil Science*, 151: 19–23.

Greenwood, D.J. (1993). The changing scene of British soil science. *Journal of Soil Science*, 44: 191–207.

Grieve, I.C., Proctor, J. and Cousins, S.A. (1990). Soil variation with altitude on Volcan Barva, Costa Rica. *Catena*, 17: 525–34.

Gupta, R.J. and Abrol, I.P. (1990). Salt-affected soils: their reclamation and management for crop production. *Advances in Soil Science*, 11: 223–88.

Hall, A.M., Mellor, A. and Wilson, M.J. (1989). The clay mineralogy and age of deeply weathered rock in north-east Scotland. *Zeitschrift für Geomorphologie*, 72: 97–108.

Hall, D.G.M., Reeve, M.J., Thomasson, A.J. and Wright, V.F. (1977). *Water Retention, Porosity and Density of Field Soils*. Soil Survey of England and Wales Technical Monograph No. 9, Harpenden.

Hallbacken, L. and Tamm, C.O. (1986). Changes in soil acidity from 1927 to 1982–1984 in a forest area of southwest Sweden. *Scandinavian Journal of Forest Research*, 1: 219–32.

Harden, J.W. (1982). A quantitative index of soil development from field descriptions: examples from a chronosequence in central California. *Geoderma*, 28: 1–28.

Harris, C. and Ellis, S. (1980). Micromorphology of soils in soliflucted materials, Okstindan, northern Norway. *Geoderma*, 23: 11–29.

Harris, J.A., Birch, P. and Short, K.C. (1989). Changes in the microbial community and physico-chemical characteristics of topsoils stockpiled during open cast mining. *Soil Use and Management*, 5: 161–8.

Harry, D.G. (1988). Ground ice and permafrost. In: M.J. Clark (ed.) *Advances in Periglacial Geomorphology*. Wiley, Chichester, 113–49.

Hartel, P.G. and Alexander, M. (1987). Effect of growth rate on the growth of bacteria in freshly moistened soil. *Soil Science Society of America Journal*, 51: 93–6.

Hasegawa, S. and Sato, T. (1987). Water uptake by roots in cracks and water movement in clayey subsoil. *Soil Science*, 143: 381–6.

Hatano, R., Iwanaga, K., Okajima, H. and Sakuma, T. (1988). Relationship between the distribution of soil macropores and root elongation. *Soil Science and Plant Nutrition*, 34: 535–46.

Hayhoe, H.N., Tarnocai, C. and Dwyer, L.M. (1990). Soil management and vegetation effects on measured and estimated soil thermal regimes in Canada. *Canadian Journal of Soil Science*, 70: 61–71.

Haynes, R.J. and Swift, R.S. (1990). Stability of soil aggregates in relation to organic constituents and soil water content. *Journal of Soil Science*, 41: 73–83.

Heathwaite, L. and Burt, T. (1992). The evidence for past and present erosion in the Slapton catchment, southwest Devon. In: M. Bell and J. Boardman (eds) *Past and Present Soil Erosion: archaeological and geographical perspectives*. Oxbow, Oxford, 89–100.

Hekstra, G.P. (1993). Ecological sustainability and the use of chemicals: is ecotoxological risk assessment doing its job properly? An introduction to chemical time bombs. *Land Degradation and Rehabilitation*, 4: 207–21.

Higashida, S. and Nishimune, A. (1988). Factors affecting CO_2 evolution from the lower plow layer of grassland soil. *Soil Science and Plant Nutrition*, 34: 203–13.

Hill, R.L. (1993). Tillage and wheel traffic effects on runoff and sediment loss from crop interrows. *Soil Science Society of America Journal*, 57: 476–80.

Hillel, D. (1982). *Introduction to Soil Physics*. Academic Press, New York.

Hodgson, J.M. (ed.) (1976). *Soil Survey Field Handbook*. Soil Survey of England and Wales Technical Monograph No. 5, Harpenden.

Hodgson, J.M. and Whitfield, W.A.D. (1990). *Applied Soil Mapping for Planning, Development and Conservation: a pilot study*. HMSO, London.

Hole, F.D. (1961). A classification of pedoturbation and some other processes and factors of soil formation in relation to isotropism and anisotropism. *Soil Science*, 91: 375–7.

Hole, F.D. (1981). Effects of animals on soil. *Geoderma*, 25: 75–112.

Hollis, J.M. (1991). The classification of soils in urban areas. In: P. Bullock and P.J. Gregory (eds) *Soils in the Urban Environment*. Blackwell, Oxford: 5–27.

Hopkins, D.W., Shiel, R.S. and O'Donnell, A.G. (1988). The influence of sward species composition on the rate of organic matter decomposition in grassland soil. *Journal of Soil Science*, 39: 385–92.

Hornung, M. (1985). Acidification of soils by trees and forests. *Soil Use and Management*, 1: 24–8.

Hornung, M. (1990). Nutrient losses from ecosystems. In: A.F. Harrison, P. Ineson and O.W. Heal (eds) *Nutrient Cycling in Terrestrial Ecosystems*. Elsevier, London, 75–9.

Hornung, M., Reynolds, B., Stevens, P.A. and Neal, C. (1987). Increased acidity and aluminium concentrations in streams following afforestation. In: *Proceedings of the International Symposium on Acidification and Water Pathways*. Norwegian National Committee for Hydrology, Bolkesjo, Norway, 259–68.

Hortensius, D. and Nortcliff, S. (1991). International standardization of soil quality measurement procedures for the purpose of soil protection. *Soil Use and Management*, 7: 163–6.

Howells, G.D. and Brown, D.J.A. (1986). Loch Fleet: techniques for acidity mitigation. *Water, Air and Soil Pollution*, 31: 817–25.

Hubble, G.D., Isbell, R.F. and Northcote, K.H. (1983). Features of Australian soils. In: CSIRO *Soils: an Australian viewpoint*. Academic Press, London, 17–47.

Hudson, N. (1992). *Land Husbandry*. Batsford, London.

Isbell, R.F., Reeve, R. and Hutton, J.T. (1983). Salt and sodicity. In: CSIRO *Soils: an Australian viewpoint*. Academic Press, London, 107–17.

Ives, J. and Messerli, B. (1989). *The Himalayan Dilemma: reconciling development and conservation*. Routledge, London.

Jackson, M.L. (1964). Chemical composition of the soil. In: F.E. Bear (ed.) *Chemistry of the Soil*. Reinhold, New York, 71–141.

Jakobson, K.M. and Dragun, A.K. (1991). Water and soil conservation in New Zealand. *Journal of Environmental Management*, 33: 1–16.

Jarvis, M.G. (1982). Non-agricultural uses of soil surveys. In: E.M. Bridges and D.A. Davidson (eds) *Principles and Applications of Soil Geography*. Longman, Harlow, 216–55.

Jarvis, N.J. and Leeds-Harrison, P.B. (1987). Some problems associated with the use of the neutron probe in swelling/shrinking clay soils. *Journal of Soil Science*, 38: 149–56.

Jarvis, R.A., Bendelow, V.C., Bradley, R.I., Carroll, D.M., Furness, R.R., Kilgour, I.N.L. and King, S.J. (1984). *Soils and their Use in Northern England*. Soil Survey of England and Wales Bulletin No. 10, Harpenden.

Jenkinson, D.S. and Johnston, A.E. (1977). Soil organic matter in the Hoosfield Continuous Barley Experiment. *Report of the Rothamsted Experimental Station for 1976*, Part 2: 87–101.

Jenny, H. (1941). *Factors of Soil Formation: a system of quantitative pedology*. McGraw-Hill, New York.

Jenny, H. (1946). Arrangement of soil series and soil types according to functions of soil-forming factors. *Soil Science*, 61: 375–91.

Johnson, C.E., Johnson, A.H. and Siccama, T.G. (1991). Whole-tree clear-cutting effects on exchangeable cations and soil acidity. *Soil Science Society of America Journal*, 55: 502–8.

Johnson, D.L. (1990). Biomantle evolution and the redistribution of earth materials and artifacts. *Soil Science*, 149: 84–102.

Johnson, D.L. and Watson-Stegner, D. (1987). Evolution model of pedogenesis. *Soil Science*, 143: 349–66.

Johnson, R.M. and Sims, J.T. (1993). Influence of surface and subsoil properties on herbicide sorption by Atlantic coastal plain soils. *Soil Science*, 155: 339–48.

Jones, A. (1982). X-ray fluorescence spectrometry. In: A.L. Page, R.H. Miller and D.R. Keeney (eds) *Methods of Soil Analysis, Part 2: chemical and microbiological properties*, Agronomy 9. American Society of Agronomy and Soil Science Society of America, Madison, Wisconsin, 85–121.

Jones, J.A.A. (1981). *The Nature of Soil Piping: a review of research*. Geo Books, Norwich.

Jones, J.A.A. (1987). The effects of soil piping on contributing areas and erosion patterns. *Earth Surface Processes and Landforms*, 12: 229–48.

Jones, R.D. and Schwab, A.P. (1993). Nitrate leaching and nitrate occurrence in a fine textured soil. *Soil Science*, 155: 272–82.

Jones, V.J., Stevenson, A.C. and Battarbee, R.W. (1986). Lake acidification and the land-use hypothesis: a mid-post-glacial analogue. *Nature*, 322: 157–8.

Jury, W.A., Gardner, W.R. and Gardner, W.H. (1991). *Soil Physics*. Wiley, New York.

Kalisz, P.J., Zimmerman, R.W. and Muller, R.N. (1987). Root density, abundance, and distribution in the mixed mesophytic forest of eastern Kentucky. *Soil Science Society of America Journal*, 51: 220–5.

Karathanasis, A.D. and Golrick, P.A. (1991). Soil formation on loess/sandstone toposequences in west-central Kentucky: I. Morphology and physicochemical properties. *Soil Science*, 152: 14–24.

Kaspar, T.C. and Bland, W.L. (1992). Soil temperature and root growth. *Soil Science*, 154: 290–9.

Kaune, A., Türk, T. and Horn, R. (1993). Alteration of soil thermal properties by structure formation. *Journal of Soil Science*, 44: 231–48.

Kemp, R.A. (1985). The cause of redness in some buried and non-buried soils in Eastern England. *Journal of Soil Science*, 36: 329–34.

Kemp, R.A. (1986). Pre-Flandrian Quaternary soils and pedogenic processes in Britain. In: V.P. Wright (ed.) *Paleosols: their recognition and interpretation*. Blackwell, Oxford, 242–62.

Kiefer, R.H. and Amey, R.G. (1992). Concentrations and controls of soil carbon dioxide in sandy soil in the North Carolina Coastal Plain. *Catena*, 19: 539–59.

King, D., Daroussin, J., Bonneton, P. and Nicoullaud, B. (1986). An improved method for combining map data. *Soil Use and Management*, 2: 140–5.

King, J.A. (1988). Some physical features of soil after open cast mining. *Soil Use and Management*, 4: 161–8.

Kirkby, S.J. and Sill, M. (1991). *The Atlas of Nepal in the Modern World*. Earthscan, London.

Klappa, C.F. (1983). A process–response model for the formation of pedogenic calcretes. In: R.C.L. Wilson (ed.) *Residual Deposits: surface related weathering processes and materials*. Blackwell, Oxford, 211–20.

Klemmedson, J.O. (1987). Influence of oak in pine forests of central Arizona on selected nutrients of forest floor and soil. *Soil Science Society of America Journal*, 51: 1623–8.

Klingebiel, A.A. and Montgomery, P.H. (1961). *Land Capability Classification*, Handbook No. 210. US Department of Agriculture Soil Conservation Service, Washington, DC.

Kozak, Z., Niecko, J. and Kozak, D. (1993). Precipitation of heavy metals in the Leczna-Wlodawa Lake region. *Science of the Total Environment*, 133: 183–92.

Krinsley, D.H. and Doornkamp, J. (1973). *Atlas of Quartz Sand Surface Textures*. Cambridge University Press, Cambridge.

Kuhnt, G. and Franzle, O. (1993). Assessment of pollution of groundwater by atrazine. *Land Degradation and Rehabilitation*, 4: 245–51.

Kukla, J. (1975). Loess stratigraphy of central Europe. In: K. Butzer and G.L. Isaac (eds) *After the Australopithecines*. Mouton, The Hague, 99–188.

Kutiel, P. and Inbar, M. (1993). Fire impacts on soil nutrients and soil erosion in a Mediterranean pine forest plantation. *Catena*, 20: 129–39.

Lal, R. and Stewart, B.A. (1990a). Soil degradation: a global threat. *Advances in Soil Science*, 11: xiii–xvii.

Lal, R. and Stewart, B.A. (1990b). Need for action; research and development priorities. *Advances in Soil Science*, 11: 331–6.

Landon, J.R. (ed.) (1991). *Booker Tropical Soil Manual: a handbook for soil survey and agricultural land evaluation in the tropics and subtropics*. Longman, Harlow.

Lautridou, J.P. (1988). Recent advances in cryogenic weathering. In: M.J. Clark (ed.) *Advances in Periglacial Geomorphology*. Wiley, Chichester, 33–47.

Lavado, R.S. and Taboada, M.A. (1987). Soil salinization as an effect of grazing in a native grassland soil in the Flooding Pampa of Argentina. *Soil Use and Management*, 3: 143–8.

Le Bissonnais, Y. and Singer, M.J. (1993). Seal formation, runoff and interrill erosion from seventeen Californian soils. *Soil Science Society of America Journal*, 57: 224–9.

Lee, K.E. (1983). Soil animals and pedological processes. In: CSIRO *Soils: an Australian viewpoint*. Academic Press, London, 629–44.

Lee, K.E. (1985). *Earthworms: their ecology and relationships with soils and land use*. Academic Press, Sydney.

Legros, J.P. (1992). Soils of alpine mountains. In: I.P. Martini and W. Chesworth (eds) *Weathering, Soils and Paleosols*. Elsevier, Amsterdam, 155–81.

Legros, J.P. and Cabidoche, Y.M. (1977). Les types de sols et leur répartition dans les Alpes et les Pyrénées cristallines. *Doc. Cartogr. Ecol.*, 19: 1–19.

Levine, S.J., Hendricks, D.M. and Schreiber, J.F. (1989). Effect of bedrock porosity on soils formed from dolomitic limestone residuum and eolian deposition. *Soil Science Society of America Journal*, 53: 856–62.

Lewkowicz, A.G. (1988). Slope processes. In: M.J. Clark (ed.) *Advances in Periglacial Geomorphology*. Wiley, Chichester, 325–68.

Liebig, M.A., Jones, A.J., Mielke, L.N. and Doran, J.W. (1993). Controlled wheel traffic effects on soil properties in ridge tillage. *Soil Science Society of America Journal*, 57: 1061–6.

Liengsakul, M., Mekpaiboonwatana, S., Pramojanee, P., Bronsveld, K. and Huizing, H. (1993). Use of GIS and remote sensing for soil mapping and for locating new sites for permanent cropland – a case study in the 'highlands' of northern Thailand. *Geoderma*, 60: 293–307.

Lietzke, D.A. and McGuire, G.A. (1987). Characterization and classification of soils with spodic morphology in the southern Appalachians. *Soil Science Society of America Journal*, 51: 165–70.

Lim, C.H. and Jackson, M.L. (1982). Dissolution for total elemental analysis. In: A.L. Page, R.H. Miller and D.R. Keeney (eds) *Methods of Soil Analysis, Part 2: chemical and microbiological properties*, Agronomy 9. American Society of Agronomy and Soil Science Society of America, Madison, Wisconsin, 1–12.

Limbird, A. (1985). Genesis of soils affected by discrete volcanic ash inclusions, Alberta, Canada. In: E. Fernandez Caldas and D.H. Yaalon (eds) *Volcanic Soils*. Catena Verlag, Cremlingen-Destedt.

Litaor, M.I. (1987). The influence of eolian dust on the genesis of alpine soils in the Front Range, Colorado. *Soil Science Society of America Journal*, 51: 142–7.

Livens, F.R. and Loveland, P.J. (1988). The influence of soil properties on the environmental mobility of caesium in Cumbria. *Soil Use and Management*, 4: 69–75.

Locke, W.W. (1979). Etching of hornblende grains in arctic soils: an indicator of relative age and paleoclimate. *Quaternary Research*, 11: 197–212.

Locke, W.W. (1986). Rates of hornblende etching in soils on glacial deposits, Baffin Island, Canada. In: S.M. Colman and D.P. Dethier (eds) *Rates of Chemical Weathering of Rocks and Minerals*. Academic Press, Orlando, 129–45.

Logan, T.J. (1990). Chemical degradation of soils. *Advances in Soil Science*, 11: 187–221.

Long, I.F., Monteith, J.L., Penman, H.L. and Szeicz, G. (1964). The plant and its environment. *Meteorologische Rundschau*, 17: 97–101.

Longhurst, J.W.S., Green, S.E. and Lee, D.S. (1987). *Acid Deposition in the Northern Hemisphere*. Acid Rain Information Centre, Department of Environmental and Geographical Studies, Manchester Polytechnic, Manchester.

Loveland, P.J., Hazelden, J., Sturdy, R.G. and Hodgson, J.M. (1986). Salt-affected soils in England and Wales. *Soil Use and Management*, 2: 150–6.

Lowe, J.J. and Walker, M.J.C. (1984). *Reconstructing Quaternary Environments*. Longman, Harlow.

Lundström, U.S. (1993). The role of organic acids in the soil solution chemistry of a podzolized soil. *Journal of Soil Science*, 44: 121–33.

McBride, R.A. and Bober, M.L. (1993). Quantified evaluation of agricultural soil capability at the local scale: a GIS-assisted case study from Ontario, Canada. *Soil Use and Management*, 9: 58–66.

McCaig, M. (1985). Soil properties and subsurface hydrology. In: K.S. Richards, R.R. Arnett and S. Ellis (eds) *Geomorphology and Soils*. George Allen & Unwin, London, 121–40.

McCracken, R.J. (1987). Soils, soil scientists, and civilization. *Soil Science Society of America Journal*, 51: 1395–400.

Macedo, J. and Bryant, R.B. (1987). Morphology, mineralogy, and genesis of a hydrosequence of Oxisols in Brazil. *Soil Science Society of America Journal*, 51: 690–8.

McEwen, F.L. and Stephenson, G.R. (1979). *The Use and Significance of Pesticides in the Environment*. Wiley, New York.

McGarry, D. (1989). The effect of wet cultivation on the structure and fabric of a Vertisol. *Journal of Soil Science*, 40: 199–207.

McGregor, D.F.M. and Barker, D. (1991). Land degradation and hillside farming in the Fall River basin, Jamaica. *Applied Geography*, 11: 143–56.

McIntosh, J.E. and Barnhisel, R.I. (1993). Erodibility and sediment yield by natural rainfall from reconstructed mine soils. *Soil Science*, 156: 118–26.

Mackay, J.R. (1984). The frost heave of stones in the active layer above permafrost with downward and upward freezing. *Arctic and Alpine Research*, 16: 439–46.

McKeague, J.A. (1983). Clay skins and argillic horizons. In: P. Bullock and C.P. Murphy (eds) *Soil Micromorphology, Vol. 2: soil genesis*. AB Academic Publishers, Berkhamsted, 367–87.

McKeague, J.A. and Brydon, J.E. (1970). Mineralogical properties of ten reddish brown soils from the Atlantic Provinces in relation to parent materials and pedogenesis. *Canadian Journal of Soil Science*, 50: 47–55.

McKeague, J.A. and Topp, G.C. (1986). Pitfalls in interpretation of soil drainage from soil survey information. *Canadian Journal of Soil Science*, 66: 37–44.

McKeague, J.A., Wang, C., Coen, G.M., DeKimpe, C.R., Laverdier, M.R., Evans, L.J., Kloosterman, B. and Green, A.J. (1983). Testing chemical criteria for spodic horizons on podzolic soils in Canada. *Soil Science Society of America Journal*, 47: 1052–4.

Mackie-Dawson, L.A., Mullins, C.E., Kirkland, J.A. and FitzPatrick, E.A. (1988). The determination of the macroporosity of impregnated blocks of a clay soil and its relation to volumetric water content. *Journal of Soil Science*, 39: 65–70.

Mackie-Dawson, L.A., Mullins, C.E., Goss, M.J., Court, M.N. and FitzPatrick, E.A. (1989). Seasonal changes in the structure of clay soils in relation to soil management and crop type. II. Effects of cultivation and cropping at Compton Beauchamp. *Journal of Soil Science*, 40: 283–92.

Macklin, M.G. (1986). Channel and floodplain metamorphosis in the River Nent, Cumberland. In: M. Macklin and J. Rose (eds) *Quaternary River Landforms and Sediments in the North Pennines*. BGRG/QRA, Cambridge, 19–33.

Macleod, D.A. (1980). The origin of the red Mediterranean soils in Epirus, Greece. *Journal of Soil Science*, 31: 125–36.

Macphail, R.I. (1986). Paleosols in archaeology: their role in understanding Flandrian pedogenesis. In: V.P. Wright (ed.) *Paleosols: their recognition and interpretation.* Blackwell, Oxford, 263–90.

Macphail, R.I. (1992). Soil micromorphological evidence of ancient soil erosion. In: M. Bell and J. Boardman (eds) *Past and Present Soil Erosion: archaeological and geographical perspectives.* Oxbow, Oxford, 197–215.

McRae, S.G. (1989). The restoration of mineral workings in Britain – a review. *Soil Use and Management,* 5: 135–42.

Magaritz, M. and Jahn, R. (1992). Pleistocene and Holocene soil carbonates from Lanzarote, Canary Islands, Spain: paleoclimatic implications. *Catena,* 19: 511–19.

Magnusson, T. (1992). Studies of the soil atmosphere and related physical site characteristics in mineral forest soils. *Journal of Soil Science,* 43: 767–90.

Mahaney, W.C. (1978). Late-Quaternary stratigraphy and soils in the Wind River Mountains, western Wyoming. In: W.C. Mahaney (ed.) *Quaternary Soils.* Geo Abstracts, Norwich, 223–64.

Major, D.J., Janzen, H.H., Olson, B.M. and McGinn, S.M. (1992). Reflectance characteristics of southern Alberta soils. *Canadian Journal of Soil Science,* 72: 611–15.

Mannion, A.M. (1991). *Global Environmental Change.* Longman, Harlow.

Marron, D.C. and Popenoe, J.H. (1986). A soil catena on schist in northwestern California. *Geoderma,* 37: 307–24.

Marshall, C.E. (1977). *The Physical Chemistry and Mineralogy of Soils, Vol. 2: soils in place.* Wiley, New York.

Martini, I.P. and Chesworth, W. (1992). Reflections on soils and paleosols. In: I.P. Martini and W. Chesworth (eds) *Weathering, Soils and Paleosols.* Elsevier, Amsterdam, 3–16.

Martz, L. (1992). The variation of soil erodibility with slope position in a cultivated Canadian prairie landscape. *Earth Surface Processes and Landforms,* 17: 543–56.

Matkin, E.A. and Smart, P. (1987). A comparison of tests of structural stability. *Journal of Soil Science,* 38: 123–35.

Matthews, J.A. (1980). Some problems and implications of ^{14}C dates from a podzol buried beneath an end moraine at Haugabreen, southern Norway. *Geografiska Annaler,* 62A: 185–208.

Matthews, J.A. (1985). Radiocarbon dating of surface and buried soils: principles, problems and prospects. In: K.S. Richards, R.R. Arnett and S. Ellis (eds) *Geomorphology and Soils.* George Allen & Unwin, London, 269–88.

Matthews, J.A. and Dresser, P.Q. (1983). Intensive ^{14}C dating of a buried palaeosol horizon. *Geologiska Föreningens i Stockholm Förhandlingar,* 105: 59–63.

Maugh, T.H. (1984). Acid rain's effects on people assessed. *Science,* 226: 1408–10.

Mbagwu, J.S.C. and Bazzoffi, P. (1989). Properties of soil aggregates as influenced by tillage practices. *Soil Use and Management,* 5: 180–8.

Mellor, A. (1985). Soil chronosequences on Neoglacial moraine ridges, Jostedalsbreen and Jotunheimen, southern Norway: a quantitative pedogenic approach. In: K.S. Richards, R.R. Arnett and S. Ellis (eds) *Geomorphology and Soils.* George Allen & Unwin, London, 289–308.

Melville, M.D. and Atkinson, G. (1985). Soil colour: its measurement and its designation in models of uniform colour space. *Journal of Soil Science,* 36: 495–512.

Messing, I. and Jarvis, N.J. (1993). Temporal variation in the hydraulic conductivity of a tilled clay soil as measured by tension infiltrometers. *Journal of Soil Science,* 44: 11–24.

Miles, J. (1985). The pedogenic effects of different species and vegetation types and the implications of succession. *Journal of Soil Science,* 36: 571–84.

Miller, H.G. (1985). The possible role of forests in streamwater acidification. *Soil Use and Management*, 1: 28–9.

Miller, M.D. and Jastrow, J.D. (1990). Hierarchy of root and mycorrhizal fungal interactions with soil aggregation. *Soil Biology and Biochemistry*, 22: 579–84.

Millington, A.C., Mutiso, S.K., Kirkby, S.J. and O'Keefe, P. (1989). African soil erosion – nature undone and the limitations of technology. *Land Degradation and Rehabilitation*, 1: 279–90.

Milne, R. (1987). Acid soils are harbouring Chernobyl's caesium. *New Scientist*, 115: 28.

Milnes, A.R. (1992). Calcrete. In: I.P. Martini and W. Chesworth (eds) *Weathering, Soils and Paleosols*. Elsevier, Amsterdam, 309–47.

Milnes, A.R. and Thiry, M. (1992). Silcretes. In: I.P. Martini and W. Chesworth (eds) *Weathering, Soils and Paleosols*. Elsevier, Amsterdam, 349–77.

Minderman, G. (1968). Addition, decomposition, and accumulation of organic matter in forests. *Journal of Ecology*, 56: 355–62.

Ministry of Agriculture, Fisheries and Food (1981). *Subsoiling as an Aid to Drainage*. Field Drainage Leaflet No. 10. MAFF Publications, Northumberland.

Mitchell, A.R. and Van Genuchten, M.T. (1993). Flood irrigation of a cracked soil. *Soil Science Society of America Journal*, 57: 490–7.

Moffat, A.J., Johnston, M. and Wright, J.S. (1990). An improved probe for sampling soil atmospheres. *Plant and Soil*, 121: 145–7.

Mohr, E.C.J., Van Baren, F.A. and van Schuylenborgh, J. (1972). *Tropical Soils: a comprehensive study of their genesis*. Mouton/Ichtiar Baru/Van Hoeve, The Hague/Paris/Djakarta.

Molope, M.B., Grieve, I.C. and Page, E.R. (1987). Contributions by fungi and bacteria to aggregate stability of cultivated soils. *Journal of Soil Science*, 38: 71–7.

Monroe, J.S. and Wicander, R. (1992). *Physical Geology: exploring the earth*. West Publishing Co., St Paul.

Monteith, J.L. and Unsworth, M.H. (1990). *Principles of Environmental Physics*. Arnold, London.

Moore, A.W., Isbell, R.F. and Northcote, K.H. (1983). Classification of Australian soils. In: CSIRO *Soils: an Australian viewpoint*. Academic Press, London, 253–66.

Moran, C.J., Koppi, A.J., Murphy, B.W. and McBratney, A.B. (1988). Comparison of the macropore structure of a sandy loam surface soil horizon subjected to two tillage treatments. *Soil Use and Management*, 4: 96–102.

Morgan, R.P.C. (1980). Implications. In: M.J. Kirkby and R.P.C. Morgan (eds) *Soil Erosion*. Wiley, Chichester, 253–301.

Morgan, R.P.C. (1986). *Soil Erosion and Conservation*. Longman, Harlow.

Morris, E.M. and Thomas, A.G. (1986). Transient acid surges in an upland stream. *Water, Air and Soil Pollution*, 34: 429–38.

Morrison, R.B. (1978). Quaternary soil stratigraphy – concepts, methods, and problems. In: W.C. Mahaney (ed.) *Quaternary Soils*. Geo Abstracts, Norwich, 77–108.

Moyo, S., O'Keefe, P. and Sill, M. (1993). *The Southern African Environment*. Earthscan, London.

Mücher, H.J., Slotboom, R.T. and ten Veen, W.J. (1990). Palynology and micromorphology of a man-made soil. A reconstruction of the agricultural history since late-Medieval times of the Posteles in the Netherlands. *Catena*, 17: 55–67.

Mulders, M.A. (1987). *Remote Sensing in Soil Science*. Elsevier, Amsterdam.

Mullins, C.E., Young, I.M., Bengough, A.G. and Ley, G.J. (1987). Hardsetting soils. *Soil Use and Management*, 3: 79–83.

Nadler, A. (1993). Negatively charged PAM efficacy as a soil conditioner as affected by the presence of roots. *Soil Science*, 156: 79–85.

Nahon, D.B. (1986). Evolution of iron crusts in tropical landscapes. In: S.M. Colman and D.P. Dethier (eds) *Rates of Chemical Weathering of Rocks and Minerals*. Academic Press, Orlando, 169–91.

Nelson, D.W. and Sommers, L.E. (1982). Total carbon, organic carbon and organic matter. In: A.L. Page, R.H. Miller and D.R. Keeney (eds) *Methods of Soil Analysis, Part 2: chemical and microbiological properties*, Agronomy 9. American Society of Agronomy and Soil Science Society of America, Madison, Wisconsin, 539–79.

Nelson, F.E. (1985). A preliminary investigation of solifluction macrofabrics. *Catena*, 12: 23–33.

Nesbitt, H.W. (1992). Diagenesis and metasomatism of weathering profiles, with emphasis on Precambrian paleosols. In: I.P. Martini and W. Chesworth (eds) *Weathering, Soils and Paleosols*. Elsevier, Amsterdam, 127–52.

Nesje, A., Kvamme, M. and Rye, N. (1989). Neoglacial gelifluction in the Jostedalsbreen region, western Norway: evidence from dated buried palaeopodzols. *Earth Surface Processes and Landforms*, 14: 259–70.

Newbould, P. (1985). Improvement of native grasslands in the uplands. *Soil Use and Management*, 1: 43–9.

Newman, R.H. and Tate, K.R. (1991). ^{13}C NMR characterization of humic acids from soils of a development sequence. *Journal of Soil Science*, 42: 39–46.

Nilsson, J. (1986). Soil is vulnerable too. *Acid Magazine*, 4: 24–7.

Nisbet, T.R., Mullins, C.E. and Macleod, D.A. (1989). The variation of soil water regime, oxygen status and rooting pattern with soil type under Sitka spruce. *Journal of Soil Science*, 40: 183–97.

Nortcliff, S. (1988). Soil survey, soil classification and land evaluation. In: A. Wild (ed.) *Russell's Soil Conditions and Plant Growth*. Longman, Harlow, 815–43.

Nortcliff, S. and Dias, A.C.D.C.P. (1988). The change in soil physical conditions resulting from forest clearance in the humid tropics. *Journal of Biogeography*, 15: 61–6.

Nullet, D., Ikawa, H. and Kilham, P. (1990). Local differences in soil temperature and soil moisture regimes on a mountain slope, Hawaii. *Geoderma*, 47: 171–84.

Nyberg, P. and Thornelof, E. (1988). Operational liming of surface waters in Sweden. *Water, Air and Soil Pollution*, 41: 3–16.

Oke, T.R. (1987). *Boundary Layer Climates*. Methuen, London.

Oldeman, L.R. and van Engelen, V.W.P. (1993). A world soils and terrain digital database (SOTER) – an improved assessment of land resources. *Geoderma*, 60: 309–25.

Ollier, C. (1984). *Weathering*. Longman, Harlow.

Olson, C.G., Ruhe, R.V. and Mausbach, M.J. (1980). The terra rossa limestone contact phenomena in karst, southern Indiana. *Soil Science Society of America Journal*, 44: 1075–9.

Olson, G.W. (1981a). *Soils and the Environment: a guide to soil surveys and their applications*. Chapman and Hall, New York.

Olson, G.W. (1981b). Archaeology: lessons on future soil use. *Journal of Soil and Water Conservation*, 36: 261–4.

Ormerod, S.J., Donald, A.P. and Brown, S.J. (1989). The influence of plantation forestry on the pH and aluminium concentrations of upland Welsh streams: a re-examination. *Environmental Pollution*, 62: 47–62.

Page, F. and Guillet, B. (1991). Formation of loose and cemented B horizons in podzolic soils: evaluation of biological actions from micromorphological features, C/N values and ^{14}C datings. *Canadian Journal of Soil Science*, 71: 485–94.

Parlange, M.B., Steenhuis, T.S., Timlin, D.J., Stagnitti, F. and Bryant, R.B. (1989).

Subsurface flow above a fragipan horizon. *Soil Science*, 148: 77–86.

Parry, M. (1990). *Climate Change and World Agriculture*. Earthscan, London.

Patrick, W.H. and Mahapatra, I.C. (1968). Transformation and availability to rice of nitrogen and phosphorus in waterlogged soils. *Advances in Agronomy*, 20: 323–59.

Payton, R.W. (1992). Fragipan formation in argillic brown earths (Fragiudalfs) of the Milfield Plain, north-east England. I. Evidence for a periglacial stage of development. *Journal of Soil Science*, 43: 621–44.

Pearce, D., Barbier, E. and Markanya, A. (1990). *Sustainable Development, Economics and Environment in the Third World*. Earthscan, London.

Pearce, F. (1993). How Britain hides its acid soil. *New Scientist*, 27 Feb.: 29–33.

Peel, R.F. (1974). Insolation and weathering: some measures of diurnal temperature changes in exposed rocks in the Tibesti region, central Sahara. *Zeitschrift für Geomorphologie Supplementband*, 21: 19–28.

Pennock, D.J. and de Jong, E. (1990). Rates of soil redistribution associated with soil zones and slope classes in southern Saskatchewan. *Canadian Journal of Soil Science*, 70: 325–34.

Percival, C.J. (1986). Paleosols containing an albic horizon: examples from the Upper Carboniferous of northern England. In: V.P. Wright (ed.) *Paleosols: their recognition and interpretation*. Blackwell, Oxford, 87–111.

Piccolo, A. and Mbagwu, J.S.C. (1990). Effects of different organic waste amendments on soil microaggregate stability and molecular sizes of humic substances. *Plant and Soil*, 123: 27–37.

Pidgeon, J.D. and Thorogood, P.J. (1985). Management of difficult soils for intensive winter cereals in central England: a farm case study. *Soil Use and Management*, 1: 124–6.

Ping, C.L. (1987). Soil temperature profiles of two Alaskan soils. *Soil Science Society of America Journal*, 51: 1010–18.

Pitty, A.F. (1979). *Geography and Soil Properties*. Methuen, London.

Pollard, W.H. (1988). Seasonal frost mounds. In: M.J. Clark (ed.) *Advances in Periglacial Geomorphology*. Wiley, Chichester, 201–29.

Poore, D. (1989). *No Timber Without Trees: sustainability in the tropical rainforest*. Earthscan, London.

Porcella, D.B. (1988). An update on the Lake Acidification Mitigation Project (LAMP). *Water, Air and Soil Pollution*, 41: 43–51.

Potter, K.N., Horton, R. and Cruse, R.M. (1987). Soil surface roughness effects on radiation reflectance and soil heat flux. *Soil Science Society of America Journal*, 51: 855–60.

Preece, R.C. (1992). Episodes of erosion and stability since the Late-glacial: the evidence from dry valleys in Kent. In: M. Bell and J. Boardman (eds) *Past and Present Soil Erosion: archaeological and geographical perspectives*. Oxbow, Oxford, 175–83.

Previtali, F. (1992). Seismipedoturbations in volcanic soils in north-eastern Ecuador. *Catena*, 19: 441–50.

Pye, K. (1992). Aeolian dust transport and deposition over Crete and adjacent parts of the Mediterranean Sea. *Earth Surface Processes and Landforms*, 17: 271–88.

Rabenhorst, M.C. and Wilding, L.P. (1986). Pedogenesis on the Edwards Plateau, Texas: II. Formation and occurrence of diagnostic subsurface horizons in a climosequence. *Soil Science Society of America Journal*, 50: 687–92.

Radtke, U. and Brückner, H. (1991). Investigation on age and genesis of silcretes in Queensland (Australia) – preliminary results. *Earth Surface Processes and Landforms*, 16: 547–54.

REFERENCES

Raghavan, G.S.V., McKyes, E., Gendron, G., Borglum, B. and Le, H.H. (1978). Effects of soil compaction on development and yield of corn (maize). *Canadian Journal of Plant Science*, 58: 435–43.

Raghavan, G.S.V., McKyes, E., Taylor, F., Richard, P. and Watson, A. (1979). The relationship between machinery traffic and corn yield reactions in successive years. *Transactions of the American Society of Agricultural Engineers*, 21: 1256–9.

Raghavan, G.S.V., Alvo, P. and McKyes, E. (1990). Soil compaction in agriculture: a view towards managing the problem. *Advances in Soil Science*, 11: 1–36.

Raina, P., Joshi, D.C. and Kolarkar, A.S. (1991). Land degradation mapping by remote sensing in the arid region of India. *Soil Use and Management*, 7: 47–52.

Rautengarten, A. (1993). Sources of heavy metal pollution in the Rhine basin. *Land Degradation and Rehabilitation*, 4: 339–49.

Rebertus, R.A. and Buol, S.W. (1985). Iron distribution in a developmental sequence of soils from mica gneiss and schist. *Soil Science Society of America Journal*, 49: 713–20.

Rehfuess, K.E. (1985). On the causes of decline of Norway spruce (*Picea abies* Karst) in central Europe. *Soil Use and Management*, 1: 30–2.

Reid, P.M., Wilkinson, A.E., Tipping, E. and Jones, M.N. (1991). Aggregation of humic substances in aqueous media as determined by light scattering methods. *Journal of Soil Science*, 42: 259–70.

Retallack, G.J. (1986). The fossil record of soils. In: V.P. Wright (ed.) *Paleosols: their recognition and interpretation*. Blackwell, Oxford, 1–57.

Retallack, G.J. (1990). *Soils of the Past: an introduction to paleopedology*. Harper Collins, London.

Retallack, G.J. (1992). Paleozoic paleosols. In: I.P. Martini and W. Chesworth (eds) *Weathering, Soils and Paleosols*. Elsevier, Amsterdam, 543–64.

Reynolds, B., Neal, C., Hornung, M. and Stevens, P.A. (1986). Baseflow buffering of stream water acidity in five mid-Wales catchments. *Journal of Hydrology*, 87: 167–85.

Reynolds, B., Hornung, M. and Stevens, P.A. (1987). Solute budgets and denudation rate estimates for a mid-Wales catchment. *Catena*, 14: 13–23.

Richardson, J.L. and Edmonds, W.J. (1987). Linear regression estimations of Jenny's Relative Effectiveness of State Factors Equation. *Soil Science*, 144: 203–8.

Righi, D. and Lorphelin, L. (1987). The soils of a typical slope in the Himalayas (Nepal): their main characteristics and distribution. *Catena*, 14: 533–51.

Roberts, G. (1987). Nitrogen inputs and outputs in a small agricultural catchment in the eastern part of the United Kingdom. *Soil Use and Management*, 3: 148–54.

Roberts, N. (1989). *The Holocene: an environmental history*. Blackwell, Oxford.

Robertson-Rintoul, M.S.E. (1986). A quantitative soil-stratigraphic approach to the correlation and dating of post-glacial river terraces in Glen Feshie, western Cairngorms. *Earth Surface Processes and Landforms*, 11: 605–17.

Robinson, D.A. and Blackman, J. (1990). Some costs and consequences of soil erosion and flooding around Brighton and Hove, autumn 1987. In: J. Boardman, I.D.L. Foster and J.A. Dearing (eds) *Soil Erosion on Agricultural Land*. Wiley, Chichester, 369–82.

Robinson, D.A. and Naghizadeh, R. (1992). The impact of cultivation practice and wheelings on runoff generation and soil erosion on the South Downs: some experimental results using simulated rainfall. *Soil Use and Management*, 8: 151–6.

Rockwell, T.K., Johnson, D.L., Keller, E.A. and Dembroff, G.R. (1985). A late Pleistocene–Holocene soil chronosequence in the Ventura basin, southern California, USA. In: K.S. Richards, R.R. Arnett and S. Ellis (eds) *Geomorphology and Soils*. George Allen & Unwin, London, 309–27.

Rose, C. (1985). Acid rain falls on British woodlands. *New Scientist*, 14 November: 52–7.

Rose, J., Boardman, J., Kemp, R.A. and Whiteman, C.A. (1985). Palaeosols and the interpretation of the British Quaternary stratigraphy. In: K.S. Richards, R.R. Arnett and S. Ellis (eds) *Geomorphology and Soils*. George Allen & Unwin, London, 348–75.

Ross, G.J. (1980). The mineralogy of spodosols. In: B.K.G. Theng (ed.) *Soils with Variable Charge*. New Zealand Society of Soil Science, Lower Hutt, 127–43.

Ross, S. (1989). *Soil Processes: a systematic approach*. Routledge, London.

Rowell, D.L. (1994). *Soil Science: methods and applications*. Longman, Harlow.

Rowell, D.L. and Wild, A. (1985). Causes of soil acidification: a summary. *Soil Use and Management*, 1: 32–3.

Rudeforth, C.C. and Webster, R. (1973). Indexing and display of soil survey data by means of feature-cards and Boolean maps. *Geoderma*, 9: 229–48.

Russell, R.S. (1977). *Plant Root Systems: their function and interaction with the soil*. McGraw-Hill, London.

Sakai, C. and Kumada, K. (1985). Characteristics of buried humic horizons at the Shiiji archaeological pits. V. Palynological studies on buried humic horizons. *Soil Science and Plant Nutrition*, 31: 81–90.

Sanchez, P.A. and Benites, J.R. (1987). Low-input cropping for acid soils of the humid tropics. *Science*, 238: 1521–7.

Santos, M.C.D., St Arnaud, R.J. and Anderson, D.W. (1986). Quantitative evaluation of pedogenic changes in Boralfs (Gray Luvisols) of east central Saskatchewan. *Soil Science Society of America Journal*, 50: 1013–19.

Saull, M. (1990). Nitrates in soil and water. *New Scientist*, 15 Sept.

Schaetzl, R.J. (1991a). Distribution of Spodosol soils in southern Michigan: a climatic interpretation. *Annals of the Association of American Geographers*, 81: 425–42.

Schaetzl, R.J. (1991b). A lithosequence of soils in extremely gravelly, dolomitic parent materials, Bois Blanc Island, L. Huron. *Geoderma*, 48: 305–20.

Schaetzl, R.J., Barrett, L.R. and Winkler, J.A. (1994). Choosing models for soil chronofunctions and fitting them to data. *European Journal of Soil Science*, 45: 219–32.

Schellmann, W. (1994). Geochemical differentiation in laterite and bauxite formation. *Catena*, 21: 131–43.

Schnitzer, M. (1991). Soil organic matter – the next 75 years. *Soil Science*, 151: 41–58.

Schofield, N.J. and Bari, M.A. (1991). Valley reforestation to lower saline ground-water tables: results from Stene's farm, western Australia. *Australian Journal of Soil Research*, 29: 635–50.

Schunke, E. and Zoltai, S.C. (1988). Earth hummocks (thufur). In: M.J. Clark (ed.) *Advances in Periglacial Geomorphology*. Wiley, Chichester, 231–45.

Schwertmann, U., Murad, E. and Schulze, D.G. (1982). Is there Holocene reddening (hematite formation) in soils of axeric temperate areas? *Geoderma*, 27: 209–23.

Selby, M.J. (1993). *Hillslope Materials and Processes*. Oxford University Press, Oxford.

Shoji, S., Nanzyo, M. and Dahlgren, R.A. (1993). *Volcanic Ash Soils: genesis, properties and utilization*. Elsevier, Amsterdam.

Shotyk, W. (1992). Organic soils. In: I.P. Martini and W. Chesworth (eds) *Weathering, Soils and Paleosols*. Elsevier, Amsterdam, 203–24.

Sigua, G.C., Isensec, A.R. and Sadeghi, A.M. (1993). Influence of rainfall intensity and crop residue on leaching of atrazine through intact no-till soil cores. *Soil Science*, 156: 225–32.

Simonson, R.W. (1978). A multiple-process model of soil genesis. In: W.C. Mahaney (ed.) *Quaternary Soils*. Geo Abstracts, Norwich, 1–25.

Simonson, R.W. (1991). Soil science – goals for the next 75 years. *Soil Science*, 151: 7–18.

Simpson, K. (1983). *Soil*. Longman, Harlow.

Singer, A. (1993). Weathering patterns in representative soils of Guanxi Province, southeast China, as indicated by detailed clay mineralogy. *Journal of Soil Science*, 44: 173–88.

Singer, M.J. and Munns, D.N. (1991). *Soils: an introduction*. Macmillan, New York.

Skole, D. and Tucker, C. (1993). Tropical deforestation and habitat fragmentation in the Amazon: satellite data from 1978 to 1988. *Science*, 260: 1905–10.

Slattery, M.C. and Bryan, R.B. (1992). Laboratory experiments on surface seal development and its effect on interrill erosion processes. *Journal of Soil Science*, 43: 517–29.

Smettan, U., Jenny, M. and Facklam-Moniak, M. (1993). Soil dynamics and plant distribution of a sand dune playa microchore catena after winter rain in the Wadi Araba (Jordan). *Catena*, 20: 179–89.

Smith, K.A. and Arah, J.R.M. (1991). Gas chromatographic analysis of the soil atmosphere. In: Smith, K.A. (ed.) *Soil Analysis: modern instrumental techniques*. Marcel Dekker, New York, 287–324.

Smith, R.M. and Stewart, D.A. (1989). A regression model for nitrate leaching in Northern Ireland. *Soil Use and Management*, 5: 71–6.

Sohet, K., Herbauts, J. and Gruber, W. (1988). Changes caused by Norway spruce in an ochreous brown earth, assessed by the isoquartz method. *Journal of Soil Science*, 39, 549–61.

Soil Survey Staff (1971). *Guide for Interpreting Engineering Uses of Soils*. United States Department of Agriculture Soil Conservation Service, Washington, DC.

Soil Survey Staff (1975). *Soil Taxonomy: a basic system of soil classification for making and interpreting soil surveys*, Agriculture Handbook No. 436. United States Department of Agriculture, Washington, DC.

Soil Survey Staff (1992). *Keys to Soil Taxonomy*. Soil Management Support Services Technical Monograph No. 19. Pocahontas Press, Blacksburg, Virginia.

Soller, D.R. and Owens, J.P. (1991). The use of mineralogic techniques as relative age indicators for weathering profiles on the Atlantic Coastal Plain, USA. *Geoderma*, 51: 111–31.

Soulsby, C. (1992). Hydrological controls on acid runoff generation in an afforested catchment at Llyn Brianne, mid-Wales. *Journal of Hydrology*, 138: 431–48.

Soulsby, C. and Reynolds, B. (1992). Modelling hydrological processes and aluminium leaching in an acid soil at Llyn Brianne, mid-Wales. *Journal of Hydrology*, 138: 409–29.

Soutar, R.G. (1989). Afforestation and sediment yields in British fresh waters. *Soil Use and Management*, 5: 82–6.

Stevens, P.A. and Hornung, M. (1988). Nitrate leaching from a felled Sitka spruce plantation in Beddgelert Forest, North Wales. *Soil Use and Management*, 4: 3–9.

Stevenson, F.J. (1986). *Cycles of Soil: carbon, nitrogen, phosphorus, sulfur, micronutrients*. Wiley, New York.

Stewart, P.A. and Cameron, K.C. (1992). Effect of trampling on soils of the St James Walkway, New Zealand. *Soil Use and Management*, 8: 30–6.

Stewart, V.I. and Scullion, J. (1989). Principles of managing man-made soils. *Soil Use and Management*, 5: 109–16.

Stocking, M. (1992). *Land Degradation and Rehabilitation Research in Africa 1980–1990: retrospect and prospect*. International Institute for Environment and Development Paper No. 34, London.

Stoops, G. (1983). Micromorphology of the oxic horizon. In: P. Bullock and C.P. Murphy (eds) *Soil Micromorphology, Vol. 2: soil genesis.* AB Academic Publishers, Berkhamsted, 419–40.

Stoops, G. (1989). Relict properties in soils of humid tropical regions with special reference to central Africa. In: A. Bronger and J.A. Catt (eds) *Paleopedology: nature and application of paleosols.* Catena, Cremlingen-Destedt, 95–106.

Strahler, A.N. and Strahler, A.H. (1989). *Elements of Physical Geography.* Wiley, New York.

Strakhov, N.M. (1967). *Principles of Lithogenesis, Vol. 1.* Oliver & Boyd, Edinburgh.

Stremme, H.E. (1989). Thermoluminescence dating of the pedostratigraphy of the Quaternary period in NW Germany. *Geoderma,* 45: 185–95.

Stromgaard, P. (1992). Traditional agriculture in Zambia. *Geography Review,* 5: 18–20.

Swanson, D.K. (1985). Soil catenas on Pinedale and Bull Lake moraines, Willow Lake, Wind River Mountains, Wyoming. *Catena,* 12: 329–42.

Swift, M.J., Heal, O.W. and Anderson, J.M. (1979). *Decomposition in Terrestrial Ecosystems.* Blackwell, Oxford.

Szabolcs, I. (1986). Agronomical and ecological impacts of irrigation on soil and water salinity. *Advances in Soil Science,* 4: 189–218.

Szott, L.T., Palm, C.A. and Sanchez, P.A. (1991). Agroforestry in acid soils of the humid tropics. *Advances in Agronomy,* 45: 275–301.

Tardy, Y. (1992). Diversity and terminology of lateritic profiles. In: I.P. Martini and W. Chesworth (eds) *Weathering, Soils and Paleosols.* Elsevier, Amsterdam, 379–405.

Tardy, Y. and Roquin, C. (1992). Geochemistry and evolution of lateritic landscapes. In: I.P. Martini and W. Chesworth (eds) *Weathering, Soils and Paleosols.* Elsevier, Amsterdam, 407–43.

Tate, K.R. (1992). Assessment, based on a climosequence of soils in tussock grasslands, of soil carbon storage and release in response to global warming. *Journal of Soil Science,* 43: 697–707.

Tedrow, J.C.F. (1977). *Soils of the Polar Landscapes.* Rutgers University Press, New Brunswick.

Terwilliger, V.J. and Waldron, L.J. (1990). Assessing the contribution of roots to the strength of undisturbed, slip prone soils. *Catena,* 17: 151–62.

Thapa, G.P. and Weber, K.E. (1991). Soil erosion in developing countries: a politicoeconomic explanation. *Environmental Management,* 15: 461–73.

Thornes, J.B. (1979). Fluvial processes. In: C. Embleton and J. Thornes (eds) *Process in Geomorphology.* Arnold, London, 213–71.

Thornton, I. (1991). Metal contamination in soils of urban areas. In: P. Bullock and P.J. Gregory (eds) *Soils in the Urban Environment.* Blackwell, Oxford, 47–75.

Thow, R.F. (1963). The effect of tilth on the emergence of spring oats. *Journal of Agricultural Science, Cambridge,* 60: 291–5.

Tisdall, J.M. and Oades, J.M. (1982). Organic matter and water-stable aggregates in soils. *Journal of Soil Science,* 33: 141–63.

Trudgill, S.T. (ed.) (1986). *Solute Processes.* Wiley, Chichester.

Trudgill, S.T. (1988). *Soil and Vegetation Systems.* Clarendon Press, Oxford.

Trudgill, S.T. (1990). The sustainable use of forests. *Geography Review,* 4: 11–13.

Ugolini, F.C. (1986). Processes and rates of weathering in cold and polar desert environments. In: S.M. Colman and D.P. Dethier (eds) *Rates of Chemical Weathering of Rocks and Minerals.* Academic Press, Orlando, 193–235.

Ulery, A.L., Graham, R.C. and Amrhein, C. (1993). Wood-ash composition and soil pH following intense burning. *Soil Science,* 156: 358–64.

Valentin, C. (1991). Surface crusting in two alluvial soils of northern Niger. *Geoderma*, 48: 201–22.

Van de Graaff, W.J.E. (1983). Silcrete in western Australia: geomorphological settings, textures, structures, and their genetic implications. In: R.C.L. Wilson (ed.) *Residual Deposits: surface related weathering processes and materials*. Blackwell, Oxford, 159–66.

Vandenberghe, J. (1988). Cryoturbations. In: M.J. Clark (ed.) *Advances in Periglacial Geomorphology*. Wiley, Chichester, 179–98.

Van Vliet, L.J.P., Kline, R. and Hall, J.W. (1993). Effects of three tillage treatments on seasonal runoff and soil loss in the Peace River region. *Canadian Journal of Soil Science*, 73: 469–80.

Van Vliet-Lanoë, B. (1985). Frost effects in soils. In: J. Boardman (ed.) *Soils and Quaternary Landscape Evolution*. Wiley, Chichester, 117–58.

Van Vliet-Lanoë, B., Helluin, M., Pellerin, J. and Valadas, B. (1992). Soil erosion in Western Europe: from the last interglacial to the present. In: M. Bell and J. Boardman (eds) *Past and Present Soil Erosion: archaeological and geographical perspectives*. Oxbow, Oxford, 101–14.

Van Wesemael, B. and Veer, M.A.C. (1992). Soil organic matter accumulation, litter decomposition and humus forms under Mediterranean-type forests in southern Tuscany, Italy. *Journal of Soil Science*, 43: 133–44.

Vita-Finzi, C. (1969). *The Mediterranean Valleys*. Cambridge University Press, London.

Walker, P.H. (1962). Soil layers on hillslopes: a study at Nowra, New South Wales, Australia. *Journal of Soil Science*, 13: 167–77.

Walling, D.E. and Quine, T.A. (1991). Use of ^{137}Cs measurements to investigate soil erosion on arable fields in the UK: potential applications and limitations. *Journal of Soil Science*, 42: 147–65.

Walsh, R.P.D. and Howells, K.A. (1988). Soil pipes and their role in runoff generation and chemical denudation in a humid tropical catchment in Dominica. *Earth Surface Processes and Landforms*, 13: 9–17.

Wang, C., McKeague, J.A. and Kodama, H. (1986). Pedogenic imogolite and soil environments: a case study of spodosols in Quebec, Canada. *Soil Science Society of America Journal*, 50: 711–18.

Ward, R.C. and Robinson, M. (1990). *Principles of Hydrology*. McGraw-Hill, London.

Warren, A. (1979). Aeolian processes. In: C. Embleton and J. Thornes (eds) *Process in Geomorphology*. Arnold, London, 325–51.

Washburn, A.L. (1979). *Geocryology: a survey of periglacial processes and environments*. Edward Arnold, London.

Watson, A. (1992). Desert soils. In: I.P. Martini and W. Chesworth (eds) *Weathering, Soils and Paleosols*. Elsevier, Amsterdam, 225–60.

Webb, T.H. and Burgham, S.J. (1994). Catenary relationships of Downland soils derived from loess, South Canterbury, New Zealand. *Australian Journal of Soil Research*, 32: 1–11.

Webster, R. and Oliver, M.A. (1990). *Statistical Methods in Soil and Land Resource Survey*. Oxford University Press, Oxford.

Westerveld, G.J.W. and Van Den Hurk, J.A. (1973). Applications of soil and interpretive maps to non-agricultural land use in the Netherlands. *Geoderma*, 10: 47–66.

Whalley, W.B. (1976). *Properties of Materials and Geomorphological Explanation*. Oxford University Press, Oxford.

Whalley, W.B. and McGreevey, J.P. (1983). Weathering. *Progress in Physical Geography*, 7: 559–86.

REFERENCES

White, I.D., Mottershead, D.N. and Harrison, S.J. (1992). *Environmental Systems.* Chapman & Hall, London.

White, R.E. (1987). *Introduction to the Principles and Practice of Soil Science.* Blackwell, Oxford.

White, S.E. (1976). Is frost action really only hydration shattering? A review. *Arctic and Alpine Research*, 8: 1–6.

Wild, A. (ed.) (1988). *Russell's Soil Conditions and Plant Growth.* Longman, Harlow.

Wilhelmi, V. and Rothe, G.M. (1990). The effect of acid rain, soil temperature and humidity on carbon mineralization rates in organic soil layers under spruce. *Plant and Soil*, 121: 197–202.

Willett, I.R., Noller, B.N. and Beech, T.A. (1994). Mobility of radium and heavy metals from uranium mine tailings in acid sulphate soils. *Australian Journal of Soil Research*, 32: 335–55.

Williams, M.A.J., Dunkerley, D.L., De Deckker, P., Kershaw, A.P. and Stokes, T. (1993). *Quaternary Environments.* Arnold, London.

Williams, P.J. and Smith, M.W. (1989). *The Frozen Earth: fundamentals of geocryology.* Cambridge University Press, Cambridge.

Williams, R.D. and Cooper, J.R. (1990). Locating soil boundaries using magnetic susceptibility. *Soil Science*, 150: 889–95.

Wilson, G.V., Jardine, P.M., Luxmoore, R.J. and Jones, J.R. (1990). Hydrology of a forested hillslope during storm events. *Geoderma*, 46: 119–38.

Wilson, M.J. (1985). The mineralogy and weathering history of Scottish soils. In: K.S. Richards, R.R. Arnett and S. Ellis (eds) *Geomorphology and Soils.* George Allen & Unwin, London, 233–44.

Wilson, M.J. (1986). Mineral weathering processes in podzolic soils on granitic parent materials and their implications for surface water acidification. *Journal of the Geological Society of London*, 143: 691–7.

Wilson, M.J. and Nadeau, P.H. (1985). Interstratified clay minerals and weathering processes. In: J.I. Drever (ed.) *The Chemistry of Weathering.* Reidal, Dordrecht, 97–118.

Wintle, A.G. and Catt, J.A. (1985). Thermoluminescence dating of soils developed in Late Devensian loess at Pegwell Bay, Kent. *Journal of Soil Science*, 36: 293–8.

Wischmeier, W.H., Johnson, C.B. and Cross, B.V. (1971). A soil erodibility nomograph for farmland and construction sites. *Journal of Soil and Water Conservation*, 26: 189–93.

Wolman, M.G. (1967). A cycle of sedimentation and erosion in urban river channels. *Geografiska Annaler*, 49A: 385–95.

Wong, M.T.F., Hughes, R. and Rowell, D.L. (1990). Retarded leaching of nitrate in acid soils from the tropics: measurement of the effective anion exchange capacity. *Journal of Soil Science*, 41: 655–63.

Wood, M. (1989). *Soil Biology.* Blackie, Glasgow.

Woodward, I. (1989). Plants in the greenhouse world. *New Scientist*, 122: 1–4.

Wright, R.J., Boyer, D.G., Winant, W.M. and Perry, H.D. (1990). The influence of soil factors on yield differences among landscape positions in an Appalachian cornfield. *Soil Science*, 149: 375–82.

Wright, V.P. (ed.) (1986). *Paleosols: their recognition and interpretation.* Blackwell, Oxford.

Wright, V.P. (1992). Paleopedology: stratigraphic relationships and empirical models. In: I.P. Martini and W. Chesworth (eds) *Weathering, Soils and Paleosols.* Elsevier, Amsterdam, 475–99.

Yakowitz, H. (1988). Policy development issues with respect to contaminated soil sites. In: K. Wolf, W.J. Van den Brink and F.J. Colon (eds) *Contaminated Soil '88.*

REFERENCES

Kluwer Academic Publishers, Dordrecht, 1515–26.

Young, A. (1976). *Tropical Soils and Soil Survey.* Cambridge University Press, Cambridge.

Young, A. and Goldsmith, P.F. (1977). Soil survey and land evaluation in developing countries: a case study in Malawi. *Geographical Journal*, 143: 407–38.

Young, A. and Saunders, I. (1986). Rates of surface processes and denudation. In: A.D. Abrahams (ed.) *Hillslope Processes.* George Allen & Unwin, Boston, 3–27.

Younger, A. (1989). Factors affecting the cropping potential of reinstated soils. *Soil Use and Management*, 5: 150–4.

Zhongli, D., Rutter, N. and Tungsheng, L. (1993). Pedostratigraphy of Chinese loess deposits and climatic cycles in the last 2.5 Myr. *Catena*, 20: 73–91.

Zoltai, S.C. and Vitt, D.H. (1990). Holocene climatic change and distribution of peatlands in western interior Canada. *Quaternary Research*, 33: 231–40.

Zuzel, J.F. and Pikul, J.L. (1993). Effects of straw mulch on runoff and erosion from small agricultural plots in northeastern Oregon. *Soil Science*, 156: 111–17.

INDEX

Bold page numbers indicate Figures; *Italics* indicate Tables.

Milton Keynes UK
Ingram Content Group UK Ltd.
UKHW040013071024
449327UK00011B/216